NEUROMETHODS

Series Editor
**Wolfgang Walz
University of Saskatchewan
Saskatoon, SK, Canada**

For further volumes:
http://www.springer.com/series/7657

Super-Resolution Microscopy Techniques in the Neurosciences

Edited by

Eugenio F. Fornasiero

Department of Neuro- and Sensory Physiology, University of Göttingen Medical Center, Göttingen, Germany

Silvio O. Rizzoli

Department of Neuro- and Sensory Physiology, University of Göttingen Medical Center, Göttingen, Germany

Editors
Eugenio F. Fornasiero
Department of Neuro- and Sensory Physiology
University of Göttingen Medical Center
Göttingen, Germany

Silvio O. Rizzoli
Department of Neuro- and Sensory Physiology
University of Göttingen Medical Center
Göttingen, Germany

ISSN 0893-2336 ISSN 1940-6045 (electronic)
ISBN 978-1-62703-982-6 ISBN 978-1-62703-983-3 (eBook)
DOI 10.1007/978-1-62703-983-3
Springer New York Heidelberg Dordrecht London

Library of Congress Control Number: 2014930566

© Springer Science+Business Media, LLC 2014
This work is subject to copyright. All rights are reserved by the Publisher, whether the whole or part of the material is concerned, specifically the rights of translation, reprinting, reuse of illustrations, recitation, broadcasting, reproduction on microfilms or in any other physical way, and transmission or information storage and retrieval, electronic adaptation, computer software, or by similar or dissimilar methodology now known or hereafter developed. Exempted from this legal reservation are brief excerpts in connection with reviews or scholarly analysis or material supplied specifically for the purpose of being entered and executed on a computer system, for exclusive use by the purchaser of the work. Duplication of this publication or parts thereof is permitted only under the provisions of the Copyright Law of the Publisher's location, in its current version, and permission for use must always be obtained from Springer. Permissions for use may be obtained through RightsLink at the Copyright Clearance Center. Violations are liable to prosecution under the respective Copyright Law.
The use of general descriptive names, registered names, trademarks, service marks, etc. in this publication does not imply, even in the absence of a specific statement, that such names are exempt from the relevant protective laws and regulations and therefore free for general use.
While the advice and information in this book are believed to be true and accurate at the date of publication, neither the authors nor the editors nor the publisher can accept any legal responsibility for any errors or omissions that may be made. The publisher makes no warranty, express or implied, with respect to the material contained herein.

Printed on acid-free paper

Humana Press is a brand of Springer
Springer is part of Springer Science+Business Media (www.springer.com)

Preface to the Series

Under the guidance of its founders Alan Boulton and Glen Baker, the Neuromethods series by Humana Press has been very successful since the first volume appeared in 1985. In about 17 years, 37 volumes have been published. In 2006, Springer Science + Business Media made a renewed commitment to this series. The new program will focus on methods that are either unique to the nervous system and excitable cells or which need special consideration to be applied to the neurosciences. The program will strike a balance between recent and exciting developments like those concerning new animal models of disease, imaging, in vivo methods, and more established techniques. These include immunocytochemistry and electrophysiological technologies. New trainees in neurosciences still need a sound footing in these older methods in order to apply a critical approach to their results. The careful application of methods is probably the most important step in the process of scientific inquiry. In the past, new methodologies led the way in developing new disciplines in the biological and medical sciences. For example, physiology emerged out of anatomy in the nineteenth century by harnessing new methods based on the newly discovered phenomenon of electricity. Nowadays, the relationships between disciplines and methods are more complex. Methods are now widely shared between disciplines and research areas. New developments in electronic publishing also make it possible for scientists to download chapters or protocols selectively within a very short time of encountering them. This new approach has been taken into account in the design of individual volumes and chapters in this series.

Saskatoon, SK, Canada *Wolfgang Walz*

Preface

Microscopy imaging experiments have contributed to lay the foundations for modern biological sciences. At the cellular level, the description of numerous cellular components and localization of individual proteins have been achieved through the combination of affinity-based labeling and light microscopy approaches. These techniques are widespread and offer a remarkable versatility for antibody staining but are limited by diffraction in the resolution that they can achieve. In practice, two infinitesimally small light sources cannot be distinguished when they are located closer than ~200 nm in the X–Y plane and ~500 nm in the Z-direction, limiting the information that can be obtained from biological specimens in conventional light microscopy. Through the use of an electron beam as an energy source, electron microscopy techniques overcame this limitation and have revealed some of the finest details of subcellular structures with accuracy in the Ångstrom range. Although electron microscopy is seen as a gold standard for optical resolution, it comes at the cost of considerably limited possibilities in terms of affinity labeling and live-imaging. As a matter of fact, until recently, neither light nor electron microscopies were able to achieve determination of molecular species at high resolution in living organisms.

The last two decades have witnessed the flourishing development of imaging techniques that contributed to tear apart the diffraction limit of light microscopy in what has been considered a revolution in optical sciences. Altogether these technical advancements, termed "super-resolution imaging techniques," achieve resolutions beyond the diffraction limit and have proven to combine some of the labeling flexibility of conventional light microscopy with nanometer-scale resolution of electron microscopy, providing new approaches for imaging living physiological samples. As most of the technological breakthroughs that have revolutionized biological sciences, super-resolution imaging approaches have determined a paradigm shift in the way in which experimental approaches are thought, designed, and analyzed, contributing to broaden both practical and theoretical approaches to neurobiological problems.

This book is intended as a comprehensive description of current super-resolution techniques, including the physical principles that allowed their development, some of the most recent neurobiological applications, and selected information for the practical use of these technologies. An historical perspective on light microscopy and general considerations about how the light diffraction limit has been overcome serve as an introduction for the basic principles of super-resolution imaging (Chaps. 1 and 2). Essentially, super-resolution can be achieved (a) with configurations that reveal fluorophores by "depleting" surrounding fluorescent molecules (in stimulated emission depletion, STED; Chaps. 3 and 4); (b) determining the positions of single fluorophores by "making up" the fluorescence signal as in photoactivated light microscopy (PALM) or in stochastic optical reconstruction microscopy

(STORM; Chaps. 5 and 6); and (c) using a structured pattern of illumination for exciting the fluorophores as in structured illumination microscopy (SIM; Chaps. 7 and 8). In addition to these techniques, a number of other exciting developments provide super-resolution images (at least in one dimension): near-field microscopy (Chap. 9), atomic force microscopy (Chap. 10), or super-resolution attempts by X-ray (Chap. 11) and Raman microscopy (Chap. 12). Finally, among super-resolution techniques, electron microscopy photooxidation should be acknowledged as a time-tested but still useful technique to combine fluorescence microscopy with the resolution of the electron microscope (Chap. 13). Higher resolution imaging has likewise raised new challenges for sample preparation and affinity labeling. Two specific chapters cover these issues, highlighting specific technical difficulties and providing useful practical solutions (Chaps. 14 and 15).

Synaptic transmission, which is at the basis of information transfer between nerve cells, is an active topic of research in modern neurosciences. Chemical synapses allow fast and specific neurotransmitter release and are essential for brain functioning. Whereas a number of aspects of synaptic physiology have been investigated by microscopy (e.g., synapse number, localization within neurons, and protein distribution by immunostaining) several features of pre- and postsynaptic contacts cannot be examined at the resolution of conventional microcopy techniques [1].

In the central nervous system, presynaptic terminals are characterized by a complex scaffold of cytomatrix proteins that organize a cluster of presynaptic vesicles filled with neurotransmitter. Upon depolarization, the neurotransmitter content of synaptic vesicles is released in the synaptic cleft and binds the receptors on the postsynaptic cell, mediating information transmission. Intracellularly postsynaptic receptors are held together by a dense proteinaceus meshwork called postsynaptic density. The elaborate synaptic molecular machinery has been studied with nanometer-scale resolution primarily by electron microscopy and immunogold-labeling. Nevertheless, immunostainings in electron microscopy reach a relatively low density and come at the cost of limited resolution of cellular details.

Super-resolution studies have contributed to describe the elaborate molecular architecture of chemical synapses. A STED approach has elucidated that the presynaptic cytomatrix of neuromuscular junctions in Drosophila is characterized by doughnut-shaped arrangement of the protein Bruchpilot (BRP, homologue of the presynaptic cytoskeletal structural protein CAST/ERC). In BRP deletion mutants, the T-bar structures that in neuromuscular junctions of Drosophila define active zones at the ultrastructural level are lost, calcium dynamics at the synapse are impaired, and neurotransmission is compromised [2]. These results, strengthened by a subsequent study [3], support the idea that BPR is a master regulator of presynaptic active zones by establishing a scaffold that allows T-bar assembly and proper organization of calcium channels.

Multicolor three-dimensional STORM imaging of olfactory bulb and cortex synapses allowed to describe the positioning of a number of pre- and postsynaptic scaffolding proteins with nanometer precision [4]. Among other observations, the authors found that synaptic scaffolding proteins are arranged at a precise reciprocal axial distance from each other with the neurotransmitter receptors localized in-between the pre- and postsynaptic scaffolding proteins. Interestingly, receptors such as NMDA and AMPA have a heterogeneous lateral distribution among different synapses, suggesting that subpopulations of glutamatergic synapses might cluster receptors to the center or the periphery of the contact depending on functional differences. Additionally, Bassoon and Piccolo, two presynaptic scaffolding proteins, show a defined oriented distribution with the carboxy group and the amino group of proteins, respectively, localized at ~50 and ~75 nm from the synaptic cleft, revealing a previously unknown level of organization of presynaptic active zones.

Along with compelling structural synaptic information, super-resolution microscopy has allowed for the description of key processes in dynamic synaptic function and plasticity. Synaptic vesicles have been largely elusive for conventional microscopy as single organelles due to their ~40–50 nm diameter. A number of super-resolution microscopy approaches allowed describing previously unknown aspects of the vesicle life cycle. A pioneering live-imaging STED study in this field showed that synaptotagmin (an integral vesicular protein) remains clustered at the plasma membrane upon vesicle fusion, indicating that synaptic vesicles maintain multimolecular integrity during recycling [5]. Three-dimensional iso-STED clarified that the very same synaptic vesicles can be used in spontaneous and in activity-triggered neurotransmitter release [6]. Plastic embedding and thin sectioning combined with STED imaging showed that endogenous synaptic vesicle components intermix at very low rates and that fluorescent fusion proteins currently used as reporters of synaptic vesicle localization and activity tend to diffuse and be mislocalized when compared to their native counterparts [7]. Video-rate STED imaging described the movement of single vesicles, revealing that these organelles generally display a random movement, characterized by a lower mobility in the regions that likely correspond to boutons [8]. A subsequent study [9] further investigated this process and revealed that upon endocytosis synaptic vesicles lose their mobility behavior within hours and that neuronal activity block by tetrodotoxin contributes to accelerate vesicle mobility loss. Dual color STED suggested that the process is likely to be tightly regulated and displays an additional layer of control provided by fast endosomal sorting of the vesicles primarily released upon stimulation, the so-called readily releasable pool [10].

Several studies took advantage of super-resolution imaging to describe the dynamic behavior of postsynaptic dendritic spines in living neurons in vitro and in vivo. STED microscopy of dendritic structures in organotypic hippocampal slices, where YFP was used as a volume marker, besides confirming the well-established paradigm of activity-dependent spine volume increase, revealed details that were previously unknown, such as changes in spine shape and spine head broadening that could suggest physical tightening of pre- and postsynaptic neuronal components functional as well as functional spine maturation [11]. Two-photon STED configurations have applied the same principle to thicker slices [12, 13]. Super-resolution imaging of dendritic spines has been extended to the somatosensory cortex of anesthetized intact living mice where a skull optical window allowed to image living neurons in combination with an upright STED setup [14]. Both PALM and STED approaches allowed following the dynamic reorganization of actin molecules within single dendritic spines, revealing that in dendritic spines, actin molecules are organized in subdomains and form bundles in the neck of spines, undergoing activity-dependent changes [15–18].

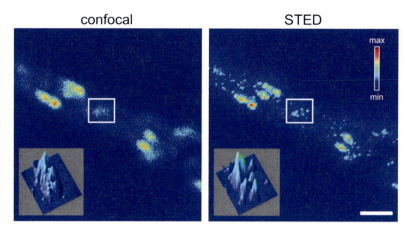

Fig. 1 Comparison of confocal and STED imaging. Primary hippocampal neurons were labeled for the synaptic vesicle marker Synapsin I and imaged with a commercially available Leica TCS STED system in confocal mode (*left*) and in STED mode (*right*). Insets: three-dimensional surface rendering of the boxed regions (820 × 760 nm). Note that in the confocal image only large spots are detected while STED imaging allows resolving smaller structures within otherwise unresolved regions. Scale bar 1 μm

We believe that the "super-resolution revolution" will not just increase the resolution in the imaging capabilities of neuroscientists (Fig. 1), but it will also contribute to change the perspective in which cellular biology processes are understood at the nanometer scale.

Göttingen, Germany *Eugenio F. Fornasiero*
Silvio O. Rizzoli

References

1. Tønnesen J, Nägerl UV (2012) Superresolution imaging for neuroscience. Exp Neurol 242: 33–40
2. Kittel RJ, Wichmann C, Rasse TM, Fouquet W, Schmidt M, Schmid A, Wagh DA, Pawlu C, Kellner RR, Willig KI, Hell SW, Buchner E, Heckmann M, Sigrist SJ (2006) Bruchpilot promotes active zone assembly, Ca^{2+} channel clustering, and vesicle release. Science (New York, NY) 312:1051–1054
3. Fouquet W, Owald D, Wichmann C, Mertel S, Depner H, Dyba M, Hallermann S, Kittel RJ, Eimer S, Sigrist SJ (2009) Maturation of active zone assembly by Drosophila Bruchpilot. J Cell Biol 186:129–145
4. Dani A, Huang B, Bergan J, Dulac C, Zhuang X (2010) Superresolution imaging of chemical synapses in the brain. Neuron 68:843–856
5. Willig KI, Rizzoli SO, Westphal V, Jahn R, Hell SW (2006) STED microscopy reveals that synaptotagmin remains clustered after synaptic vesicle exocytosis. Nature 440:935–939
6. Wilhelm BG, Groemer TW, Rizzoli SO (2010) The same synaptic vesicles drive active and spontaneous release. Nat Neurosci 13: 1454–1456
7. Opazo F, Punge A, Bückers J, Hoopmann P, Kastrup L, Hell SW, Rizzoli SO (2010) Limited intermixing of synaptic vesicle components upon vesicle recycling. Traffic (Copenhagen, Denmark) 11:800–812
8. Westphal V, Rizzoli SO, Lauterbach MA, Kamin D, Jahn R, Hell SW (2008) Video-rate far-field optical nanoscopy dissects synaptic vesicle movement. Science (New York, NY) 320:246–249

9. Kamin D, Lauterbach MA, Westphal V, Keller J, Schönle A, Hell SW, Rizzoli SO (2010) High- and low-mobility stages in the synaptic vesicle cycle. Biophys J 99:675–684
10. Hoopmann P, Punge A, Barysch SV, Westphal V, Bückers J, Opazo F, Bethani I, Lauterbach MA, Hell SW, Rizzoli SO (2010) Endosomal sorting of readily releasable synaptic vesicles. Proc Natl Acad Sci U S A 107:19055–19060
11. Nägerl UV, Willig KI, Hein B, Hell SW, Bonhoeffer T (2008) Live-cell imaging of dendritic spines by STED microscopy. Proc Natl Acad Sci U S A 105:18982–18987
12. Ding JB, Takasaki KT, Sabatini BL (2009) Supraresolution imaging in brain slices using stimulated-emission depletion two-photon laser scanning microscopy. Neuron 63:429–437
13. Takasaki KT, Ding JB, Sabatini BL (2013) Live-cell superresolution imaging by pulsed STED two-photon excitation microscopy. Biophys J 104:770–777
14. Berning S, Willig KI, Steffens H, Dibaj P, Hell SW (2012) Nanoscopy in a living mouse brain. Science (New York, NY) 335:551
15. Tatavarty V, Kim E-J, Rodionov V, Yu J (2009) Investigating sub-spine actin dynamics in rat hippocampal neurons with super-resolution optical imaging. PLoS One 4:e7724
16. Frost NA, Shroff H, Kong H, Betzig E, Blanpied TA (2010) Single-molecule discrimination of discrete perisynaptic and distributed sites of actin filament assembly within dendritic spines. Neuron 67:86–99
17. Izeddin I, Specht CG, Lelek M, Darzacq X, Triller A, Zimmer C, Dahan M (2011) Super-resolution dynamic imaging of dendritic spines using a low-affinity photoconvertible actin probe. PLoS One 6:e15611
18. Urban NT, Willig KI, Hell SW, Nägerl UV (2011) STED nanoscopy of actin dynamics in synapses deep inside living brain slices. Biophys J 101:1277–1284

Contents

Preface to the Series		*v*
Preface		*vii*
Contributors		*xv*
1	Light Microscopy and Resolution **Sinem K. Saka**	1
2	Super-Resolution Microscopy: Principles, Techniques, and Applications **Sinem K. Saka**	13
3	Foundations of STED Microscopy **Marcel A. Lauterbach and Christian Eggeling**	41
4	Application of Real-Time STED Imaging to Synaptic Vesicle Motion **Benjamin G. Wilhelm and Dirk Kamin**	73
5	Photoactivated Localization Microscopy for Cellular Imaging **Paulina Achurra, Seamus Holden, Thomas Pengo, and Suliana Manley**	87
6	Data Analysis for Single-Molecule Localization Microscopy **Steve Wolter, Thorge Holm, Sebastian van de Linde, and Markus Sauer**	113
7	Super-Resolution Fluorescence Microscopy Using Structured Illumination **Kai Wicker**	133
8	Application of Three-Dimensional Structured Illumination Microscopy in Cell Biology: Pitfalls and Practical Considerations **Daniel Smeets, Jürgen Neumann, and Lothar Schermelleh**	167
9	Scanning Near-Field Optical Microscopy for Investigations of Bio-Matter **Christiane Höppener**	189
10	Atomic Force Microscopy of Living Cells **David Alsteens and Yves F. Dufrêne**	225
11	X-Ray Microscopy for Neuroscience: Novel Opportunities by Coherent Optics **Tim Salditt and Tanja Dučić**	257
12	Nonlinear Optics Approaches Towards Subdiffraction Resolution in CARS Imaging **Klaus-Jochen Boller, Willem P. Beeker, Carsten Cleff, Kai Kruse, Chris J. Lee, Petra Groß, Herman L. Offerhaus, Carsten Fallnich, and Jennifer L. Herek**	291

13 Photooxidation Microscopy: Bridging the Gap Between Fluorescence and Electron Microscopy 325
Annette Denker and Silvio O. Rizzoli

14 Requirements for Samples in Super-Resolution Fluorescence Microscopy 343
Marko Lampe and Wernher Fouquet

15 Probing Biological Samples in High-Resolution Microscopy: Making Sense of Spots 369
Felipe Opazo

Index *387*

Contributors

PAULINA ACHURRA • *Laboratory of Experimental Biophysics (LEB), Ecole Polytechnique Federale de Lausanne (EPFL), Lausanne, Switzerland*
DAVID ALSTEENS • *Institute of Condensed Matter and Nanosciences, Université Catholique de Louvain, Louvain-la-Neuve, Belgium*
WILLEM P. BEEKER • *Laser Physics and Nonlinear Optics Group, MESA+ Research Institute for Nanotechnology, University of Twente, Enschede, The Netherlands*
KLAUS-JOCHEN BOLLER • *Laser Physics and Nonlinear Optics Group, MESA+ Research Institute for Nanotechnology, University of Twente, Enschede, The Netherlands*
CARSTEN CLEFF • *Institut für Angewandte Physik, Westfälische Wilhelms-Universität, Münster, Germany*
ANNETTE DENKER • *Biomedical Genomics, Hubrecht Institute, Utrecht, The Netherlands*
TANJA DUČIĆ • *Institut für Röntgenphysik, Universität Göttingen, Göttingen, Germany*
YVES F. DUFRÊNE • *Institute of Condensed Matter and Nanosciences, Université Catholique de Louvain, Louvain-la-Neuve, Belgium*
CHRISTIAN EGGELING • *MRC Human Immunology Unit, Weatherall Institute of Molecular Medicine, University of Oxford, Oxford, UK*
CARSTEN FALLNICH • *Institut für Angewandte Physik, Westfälische Wilhelms-Universität, Münster, Germany*
EUGENIO F. FORNASIERO • *Department of Neuro- and Sensory Physiology, University of Göttingen Medical Center, Göttingen, Germany*
WERNHER FOUQUET • *Leica Microsystems CMS GmbH, Mannheim, Germany*
PETRA GROß • *Institut für Angewandte Physik, Westfälische Wilhelms-Universität, Münster, Germany*
JENNIFER L. HEREK • *Optical Sciences Group, MESA+ Research Institute for Nanotechnology, University of Twente, Enschede, The Netherlands*
SEAMUS HOLDEN • *Laboratory of Experimental Biophysics (LEB), Ecole Polytechnique Federale de Lausanne (EPFL), Lausanne, Switzerland*
THORGE HOLM • *Lehrstuhl für Biotechnologie und Biophysik, University of Würzburg, Würzburg, Germany*
CHRISTIANE HÖPPENER • *Institute of Optics, University of Rochester, Rochester, NY, USA; Institute of Physics, University of Münster, Münster, Germany*
DIRK KAMIN • *Department of NanoBiophotonics, Max Planck Institute for Biophysical Chemistry, Göttingen, Germany*
KAI KRUSE • *Laser Physics and Nonlinear Optics Group, MESA+ Research Institute for Nanotechnology, University of Twente, Enschede, The Netherlands*
MARKO LAMPE • *Advanced Light Microscopy Core Facility, EMBL Heidelberg, Heidelberg, Germany*
MARCEL A. LAUTERBACH • *Wavefront Engineering Microscopy Group, Neurophotonics Laboratory, University Paris Descartes, Sorbonne Paris City, Paris, France*

CHRIS J. LEE • *Laser Physics and Nonlinear Optics Group, MESA+ Research Institute for Nanotechnology, University of Twente, Enschede, The Netherlands*

SEBASTIAN VAN DE LINDE • *Lehrstuhl für Biotechnologie und Biophysik, University of Würzburg, Würzburg, Germany*

SULIANA MANLEY • *Laboratory of Experimental Biophysics (LEB), Ecole Polytechnique Federale de Lausanne (EPFL), Lausanne, Switzerland*

JÜRGEN NEUMANN • *Department of Biology, Ludwig Maximilians University Munich, Martinsried, Germany*

HERMAN L. OFFERHAUS • *Optical Sciences Group, MESA+ Research Institute for Nanotechnology, University of Twente, Enschede, The Netherlands*

FELIPE OPAZO • *Department of Neuro- and Sensory Physiology, University of Göttingen Medical Center, Göttingen, Germany*

THOMAS PENGO • *Laboratory of Experimental Biophysics (LEB), Ecole Polytechnique Federale de Lausanne (EPFL), Lausanne, Switzerland*

SILVIO O. RIZZOLI • *Department of Neuro- and Sensory Physiology, University of Göttingen Medical Center, Göttingen, Germany*

SINEM K. SAKA • *Department of Neuro- and Sensory Physiology, University of Göttingen Medical Center, Göttingen, Germany*

TIM SALDITT • *Institut für Röntgenphysik, Universität Göttingen, Göttingen, Germany*

MARKUS SAUER • *Lehrstuhl für Biotechnologie und Biophysik, University of Würzburg, Würzburg, Germany*

LOTHAR SCHERMELLEH • *Department of Biology, Ludwig Maximilians University Munich, Martinsried, Germany; Department of Biochemistry, University of Oxford, Oxford, UK*

DANIEL SMEETS • *Department of Biology, Ludwig Maximilians University Munich, Martinsried, Germany; Department of Biochemistry, University of Oxford, Oxford, UK*

KAI WICKER • *Institute of Physical Chemistry, Friedrich-Schiller-University, Jena, Germany*

BENJAMIN G. WILHELM • *Department of Neuro- and Sensory Physiology, University of Göttingen Medical Center, Göttingen, Germany*

STEVE WOLTER • *Lehrstuhl für Biotechnologie und Biophysik, University of Würzburg, Würzburg, Germany*

Chapter 1

Light Microscopy and Resolution

Sinem K. Saka

Abstract

Galileo Galilei invented the first microscope "occhiolino," by combining a concave and a convex lens in 1600s. Robert Hooke and Anton van Leeuwenhoek later modified it to look at living things. Since then, light microscopy has gained immense popularity and has been pushing the limits of optical technology. The race to improve the power of seeing continued by introduction of new techniques, more and more powerful lenses, better optical corrections, stronger light sources, higher-sensitivity detectors, and assembly of cutting-edge systems. However, the wish for a better peek at the cellular world hit a wall in the 1990s, already envisioned by Ernst Abbe in 1873. Diffraction, which enables an image to be formed in the first place, is also a barrier that obscures the finer details of cells and biomolecules. This chapter presents a discussion of resolution, briefly introducing some of the fundamental concepts such as diffraction and contrast.

Key words Numerical aperture, Diffraction, Point spread function, Image formation, Airy disk, Resolution limit

1 Introduction

In 1590, two Dutch lens grinders, father and son Janssens, have combined two lenses in a tube, creating one of the first compound light microscopes in history [1]. Their simple invention was later modified by Galileo Galilei who named the new device "occhiolino" (small eye). It would then be renamed "microscope" and be known by this name. In 1665, Robert Hooke used a refined multi-lens (compound) microscope and made a discovery that generated a completely different perspective on the living world and laid the basis for the scientific fields of today; he saw "cells" for the first time [2]. Anton van Leeuwenhoek, in parallel, designed a single-lens microscope with higher magnification power (up to 275×) than the compound microscope and observed microorganisms among many other cells, thus becoming the father of microbiology [3]. With those exciting initial observations, microscopy opened a window to the small world beyond human vision and eventually became an indispensible component of cell biology. Technology was pushed to its limits to get the best possible images of the cellular/subcellular world. Introduction of the green fluorescent protein (GFP [4, 5]) made a huge breakthrough, rendering fluorescence microscopy one of the most important tools for cell biology (most of the publications in biological sciences now include fluorescence micrographs).

Many subcellular details and localizations of important constituents have been discovered, thanks to the refined objective and microscope designs. However, light microscopy has not been the method of choice for making observations at the molecular level, as the resolution it offers is not high enough to distinguish fine subcellular structures. When light waves interact with objects (such as the lens itself) they spread, causing the fine structures to be observed as "unresolvable blobs." This phenomenon, known as diffraction, limits the resolution that can be obtained with light microscopes to approximately half of the wavelength of the light used, which gives ~200 nm at best. In this regard, electron microscopy has a much higher resolving power, since the accelerated electrons used for imaging have extremely short wavelengths compared to light. However, despite the higher resolution electron microscopy offers, light microscopy has unique advantages for biological imaging. Compared to electron microscopy, sample preparation and handling is easier and less artifactual, and it allows highly specific labeling for multiple targets, even in living cells in their normal environment. Therefore any attempt to increase the resolving power of light microscopy is valuable and creates an immense potential for biology.

In the last decade, important studies have been undertaken to find ways to circumvent the diffraction problem and improve the capabilities of light microscopy. These endeavors resulted in the rise of the field of high-resolution/super-resolution optical microscopy and emergence of many high-resolution techniques. The next sections briefly present the basic concepts to understand the limitations that should be overcome by these techniques to attain high resolution.

2 Wavelength and the Numerical Aperture

Optical microscopy operates in the visible range of the electromagnetic radiation spectrum. The visible range, which is determined based on the response of the human eye, corresponds to wavelengths from ~400 to 750 nm. Below the visible light (200–400 nm) is the ultraviolet (UV) and above (750–1,000 nm) is the infrared part of the spectrum. In essence, light is a form of energy that displays both particle and wave properties. The particle properties of light are embodied in its quanta, the discrete units of energy called photons. Photons can be measured by use of cameras or detectors. The wave nature of the light, on the other hand, is represented in the propagation of the photons oscillating at a specific wavelength (λ). The wavelength of light is an important property for imaging. First, it indicates the energy level of the photons; shorter wavelengths are more energetic compared to longer ones. This is why the light of shorter wavelengths, such as UV radiation, can damage the cells and macromolecules. Second, in fluorescence microscopy it specifies the color of the light and affects the choice of fluorophores and filters to be used for spectral selection (selection of photons only from a certain part of the spectrum while avoiding the others). Finally, the wavelength of the light used for imaging is one of the major determinants of the resolution obtained with a microscope, as explained below.

In a microscope, light is collected through the objective lens and directed into the light path of the setup. The objective is the key element in image formation. Its properties are important for the magnification and resolution of the resulting image. The objective captures the light coming from the specimen through its aperture (the aperture is an opening in an otherwise opaque material). In a microscope the aperture serves the purpose of letting a fraction of the incoming light enter the light path, while eliminating the stray light coming from outside of the focused specimen area. The aperture angle shows how much of the spherical wave front can be collected (wave front is the surface of the light waves having the same phase; when the light originates from a point source the wave front is spherical). A wider aperture can collect more of the incoming light. Similarly, the refractive index of the imaging medium affects the light collection. When the light passes from one medium to another it changes direction, a phenomenon known as refraction. Refraction index of a medium describes how the propagation of light will be in that medium. The medium affects the velocity of light, so if light of wavelength λ propagates in medium of refraction index n, the velocity decreases and the wavelength becomes λ/n. The decrease in velocity is compensated by a shortening of the optical path length, which is realized by refraction, which causes a change in the propagation angle of light. Imaging media of higher refractive indices enable collection of light from wider angles. Since objectives are manufactured to be used with media (air, water, oil) of certain refractive indices, the refraction index and aperture angle are brought together to define the light gathering ability of an objective with a single parameter, the numerical aperture (NA):

$$NA = n \bullet \sin \alpha \qquad (1)$$

where n is the refractive index of the medium between the lens and the cover slip (usually air, water, glycerol, or oil) and α is the half angle over which the objective can gather light from the specimen

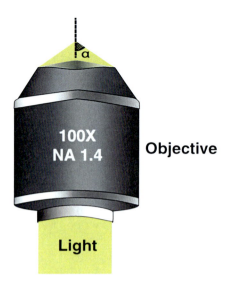

Fig. 1 Numerical aperture. The drawing depicts a common oil immersion objective with semi-aperture angle α of 68°. When a medium with refractive index of $n=1.51$ is used, the numerical aperture is 1.40

(i.e., half of the angular aperture of the objective) (Fig. 1). Like the wavelength, the numerical aperture is an important determinant of optical resolution as will be further discussed in the next section.

3 Resolution and Diffraction

Resolution of a microscope is defined as the smallest distance at which two point-like objects could be discriminated. Optically, the resolution that a microscope can offer is limited by diffraction. Diffraction can be defined as the spreading of light waves as a result of their interaction with an object. In a microscope the light (transmitted light or fluorescence light) is diffracted in the aperture of the objective lens or at the specimen itself, resulting in a redistribution of the light waves. For a point object, diffraction of the light causes production of an image that is broadened into a diffraction pattern, much larger than the size of the original object. This diffraction pattern is called "the Airy disk pattern," named after Sir George Airy. It consists of a central diffraction disk (known as the Airy disk and containing 84 % of the light coming from the point source) and a series of diffraction rings separated by zero intensity minima [6] (Fig. 2). The size of the Airy disk is related to the wavelength of light, the aperture angle of the objective and the refractive index of the imaging medium. Since the objective NA and the wavelength of the light used determine the resolution for every diffraction-limited system, for convenience a dimensionless

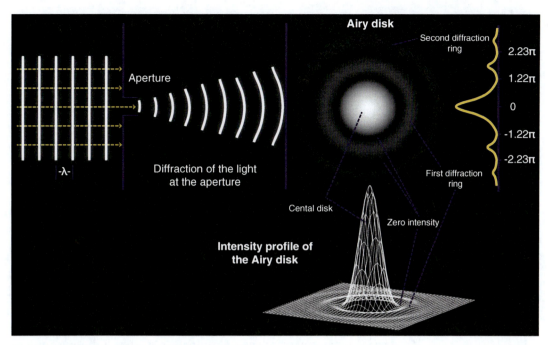

Fig. 2 The diffraction pattern and the Airy disk. The image of a point-like object is viewed as a diffraction pattern in a microscope. The diffraction pattern consists of a central diffraction disk (also called "zero order maxima") and the surrounding diffraction disks. At the minima between the disks, the intensity of the light is zero. The central disk is called the Airy disk and the diffraction pattern is known as the Airy pattern. The graph on the *right* shows the normalized intensity distribution (point spread function) of the diffraction pattern

unit is employed to express the distances in optical microscopy. This so-called optical unit (o.u.) is defined as:

$$1 \text{ o.u.} = \frac{\lambda}{2\pi NA_{objective}} \quad (2)$$

where λ is the wavelength. For transmitted light, λ is equal to the average illumination wavelength, whereas it is the excitation wavelength for fluorescence. When the distance is normalized in this way, the radius of the Airy disk is equivalent to 1.22π (Fig. 2). If two point-like objects (like two points in a specimen) are very close to each other, their diffraction patterns overlap. Depending on the degree of overlap, it might not be possible to separate these spots as different entities; hence, they cannot be optically resolved.

In 1896, Lord Rayleigh defined the resolution limit mathematically (the Rayleigh criterion [7]) based on the Airy disk pattern. The Rayleigh criterion states that the minimum distance between two objects that are still resolved is reached when the central disk of the Airy pattern from one coincides with the first diffraction minimum of the other. This distance is equal to the radius of the central disk of the Airy pattern [8]. Accordingly,

$$d_{min} = 1.22\pi \text{ o.u.} = \frac{0.61\lambda}{NA_{objective}} \qquad (3)$$

Although the Rayleigh criterion gives a simple formulation of the resolution for point objects and is very commonly applied for telescopes, a very close but slightly narrower approximation is more frequently used for microscopes, the Abbe limit.

In 1873, Ernst Abbe who was collaborating with Carl Zeiss in Jena, Germany, put forward a theory for image formation in a microscope. He used a periodic object, a diffraction grating (a series of lines spaced very close together, approximately as close as the wavelength of light), and observed the diffraction pattern at the rear focal plane of the objective lens (the diffraction plane), as well as how the image is formed at the image plane. Each space on the grating behaves as an independent source of light waves (Fig. 3a). A certain amount of light does not interact with the grating and goes through it non-diffracted. This non-diffracted (direct) light gives rise to the 0th diffraction order, which is observed as the central spot in the diffraction plane. At the image plane, these non-deviated rays create an even illumination. Other light rays interacting with the grating are diffracted, radiating outward at different angles. The rays diffracted at the same unique angle will form higher order diffraction spots (first, second, third…), flanking the 0th order central spot at the diffraction plane. The light waves forming this diffraction pattern go on and interfere (superimpose and recombine into a wave of higher or lower amplitude) to form an image of the grating in the image plane. For any image to be formed, it is necessary that at least two different diffraction orders are collected. This means that the image formation depends on the interference between non-diffracted light (the 0th order) and diffracted light (1st and higher orders [6, 9]). The objective drives the interference by focusing the incoming light at the focal point.

The diffraction pattern formed at the diffraction plane is actually an inverse transform (Fourier transform) of the resulting image. There is a reciprocal relationship between the diffraction pattern and the image such that for a periodic object like the grating, the distance between the spots in the diffraction pattern is inversely proportional to the spacing of the grating (equivalent to the object size) and the resulting image. Accordingly, the finest details of the object with small separation distances (high spatial frequency) diffract light at larger angles creating diffraction spots that are farthest from each other in the diffraction plane. On the other hand, the more coarse features that are well separated from each other (low spatial frequency) in the object form the more central spots in the diffraction plane that are close to each other [6]. When the lens geometry is considered, it can be inferred that the lens is only able to collect the light scattered up to a certain angle. Therefore, while the more central diffraction spots are transformed into the image,

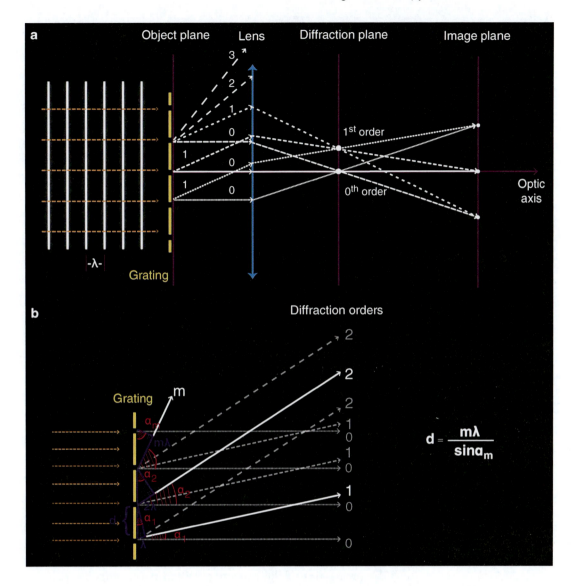

Fig. 3 Abbe's theory of optical image formation. (**a**) The paths of the diffracted light rays coming from a diffraction grating. Light rays of the same diffraction order (having the same scattering angle) form the diffraction spots at the diffraction plane (rear focal plane of the objective lens). The diffracted light from different orders (here only 1st order of diffraction is shown for simplicity) interferes with the non-diffracted light (0th order) at the image plane. For an image to be formed, at least two different diffraction orders need to be collected. (**b**) The relationship between the grating spacing (d), scattering angle (α_m), wavelength (λ), and order of diffraction (m)

numerical aperture limits the gathering of light from higher diffraction orders. However since the high diffraction orders arise from the smaller features in the object, they essentially carry the high-resolution information. Thus, it can be stated that the more diffraction orders transformed into the final image the higher the resolution is. Figure 3b shows the relationship between the grating spacing, scattering angle, and wavelength. Here, the minimum period of the grating can be thought to be equivalent to the minimum distance between two adjacent resolvable objects (resolution). For an imaging medium with refractive index of n, the minimum resolvable distance d is:

$$d = \frac{\lambda}{NA_{condenser} + NA_{objective}} \quad (4)$$

In this case, the diffraction order m described in Fig. 3c is equal to 1. In systems where the condenser NA is bigger than the objective NA or illumination occurs through the objective as in fluorescence microscopy,

$$NA_{condenser} + NA_{objective} = 2NA_{objective} \quad (5)$$

Therefore,

$$d = Abbe\ Resolution_{x,y} = \frac{\lambda}{2NA_{objective}} \quad (6)$$

Equation 6 gives the resolution in the lateral image plane. In comparison, the axial resolution is conventionally lower. When three dimensions are considered, the diffraction pattern can be thought to be rather elliptical and axially elongated, as only a part of the spherical wave front can be collected by the limited aperture angle of the objective. So a second equation defines the axial resolution [8, 10]:

$$Abbe\ Resolution_z = \frac{2\lambda n}{NA_{objective}^2} \quad (7)$$

The practical outcome of these equations is that all the point objects in a specimen are imaged as disks whose resolution is dependent on the wavelength of the light, the imaging medium, and the properties of the objective. When two points in the specimen are closer to each other than this resolution limit, their centers overlap and they would not be distinguished as different entities. Consequently, even when visible light of lower wavelengths and the objectives of highest numerical aperture are employed, the resolution does not reach beyond 200 nm in the x, y-plane. On the third dimension, it is worse; the axial resolution limit is in the order of 500 nm. For years, this has been a source of disappointment for cell biologists as even under the best imaging conditions many interesting phenomena at the subcellular level are lost in a diffraction-limited blob. In most cases, the actual resolution is poorer than these theoretical limits, due to optical aberrations, sample and preparation-related complications or noise. For a particular microscope, the intensity distribution of the obtained diffraction pattern defines the point spread function (PSF; the graph in Fig. 2 shows an example intensity distribution for the Airy pattern). In practical terms, the narrower this distribution is the higher the resolution. A microscope's PSF can be calculated by imaging fluorescent beads of known sizes. Actually, for the dimension of interest (lateral or axial) the full width of the microscope's PSF maxima at the point corresponding to the half of its maximum height (abbreviated as FWHM, full width as half maximum) is a good estimate of the practical resolution, which can expectedly be lower than the theoretical Abbe limit [11].

4 Other Limiting Factors: Contrast and Noise

Contrast and resolution are interlinked concepts. The resolution definitions given above, represent the theoretical resolution that depends on the optical elements of a microscope. However, several other factors affect the effective resolution or the final visibility of close objects. These factors include various steps of the imaging process involving the optical system, detector, image acquisition, and image digitization, as well as the properties of the sample and experimental limitations. Unlike theoretical resolution, contrast is a product of the whole imaging process and therefore is a determining factor for the effective resolution.

In general terms, contrast is proportional to the ratio of signal variation to the average signal intensity [8]. For two close objects (of equal intensities) with partially overlapping images, contrast is defined as the difference between the lowest intensity (the dip between their Airy disks) and highest intensity found in these images [11]. If the Airy disk peak is normalized to 1, the highest possible contrast is equal to 1 and is achieved when the separation between the objects is bigger than the Rayleigh limit. If the objects are closer, the contrast decreases as the two PSFs start to overlap. The distance at which the contrast becomes zero and the peaks of Airy disks cannot be distinguished is called "contrast cut-off distance." The relationship of contrast to the distance between two objects (separation distance) is represented by a function called the contrast transfer function. Plotting the contrast as a function of the separation distance allows expressing the resolution in terms of the peak-to-peak distance at a certain contrast value. For example, the separation distance given by the Rayleigh criterion corresponds to a contrast value of 26.4 %. This means that the difference between the dip and the peak is equal to the 26.4 % of the peak value. Alternatively, a third resolution definition, the Sparrow criterion, states that resolution is equal to the contrast cut-off distance. Essentially, however high the theoretical resolution of a microscope is, the resulting images would not be of much use if the contrast is low.

One important factor that reduces the contrast of the final image is the noise. Noise could be defined as the variation observed for repeated measurements of signal. It presents an uncertainty in the intensity quantifications, decreasing the accuracy of any particular measurement [11]. The sources of noise could be intrinsic, such as variation in photon counts, or extrinsic as for electronic noise (due to random fluctuations in the electrical signal) or variation in the illumination intensity. Fluorescence microscopy generally operates with small numbers of photons, typically making the intrinsic noise the dominant factor.

Statistically, repeating the same observation many times and averaging these observations reduces the noise. If n photons are collected per observation, the probability that any particular observation will lie in the range between $n + \sqrt{n}$ and $n - \sqrt{n}$ is 63 % [8]. Digital images are divided to a finite number of picture elements called pixels. Each pixel contains optical information from a certain area of the imaged specimen. For a pixel that is formed by detection of n photons, signal-to-noise ratio will be $n / \sqrt{n}$, which is equivalent to $\sqrt{n}$. It can be seen that if only a small number of photons are collected for a pixel or image, the associated uncertainty will be very high. This is especially important to consider for high-resolution imaging, because in many cases, resolution improvement is linked to detection of photons from a smaller area or volume, hence images are generated smaller number of photons and have a decreased signal-to-noise ratio. In such cases, although a theoretical resolution increase is envisioned, the final image contrast could be diminished by the noise, negating the advantages of improved resolution.

5 Super-Resolution Microscopy

Abbe's limit of resolution is a physical law that arises from the nature of light. In that sense it is a real barrier that is seemingly impossible to overcome. On the other hand, it derives from some assumptions, such as the standard design of light collection by the objective and use of uniform excitation light. It also relies on single photon fluorescence with linear absorption and emission properties. Super-resolution microscopy is a collective term for light microscopy techniques that offer resolution that is beyond the diffraction limit. To overcome the diffraction limit, super-resolution techniques emerging in the last 2 decades have circumvented these assumptions in different ways. One way is to modulate the PSF by exploiting saturation-based nonlinear optical effects and patterned illumination. Stimulated emission depletion (STED) microscopy [12] and (saturated) structured illumination microscopy (SIM [13, 14]) are examples of such modulations. Another method is to change the traditional objective geometry in order to collect more of the spherical wave front. 4Pi microscopy [15] and I^{5}M [16] employ opposing lenses that allow collection of information from both sides of the sample. Finally, a number of techniques, including stochastic optical reconstruction microscopy (STORM [17]), photoactivation localization microscopy (PALM [18]), fluorescence photoactivation localization microscopy (FPALM [19]), are based on localization of single fluorophores. These techniques rely on photoswitching/photoactivation of fluorophores to enable sequential collection of emission from active single fluorophores.

These main methodologies make it possible to attain resolution well beyond the diffraction limit of light. However, one should keep in mind that resolution is not everything by itself and, contrast and noise should also be taken into consideration to achieve more informative and higher-resolution images. Each super-resolution technique offers different advantages, but also poses various challenges, as will be discussed in the next chapter.

References

1. Karlsson RC (2009) Milestone 1: The beginning. Nature Cell Biology 11:S6. doi:10.1038/ncb1938
2. Hooke R (1665) Micrographia: or, some physiological descriptions of minute bodies made by magnifying glasses. With observations and inquiries thereupon. John Martyn and James Allestry, London
3. van Leeuwenhoek A (1977) The select works of anthony van leeuwenhoek: containing his microscopical discoveries in many of the works of nature. Arno Press, New York
4. Chalfie M, Tu Y, Euskirchen G et al (1994) Green fluorescent protein as a marker for gene expression. Science 263:802–805
5. Tsien RY (1989) Fluorescent probes of cell signaling. Annu Rev Neurosci 12:227–253. doi:10.1146/annurev.ne.12.030189.001303
6. Murphy DB, Davidson MW (2012) Fundamentals of light microscopy and electronic imaging, 2nd edn. Wiley-Blackwell, Hoboken
7. Rayleigh L (1896) XV. On the theory of optical images, with special reference to the microscope. Philos Mag 42:167–195. doi:10.1080/14786449608620902
8. Pawley JB (2006) Handbook of biological confocal microscopy, 3rd edn. Springer, New York
9. Abbe E (1873) Beiträge zur Theorie des Mikroskops und der mikroskopischen Wahrnehmung. Archiv f mikrosk Anatomie 9:413–418. doi:10.1007/BF02956173
10. Linfoot EH, Wolf E (1953) Diffraction images in systems with an annular aperture. Proc Phys Soc B 66:145–149. doi:10.1088/0370-1301/66/2/312
11. Stelzer EHK (1998) Contrast, resolution, pixelation, dynamic range and signal-to-noise ratio: fundamental limits to resolution in fluorescence light microscopy. J Microsc 189:15–24
12. Hell SW, Wichmann J (1994) Breaking the diffraction resolution limit by stimulated emission: stimulated-emission-depletion fluorescence microscopy. Opt Lett 19:780–782. doi:10.1364/OL.19.000780
13. Heintzmann R, Cremer CG (1999) Laterally modulated excitation microscopy: improvement of resolution by using a diffraction grating. In: Altshuler GB, Benaron DA, Ehrenberg B, et al. (eds) SPIE Proceedings. SPIE, pp 185–196
14. Gustafsson MGL (2005) Nonlinear structured-illumination microscopy: wide-field fluorescence imaging with theoretically unlimited resolution. Proc Natl Acad Sci U S A 102:13081–13086. doi:10.1073/pnas.0406877102
15. Hell S, Stelzer EHK (1992) Properties of a 4Pi confocal fluorescence microscope. J Opt Soc Am A 9:2159. doi:10.1364/JOSAA.9.002159
16. Gustafsson MG, Agard DA, Sedat JW (1999) I5M: 3D widefield light microscopy with better than 100 nm axial resolution. J Microsc 195:10–16. doi:10.1046/j.1365-2818.1999.00576.x
17. Rust MJ, Bates M, Zhuang X (2006) Sub-diffraction-limit imaging by stochastic optical reconstruction microscopy (STORM). Nat Methods 3:793–796. doi:10.1038/nmeth929
18. Betzig E, Patterson GH, Sougrat R et al (2006) Imaging intracellular fluorescent proteins at nanometer resolution. Science 313:1642–1645. doi:10.1126/science.1127344
19. Hess ST, Girirajan TPK, Mason MD (2006) Ultra-high resolution imaging by fluorescence photoactivation localization microscopy. Biophys J 91:4258–4272. doi:10.1529/biophysj.106.091116

Chapter 2

Super-Resolution Microscopy: Principles, Techniques, and Applications

Sinem K. Saka

Abstract

Diffraction sets a physical limit for the theoretically achievable resolution; however, it is possible to circumvent this barrier. That's what microscopists have been doing in recent years and in many ways at once, starting the era of super-resolution in light microscopy. High-resolution approaches come in various flavors, and each has specific advantages or disadvantages. For example, near-field techniques avoid the problems associated with the propagation of light by getting very close to the specimen. In the far-field, the strategies include increasing the light collecting capability, sharpening the point spread function or high-precision localization of individual fluorescent molecules. In this chapter, the major super-resolution approaches are introduced briefly, together with their pros and cons, and exemplar biological applications.

Key words Superresolution, NSOM, AFM, TIRF, STED, SIM, I^{5}M, 4Pi, Localization microscopy, STORM, PALM, FPALM

1 Near-Field Techniques

Light microscopy can work in two different modes depending on how far the light is allowed to propagate: far-field light propagates through space in an unconfined manner and is the typical light in conventional microscopy, while near-field light consists of a non-propagating field that exists very close to the surface of an object, at a distance smaller than the wavelength of light used for imaging. Near-field microscopy, which is in general limited to imaging of surfaces, has attained nanoscale resolution earlier than many far-field techniques. However, it has not been able to provide an increased resolution for most common biological applications, since the biological specimen are often too thick for near-field imaging. Still, some techniques like scanning near-field optical microscopy, atomic force microscopy (AFM), and total internal reflection microscopy (TIRF) have proven to be very useful for particular questions. Hence they will be briefly discussed below.

1.1 Near-Field Scanning/Scanning Near-Field Optical Microscopy

Near-field in optics is a general term for configurations that involve illumination or detection of the specimen through an element with subwavelength (smaller than the wavelength of the light used for imaging) size, located at a subwavelength distance from the specimen. This concept allows obtaining resolution substantially below the wavelength of light, because light is not allowed to diffract after leaving the aperture. The simplest way to design a setup for near-field microscopy is to avoid using lenses. Instead, an aperture smaller than the wavelength of light is positioned very close to the surface (at a distance smaller than the wavelength itself). In such a system, resolution depends on the aperture size and the distance from the sample, but not on the wavelength.

The theory of near-field scanning optical microscopy (NSOM) was proposed by Synge in 1928 [1]. The proposal involved placing an opaque plate with a hole (aperture) of 10 nm diameter in front of a strong light source. In this configuration, he envisioned local illumination of a thin biological section, which should be closer to the hole than the diameter of the hole itself. Transmitted light was to be detected point by point by a detector. Unfortunately, as he was also aware of the technology at the time was not advanced enough to build four critical components of such a microscope [1]. First, the light sources of the time were not strong enough. Second, it was very hard to move the section in very small regular increments. Third, getting a thin biological section with a flat surface was quite difficult at the time. Finally, constructing an opaque plate with a very small hole was also quite complicated. Therefore, it took several decades to develop the technology necessary for this design. Initially, the approach was demonstrated to achieve subwavelength resolution (reaching $\lambda/60$) in near-field at the microwave spectrum [2]. Then with the development of techniques in aperture fabrication, several groups independently applied it for shorter wavelengths [hence the different abbreviations NSOM and SNOM (scanning near-field optical microscopy)] using different aperture constructions [3–6]. The "aperture" here is the nano-sized opening at the end of a metal-coated optical fiber. Although theoretically arbitrarily small apertures can be created to increase the resolution, one limiting factor is the penetration depth of light into the metal aperture, which restricts the achievable resolution to 30–50 nm [7]. Alternatively, apertureless microscopes have also appeared. The apertureless NSOM (ANSOM) uses a sharp nano-sized mechanical probe tip (similar to that of an atomic force microscope as detailed below). The tip creates a local change, for example in the effective fluorescence, when it comes close to a point in the specimen surface, and this change is detected in the far-field. Since the effect is limited to the area interacting with the probe, objects much closer than the diffraction limit are discernible [8]. With apertureless setups, it is possible to reach below 10 nm resolution [7].

Using NSOM it is possible to obtain a resolution of at least ~80 nm and it also allows combining optical and topographical (force-based) imaging simultaneously. To keep the distance between the probe and the sample surface constant, NSOM uses a shear-force feedback control mechanism. For that, the probe tip is let to oscillate with an amplitude of less than 1 nm. When the tip is too close to the sample, the shear forces cause a change in the oscillation, which is detected through an electronic system. However, the use of this system requires the surface to be dry. Although there are protocols for structure-preserving drying processes, this creates a problem for biological specimens and fluorophores that might need an aqueous environment for desired fluorescence effects and are subject to increased photobleaching due to direct contact with oxygen [9]. These reasons, as well as the applicability of the method only to surfaces, prevented its extensive use for biology. However, in the last decade there have been successful attempts to improve the tip-sample distance regulation in liquid [10] and various cellular structures have been imaged in aqueous medium. To name a few, fluorescently labeled plasma membranes of fibroblasts or transmembrane proteins like HLA class 1 molecules have been visualized with a resolution of ~40 nm [11]. Similarly, T-cell receptor domains and aggregates of various properties have been imaged with 50 nm resolution using quantum dots for labeling [12].

1.2 Atomic Force Microscopy

Like NSOM, AFM [13] is a scanning probe microscopy technique, in which the surface of the sample is scanned with a very fine tip. While NSOM offers a direct observation of the nanostructures, AFM can provide either direct or indirect topographical images and measurements of atomic interactions. The AFM tip is positioned with high-precision in respect to the sample and is mounted to a flexible cantilever that deflects as a result of the force on the tip (Fig. 1a–c). The deflection of the cantilever is a quantitative measure of the force. To detect the deflection, a laser beam is reflected from the end of the cantilever into a detector. Through scanning, the interaction or the force between the tip and the surface is measured in a continuous feedback loop. The feedback serves to keep some of the parameters constant, allowing the measurement of the desired interactions.

AFM can be operated in four different primary modes that are distinguished by the way the tip is moved over the specimen. In the contact (static) mode [15], deflection is used as a feedback parameter by keeping the force constant. This mode is preferred especially for hard surfaces and single molecules, as the risk of damage is high for biological samples. In the non-contact mode [16], the cantilever oscillates at a small distance (1–10 nm) above the surface during the scan, without touching it. By keeping a constant distance, the changes in the amplitude, the frequency or the phase of the

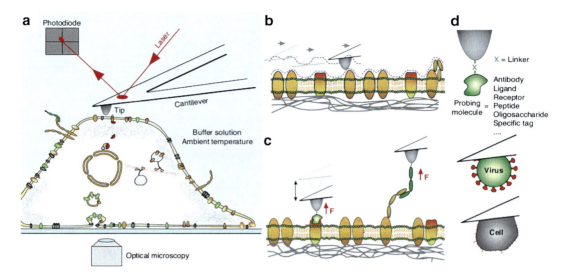

Fig. 1 Atomic force microscopy, for structural and functional imaging. (**a**) Use of AFM to probe the surface of a cell, in combination with modern optical microscopy techniques. (**b**) In the imaging mode, AFM scans the cell surface with nanometer resolution. (**c**) In force spectroscopy, the small interaction force acting between the AFM tip and the cell surface is measured while the tip approaches the cell and is retracted from the cell. (**d**) Various modalities may be used depending on whether the tip is functionalized with biomolecules or viruses or by a full cell. Reprinted with permission from [14]

cantilever are detected and used for feedback control. Attractive forces acting between the tip and the sample are sensed and topographic images are constructed. Another mode is the tapping mode [17] in which the tip briefly contacts the surface and then lifts off alternately. In this mode, the cantilever oscillates at higher amplitudes than the non-contact mode, which is advantageous to prevent trapping of the tip in thin liquid layer on the surfaces. It is especially used to image larger areas, which might have greater topographical variations. For biological samples, the last two modes are preferred, as the damage to the sample is greatly reduced. The fourth primary mode is the torsional resonance mode, where the cantilever oscillates in a twisting motion around its long axis. The twisting motion is changed upon the lateral forces the tip encounters and these changes are sensed by the detector [18]. There are also secondary imaging modes that are derived from the primary modes for detection of a specific type of interaction, such as friction forces between the probe tip and sample surface, conductivity or elasticity of the surface, electrostatic surface potential, temperature, magnetic force, or electric field gradients.

The biggest advantage of AFM is the possibility to attain single protein resolution in aqueous solutions. The resolution is in theory limited only by the apical probe geometry and the sample geometry (tip-sample dilation), but is also affected by the thermal noise and the softness of the sample. For example, for soft and dynamic cell surfaces, such as those of mammalian cells, the typical resolution is 50–100 nm [14]. Although AFM offers high resolution and the possibility to probe various interactions, it has its own limitations. Since it is a surface scanning technique, only the top surface of the sample can be imaged at high resolution and the sample has to be immobilized on a support. For accurate results, effective vibration isolation is necessary. For biological samples, there is also the inherent risk of sample deformation by the tip. Live imaging is possible, but is typically rather slow (1 image/min). There are attempts to tackle this challenge by advancing the instrumentation, like incorporating high-speed scanners. Despite these disadvantages, AFM is a valuable tool for nanoscale resolution especially for in vitro characterization of biomolecules and has been employed for several biological questions. It has recently been used to observe the motion of single Kinesin-1 dimers along microtubules with enough resolution to observe both kinesin heads binding 8 nm apart to the same protofilament [19]. In another in vitro study, the mechanism of actin remodeling by a neuronal actin binding protein, drebrin A, has been investigated with subnanometer resolution [20]. Performing contact mode AFM imaging, nuclear pore complexes have been visualized together with ribonucleoprotein particles traversing the nuclear envelope [21]. AFM has also been used to study the conformation of various amyloid molecules in reconstituted membranes. The tested amyloid peptides (including α-synuclein, Aβ(1–40), amylin, and more) were found to form morphologically compatible ion-channel-like structures and elicit single ion-channel currents in reconstituted membranes [22].

The AFM-tip can be biologically modified to image-specific molecules in an approach called molecular recognition mapping (MRM [23]). For this purpose, various chemical groups can be adsorbed or immobilized on the tip (Fig. 1d). Molecules of interest, such as receptors, antibodies, ligands, tags, or even whole cells, can be conjugated to the chemically modified tip, usually through a linker molecule. By using modified tips, specific interactions can be probed [24] and interaction maps can be produced. This approach, for example, has been used to study the organization and rigidity of protein domains containing GPI-anchored proteins on neuronal membranes [25].

Apart from its use for imaging, AFM is also an important tool for force spectroscopic measurements and nano-manipulations. For example, the AFM tip could be employed to seize a single protein. By retracting the cantilever, the protein could be forced to unfold bond by bond and the amount of retraction could be measured to calculate the relative strengths of the chemical bonds [26, 27].

1.3 Total Internal Reflection Microscopy

When light encounters the interface between two media of different refractive indices, it bends (refraction). The refraction angle is determined by the refractive indices of the two media and the

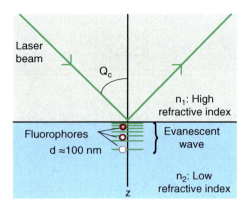

Fig. 2 The principle of TIRF microscopy. When a light beam arrives at an interface between media of different refractive indices (such as from glass into water), at an angle higher than the critical angle, it goes under total internal reflection. This generates an evanescent electromagnetic field at the other side of the interface at an intensity that decreases exponentially with the distance in z. TIRF makes use of this principle for excitation of only a subset of molecules that are very close (typically ≤100 nm) to the interface

incidence angle of the light. Refraction is usually accompanied by a small amount of reflection from the interface between the two media. If the light reaches the interface at an angle high enough, termed the critical angle, it refracts parallel to the interface. As the incidence angle increases toward the critical angle, the proportion of the refraction decreases and more light is reflected. When the angle is higher than the critical angle, the light coming through the first medium (of higher refractive index, such as glass) does not propagate in the second medium (of lower refractive index, such as an aqueous solution) and is instead completely reflected back from the interface (total internal reflection). In such a case, although the light itself stays in the first medium, an electromagnetic field, called "evanescent field," is created in the second medium at close proximity of the interface. The evanescent field can only extend a few hundred nanometers in the z-direction, since the intensity decreases exponentially with the distance from the interface. TIRF microscopy (also called evanescent wave microscopy) exploits this principle to excite fluorophores that are in close vicinity of the specimen-glass interface [28]. The excitable region is typically restricted to a depth of less than 100 nm. To use the total internal reflection effect, TIRF setups provide laser excitation at an angle greater than the critical angle (Fig. 2).

As the physical principle suggests, the main characteristic of TIRF is restricting the fluorescence emission only to a very shallow region (≤100 nm) close to the coverslip. This greatly reduces the background and provides a high signal-to-noise ratio. Although the lateral resolution is not altered, it is possible to achieve high axial resolution (typically ≤100 nm, but can be decreased down to 10 nm) on large fields of view without scanning. Thanks to the immense reduction in the background, high sensitivity, and minimized photobleaching, it is also a popular technique for single particle tracking approaches in the near-field [29]. TIRF is easily applied to live cell imaging and has been commonly employed for studies of dynamics near the surface. For instance, the inward movement of single clathrin-coated pits accompanied by dynamin recruitment and actin assembly has been visualized using TIRF [30].

Improvements in the TIRF methodology have also been important for advancement of other super-resolution techniques that rely on single molecule detection. By making use of the TIRF technique, it is possible to collect photons from single emitters, as the detection area is highly restricted. Provided that a sufficient number of photons are collected from single molecules, it is possible to find the localization of each molecule with very high precision by determining the centers of their PSFs. Thus, sub-diffraction localization information can be obtained from a diffraction-limited imaging process. The first of these TIRF-based single-molecule localization methods is "fluorescence imaging with one-nanometer accuracy" (FIONA [31]). Among its applications is the achievement of 10 nm axial resolution in living cells using TIRF and single-molecule localization for visualization of clathrin and its adaptor AP-2 in clathrin-coated pits [32]. The motion of secretory granules and the variations in their plasma membrane attachment have been visualized by TIRF as well, with detection of very small (<20 nm) axial displacements [33].

Although FIONA and its derivatives provide localization information with very high accuracy, they can only be used when the fluorophores are distant enough from each other. This is an important limitation for the wide-range biological applicability of these techniques. In this regard, one step forward was taken by incorporation of photoswitching or photoconversion of fluorophores, so that it is possible to detect individual fluorophores even when multiple emitters overlap. Super-resolution techniques that use this principle are explained below in more detail.

2 Far-Field Techniques

Despite the impressive resolution attainable by near-field techniques, the imaging power is limited to only a very thin section at (or close to) the specimen surface. Unfortunately, most cellular processes take place deeper in the cells. Far-field microscopy techniques enable detection at much larger distances than the wavelength of light, making them more easily applicable to biological questions.

2.1 Confocal Laser Scanning Microscopy

The confocal microscope has been invented and patented by Marvin Minsky in 1957 [34]. Although the main principles of his original design are still valid, several technological advancements have been incorporated into it over the decades. A modern confocal system consists of lasers for point-like and strong illumination of specimen, an illumination pinhole and an objective lens for obtaining a point-like focused scanning laser beam (point-like focal spot) and a detection pinhole for formation of a point-like image on a sensitive electronic photon detector. In this system, the point of focus on the object and the image are optically aligned to each other (conjugated together) along the path of light, hence the origin of the name "confocal" (having the same focus).

The main achievement of confocal microscopy is the elimination of the stray light and out-of-focus blur as the pinhole blocks the light emitted from the areas outside the focus. This gives clearer images and higher contrast, as well as the possibility to create optical sections. In a confocal microscope the contrast cut-off distance (the distance at which the contrast becomes zero and the intensity peaks cannot be distinguished) is smaller than for wide-field [35]. At the level of PSF, the higher contrast of the confocal microscope originates from a decrease in the height of the first diffraction ring in comparison to the central disk. It is especially useful for better separation of dim objects that are further away than the diffraction limit [36].

Theoretically, it is possible to achieve resolution beyond the diffraction limit in fluorescence imaging, by decreasing the pinhole size significantly lower than the Airy disk size. The attainable lateral and axial resolution improvement lies in the range of $\sqrt{2}$ *fold* [36, 37]. However, this is not very practical for the imaging of most biological specimens, as it causes a substantial decrease in signal-to-noise ratio due to the massive rejection of the emission signal by the small pinhole. Therefore, the lateral resolution improvement obtained with a confocal laser scanning microscopy (CLSM) is usually little or none, although the gain in contrast gives rise to crisper images. Nevertheless, the concept of the scanning confocal microscope has been used as a starting point for many high-resolution techniques.

2.2 Super-Resolution Microscopy Techniques

The techniques employed to attain high-resolution in the far-field differ in principle of operation, instrumental design, sample requirement and imaging capabilities. It is possible to classify them based on different aspects. Here, a basic classification is made according the main resolution-improvement strategies.

One general method to attain sub-diffraction resolution is to engineer/modulate the PSF. Techniques such as stimulated emission depletion (STED) microscopy [38] and structured illumination microscopy (SIM [39, 40]) modulate the geometry of the excitation beam by using non-linear optical effects and high-intensity lasers.

Another general method exploits the photoswitching/photoconversion properties of the fluorescent probes and spot localization at high-precision in the absence of nearby diffracting spots. This method led to a set of super-resolution techniques known as single-molecule localization methods, including stochastic optical reconstruction microscopy (STORM), photoactivation localization microscopy (PALM), fluorescence photoactivation localization microscopy (FPALM) [41–43].

A third strategy is to reconfigure the conventional objective geometry and light collection. The resolution improvement is achieved by increasing the fraction of the wavefront collected by the objective. 4Pi microscopy [44] and I^5M [45] are the techniques that successfully apply this method to reach impressive levels of resolution especially in the axial dimension. In many cases, newer techniques arise by a modification or combination of these three concepts.

2.2.1 4Pi Microscopy

Most of the current super-resolution techniques (as explained below) give a higher priority to enhancing the lateral resolution. In contrast, 4Pi microscopy aims to increase the axial resolution, which is conventionally much lower than the lateral counterpart. For a single lens, the angle at which the spherical wavefront is collected is theoretically limited to 2π (where a full sphere encompasses an angle of 4π). However, in reality this requires a 90° semi-aperture angle, which is not attainable. By using two opposing lenses, this angle approaches the limit of 4π (hence the name 4Pi). If a spherical focal plane can be employed to collect the spherical wavefront, the image would not be as stretched as obtained with a planar focus plane [44]. 4Pi microscopy increases the aperture angle by placing two objectives opposite to each other, so that they have a common focus and the spherical segments of the wavefront can be collected from both sides, and interfered constructively. This gives rise to a 3–7-fold narrower main peak in the axial direction of the PSF. Thus axial resolution can be increased to 80–160 nm [46]. The lateral resolution however, stays the same (although crisper image details could be obtained, as the optical sections are thinner than in a conventional confocal microscope). Lateral imaging is performed through scanning. The specimen is generally placed in between two coverslips to maintain optical accessibility from both sides.

4Pi setups can differ in design depending on which beams are interfered [44]. Type A microscopes exploits only the interference of the excitation light and the emerging fluorescence is collected by a single lens; whereas Type B use only the interference of the emission light for coherent detection of the fluorescence. A third variant, Type C (Fig. 3a, b) combines A and B using two lenses both for excitation and emission [49].

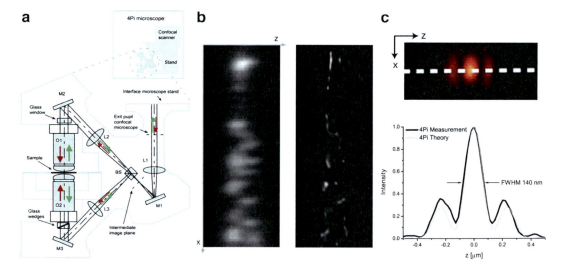

Fig. 3 4Pi microscopy. (**a**) Sketch of a 4Pi microscope (type C). Excitation light coming from the microscope stand is divided into two by the beam splitter (BS) and both beams are focused onto the same sample spot by the opposing objective lenses O1 and O2. The intermediate optical system consisting of the lenses (L1, L2, and L3) and the mirrors (M1, M2, and M3) maintain the beam paths for point scanning with two objective lenses. Fluorescence collected by both lenses meet at BS, and led back to the microscope stand. The movable glass wedges and the glass window are used to adapt the system for dispersion of light at a range of wavelengths. A standard confocal microscope unit is used for scanning. Reprinted with permission from [46]. (**b**) Confocal (*left*) and restored 4Pi-confocal (*right*) XZ-images of microtubules in a cell. The 4Pi image has an axial resolution in the 100-nm range. Scale bar is 400 nm. Reprinted with permission from [47]. (**c**) XZ-image of a 100 nm fluorescent bead taken by a type A 4Pi-confocal microscope. The line profile along the dashed white line displays side lobes of 35 %, which can be removed through deconvolution. The measured FWHM is very close to the theoretical value. Reprinted with permission from [48]

Interference of the two equal wavefronts of opposite directions generates a standing wave that does not propagate and stays in a constant position. This kind of wave results in the formation of two smaller secondary maxima next to the main PSF peak (more explicitly below and above according to the specimen plane). These secondary interference maxima are called "side lobes" and can cause ghost images if not removed properly (Fig. 3c). To remove the side lobes, a deconvolution procedure has to be applied (deconvolution is the mathematical reversion of the optical distortion that takes place in a microscope, in order to improve the image quality). The higher these side lobes are, the harder it gets to remove them by deconvolution. As a general rule, if they are higher than 50 % of the main PSF peak, it is not possible to get unambiguous data [48]. The height of the side lobes decreases with increasing semi-aperture angle. With most available immersion lenses the semi-aperture angle would be below 68°, giving primary side lobes higher than 60 % of the main peak [50]. Therefore, in order to suppress the side lobes, other optical modifications are generally used, including addition of a confocal pinhole to decrease the fluorescence detection from the side lobes or using two-photon excitation [51], which decreases the height of side lobes relative to the main peak. Two-photon excitation also enhances the effect of the confocal pinhole, as longer excitation wavelengths move the side lobes away from the focal plane [44]. In general Type C setups are better at lowering the side lobes. The best results are therefore obtained by using a confocal Type C 4Pi microscope with two-photon excitation [46, 49].

4Pi microscopy has been used for mammalian cells to visualize in dual-color the Golgi stacks [48] and nuclear pore complex components [52] with 110–130 nm axial resolution. For live imaging, there are a few compromises. First, the specimen needs to be mounted in aqueous medium, which creates a refractive index mismatch between the mounting and the immersion medium used for high numerical aperture oil-immersion lenses. This mismatch deforms the spherical detection of the wavefronts and causes higher side lobes. Water-immersion lenses need to be used to prevent the mismatch. However, these lenses have smaller aperture angles compared to oil-immersion lenses, consequently enhancing the side lobes [53]. To account for that, it is necessary to apply multi-photon excitation. Second, as in some other super-resolution techniques, the focal volume is reduced to improve the resolution. This gives rise to a reduction in total fluorescence and a decrease in signal-to-noise ratio. Therefore, the recording time should typically be increased to reduce the noise, in order to compensate for the decreased volume. To counterweigh this problem 4Pi setups with parallelized recording from multiple foci have been designed. This design is called "multifocal multi-photon microscope" (MMM-4Pi [54, 55]). This modification, together with confocal pinholes and image restoration, has been useful to obtain 3D images of the mitochondrial network in living yeast at an equilateral resolution of ~100 nm [56]. Live mammalian cells have also been imaged with a similar 3D resolution, using a GFP-labeled Golgi marker [57].

In addition to imaging at high resolution, the shape of the 4Pi PSF has offered a new tool to quantify the thickness of cellular structures (varying between 75 and 500 nm) with high precision. For this quantification the ratio of the average value of the two primary minima to the average value of the two side maxima is calculated. The smaller the ratio (the higher the side lobes relative to the primary minima), the thinner is the object [53]. This measurement was used to quantify the average diameter of the tubules of the mitochondrial network [56].

In order to reach super-resolution in 3D, 4Pi has also been employed in combination with super-resolution techniques of high lateral performance such as STED microscopy [58], as will be detailed below.

2.2.2 I⁵M

I⁵M is technically related to 4Pi. It is a combination of two kinds of interference-based approaches, image interference microscopy (I²M) and incoherent interference illumination (I³) [59]. I² and I³M can be said to be wide-field counterparts of type B and type A confocal 4Pi microscopy, respectively. In the I²M mode, emission beams are collected from two opposing objective lenses and after traveling over different paths with equal distances they are superimposed on one CCD camera. The image is formed through interference of the two fluorescent beams. In the I³ mode, the sample is illuminated from both sides using an incoherent light source such as a standard arc lamp, in contrast to a laser as in 4Pi. The illumination light is polarized perpendicular to the axial direction, so light beams coming from opposite directions interfere at the focal plane creating standing waves [53] that excite a narrower area. Emitted fluorescence is detected by a CCD camera. By combining I²M and I³, I⁵M attains high axial resolution through interference in both excitation and emission detection, in a similar fashion to type C confocal 4Pi microscopy, but operating in the wide-field context with single-photon illumination [60].

The FWHM of the I⁵M PSF is typically comparable to that obtained by 4Pi, ~7 times narrower than conventional wide-field microscopy. However, compared to 4Pi, the side lobes are more pronounced in I⁵M. On the other hand, I⁵M is faster and has a lower-cost, and can obtain more photons as single-photon excitation is used. Still, the difficulty of removing comparatively higher side lobes has been limiting widespread biological applications of I⁵M.

The compromises for live 4Pi imaging are also valid for I⁵M; however, the generally elevated side lobes in I⁵M cannot be as easily compensated. Therefore, so far significant resolution improvement by I⁵M has only been demonstrated with oil immersion lenses, limiting its use to fixed samples [53]. Under favorable conditions, better than 100 nm axial resolution has been experimentally verified imaging filamentous actin in fixed mammalian cells [45].

Like 4Pi, for 3D resolution improvement, I⁵M has also been combined with other super-resolution techniques, for example, SIM (I⁵S [61]), giving rise to ~100 nm resolution in all dimensions.

2.2.3 Stimulated Emission Depletion Microscopy

STED microscopy exploits the physical phenomenon known as "stimulated emission" to modulate the effective PSF. If an electron in excited state encounters a photon at an energy level similar to the difference between the electron's energy and the ground energy, the photon will stimulate the electron to fall into the ground energy state. This is realized through emission of a photon with the same wavelength as the incoming one. The excited molecule is therefore not allowed to undergo spontaneous emission of a fluorescence photon, but is subject to STED and will emit a photon at the wavelength of the light used for the STED effect. These more red-shifted photons can be separated from the

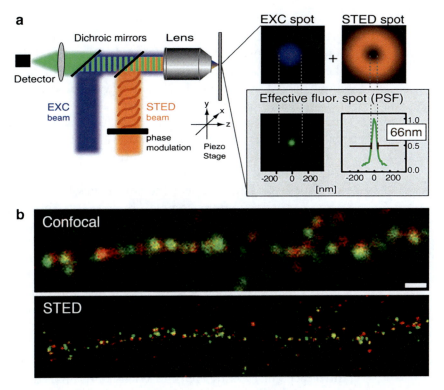

Fig. 4 STED microscopy. (**a**) Principle of operation. The excitation beam (*blue*) is superimposed with a doughnut-shaped depletion beam (*orange*), resulting in the de-excitation of all fluorophores except those located in the center of the depletion beam. This yields an effective fluorescent spot of sub-diffraction size (*green, below*). Reprinted with permission from [63]. (**b**) Synaptotagmin 1 (*red*) and synaptophysin (*green*) colocalize well on the plasma membrane of neurons, after vesicular fusion, as visualized by STED microscopy. Note the precision of the full or partial colocalizations in the two-color STED image over the confocal counterpart. Scale bar is 1 μm. Reprinted with permission from [64]

fluorescence emission by use of filters. Fluorescent dyes can thus be "de-excited" by additional irradiation with a red-shifted depletion beam: STED. The STED microscope functions like an improved confocal laser-scanning microscope, in which a conventional excitation beam is used to turn on the fluorophores in a diffraction-limited spot. The additional feature is the introduction of a second (STED) beam that is modulated by a phase plate into a doughnut-like shape. By alignment of the two beams, the emission of fluorescence is only allowed from the center of the excited spot where the depletion beam has zero intensity, so a much smaller focal spot can be obtained [38, 62], resulting in images with substantially higher resolution (Fig. 4). The image is generated through scanning of the chosen area. The obtained improvement in resolution is theoretically not limited by a physical barrier. In principle if the intensity of the de-excitation beam is high enough, the focal spot can be decreased to molecular size [65].

Since STED was the first of the many following diffraction-limit breaking strategies, and has been used to address biological questions for over a decade, it is worthwhile to go briefly through the key studies that contributed to the development of the technique. The development timeline of STED microscopy, as detailed below, is a good example of the typical steps by which a super-resolution technique emerges in biology. New techniques are put forward first by demonstration of the use of an optical/physical phenomenon to obtain high-resolution with a proof-of-principle study, followed by application to a simple, fixed biological specimen such as bacteria. Then, the advantages of the technique are demonstrated on more complex samples, to answer previously unresolvable biological questions. Next, the technique is adapted to the needs of biology by making implementations enabling multicolor imaging, reduced sample damage, live imaging, high temporal resolution, broader fluorophore selection, further-improved or 3D super-resolution, as well as easier operation and lower costs. In the course of development, the commercialization of the specialized equipment is also important for the new technique to become widely used.

The theoretical principles of STED microscopy were first introduced in 1994 by Hell and Wichmann [38]. Later in a proof-of-principle study the proposed setup, which featured an axially modulated PSF, was used on living yeast and *E. coli* cells [62]. A fluorescent spot of 18-times smaller focal volume was obtained. This was followed by implementation of STED in a 4Pi fashion, giving rise to an axial resolution of 33 nm. The focal spot size of $\lambda/23$ was the lowest obtained in far-field microscopy by 2002 [66]. Until this point, solely membrane dyes were used for labeling. Since the resolution is highly dependent on the high excitation intensity it was questionable if similar sub-diffraction resolution could be obtained with other types of labeling. Therefore the next important study applied STED to a conventional immunofluorescence preparation and microtubules of mammalian cells were imaged at an axial resolution of ~50 nm, with a 4Pi–STED setup. The resolution improvement obtained with STED was until this time point largely axial (along the optical axis) due to radial polarization of the STED beam and was also accompanied with higher side lobes, which required extra image processing. In a 2005 study, the phase plate that modulates the STED beam was modified into a doughnut-shaped composite to obtain sub-diffraction resolution in the focal plane, yielding a lateral resolution of 16 nm [67]. With this version of the setup it was experimentally proven that the resolving power increases with the square root of the saturation level of STED. Therefore, for calculation of resolution in such microscopes the Abbe equation [68] is replaced with the following:

$$\Delta r \approx \frac{\lambda}{2\sin\alpha \sqrt{1+\frac{I}{I_{sat}}}} \qquad (1)$$

where λ and α denote the wavelength and aperture angle of the lens, respectively. I is the maximal focal intensity applied for STED and I_{sat} is a characteristic value at which the fluorescence excitation is reduced to half [63, 67]. Here it could be seen that when I/I_{sat} is increased toward infinity, resolution (Δr) approximates to zero; hence resolution is now independent of diffraction.

After this point, STED was applied to biological questions. One of the best-known applications was performed to answer a neuroscience question: do the components of synaptic vesicles diffuse on the plasma membrane or do they remain together after exocytosis? Synaptic vesicles are ~40 nm in diameter and are densely packed with proteins making them or their constituents very hard to resolve. Using STED, it was possible to show that the vesicle membrane protein Synaptotagmin 1 is found in patches of 70 nm on the plasma membrane after fusion of the vesicle [63]. In another study STED was used to visualize the active zone protein Bruchpilot in Drosophila synapses and neuromuscular junctions. Surprisingly, the protein was found to form ring-shaped structures, which could not be resolved with confocal microscopy [69].

In 2006, fluorescent proteins (following membrane dyes and organic-dye conjugated antibodies) were shown to be compatible with STED [70]. These studies continued with introduction of two-color STED, which enabled simultaneous super-resolution imaging in multiple colors. For that it was necessary to use two pulsed laser diodes for excitation at different wavelengths and two amplifiers for two STED beams of different wavelengths. A lateral resolution of 30 and 65 nm was obtained for fluorescent beads imaged in the green and the red color channels, respectively [71]. 3D imaging with the STED technique was also made possible by introduction of isoSTED [58]. This setup used two different STED beams of the same wavelength orthogonal to each other so that the STED effect is obtained both laterally and axially. It also had a 4Pi-like organization, using two opposing lenses to collect more of the wavefront. As a result, a spherical focal spot was obtained with ~40 nm resolution in each dimension. By making optical stacks, high-resolution 3D images of mitochondria (immunolabeled against Tom20) were obtained. The method was also used with the addition of a second color, displaying a homogeneously distributed mitochondrial matrix protein together with the outer membrane protein Tom20, which was found to form distinct clusters at the organelle's boundary. In parallel, a more economical and easy-to-use laser source, continuous wavelength (CW) laser, was used instead of the more expensive and complicated pulsed-laser [72]. This development increased the accessibility of the technique for more laboratories, as well as the speed of image acquisition.

STED was also applied to live, deep tissue imaging. Neurons were filled with YFP and the morphological changes of dendritic spines (with structures ≤40 nm) were imaged in time-lapse after stimulation [73]. The fact that the technique requires scanning makes it relatively slow. However, when small fields of view are taken, it is possible to make video-rate (28 frames per second) movies with a fast scanner and a fluorescent label with high photon yield. This was demonstrated by imaging synaptic vesicles in living neuronal preparations. It was possible to track these small organelles and identify different pools based on their speed [74, 75]. Another biologically useful effort was combining advantages of STED and electron microscopy in a correlative fashion to obtain specific protein localization information in the morphological context of the cell [76]. Of course, this kind of correlation microscopy requires the fixation and embedding conditions to be optimized to get the best of both techniques. Finally, STED was shown to be applicable to multicellular living organisms. Fluorescent proteins were expressed in neuronal cells of *C. elegans* and in the brain of an adult mouse [77, 78]. In both cases, axonal or dendritic structures in neurons were imaged with 110 and 67 nm resolution, respectively, for *C. elegans* and mouse brain.

As could be seen from all these applications, STED microscopy has undergone various modifications to satisfy the needs of biology. Every adaptation though, as one would expect, has the risk of complicating the instrumentation. Also economically, a STED setup is rather costly. Still, its key advantage is the fact that it does not require image processing and the super-resolution image is obtained right during acquisition. Fifty to 70 nm resolution is commonly reached, while it is possible to reach much higher resolution under optimized conditions. The fluorophore choice depends on the capabilities of the setup used, and is still limited to bright and rather photostable fluorophores whose spectral properties match that of the available lasers for a particular setup.

2.2.4 Structured Illumination Microscopy

SIM is a wide-field (non-scanning) technique. It is sometimes also called "poor man's confocal microscope," referring to its immense optical sectioning ability without the use of an expensive confocal system [79]. In a conventional SIM configuration, a periodic diffraction grating of known pattern is inserted in the illumination beam path and is projected onto the sample [39]. This causes the formation of interference patterns (referred to as "moiré fringes") in the emission (Fig. 5a). Moiré patterns are generated by multiplication of the initial pattern (of known periodicity) with the

Super-Resolution Microscopy: Principles, Techniques, and Applications 29

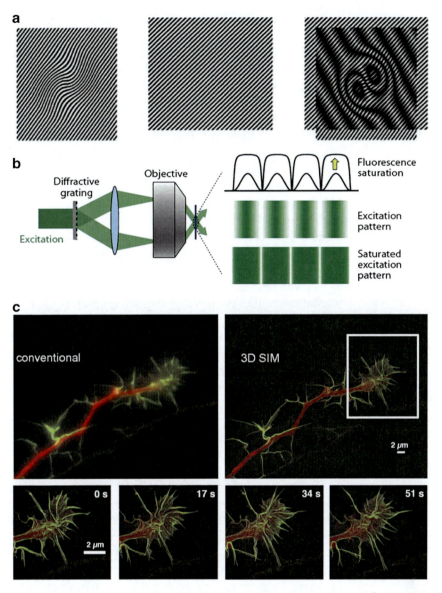

Fig. 5 Structured illumination microscopy. (**a**) Resolution extension through the moiré effect. When a known regular illumination pattern is applied to a sample, moiré fringes are generated at a significantly lower spatial frequency than that of the sample and imaged by the microscope. Multiple images that resulted from scanning and rotating the excitation pattern are then used to reconstruct the sample structure. Reprinted with permission from [40]. (**b**) The principle of SIM and SSIM. A diffractive grating placed in the excitation path causes the single light beam to split. The interference of these after passing the objective creates a pattern with alternating peaks and zero points. For SSIM, the strong excitation light creates sharper transitions, as it saturates the fluorescence emission at the peaks, while not changing the zero points. Reprinted with permission from [80]. (**c**) Live two-color 3D SIM imaging of cultured hippocampal neurons. To reveal the actin structure in the growth cone, actin was labeled with tdTomato-LifeAct (pseudocolored in *green*) and cytosol was labeled with GFP (shown in *red*). Live images were collected for 20 time points, with each time point acquired within 17 s. *Top panel* shows the maximum intensity projections (along the z-axis through the entire 3 μm volume thickness) for the conventional and SIM images for the first time point. Magnified images of the boxed region at different time points are shown in the *bottom panels*. Reprinted with permission from [81]

unknown sample periodicity. The resulting interference pattern can be used to back-calculate the spatial information from the sample. For that, it is necessary to collect a number of images by applying the illumination pattern in different orientations (a minimum of 3 per position). Because the moiré fringes are much coarser than both the original illumination pattern and the sample information, they are easily detected and can be used to gain spatial information that is normally unresolvable. In the case of patterned illumination, Abbe's limit for detection is extended by addition of the spatial frequencies in the pattern. Therefore, the resolution improvement is proportional to the spatial frequency of the illumination pattern. Here, the maximum spatial frequency is obtained by summing the highest spatial frequency in the patterned illumination and highest frequency detected [82]. Since the pattern itself is also diffraction limited, the lateral resolution increase by this summation is maximum twofold [83].

Nevertheless, by using optical nonlinearity near saturation the resolution could be further improved [84]. This is achieved by saturated structured illumination microscopy (SSIM [40]). When illumination is strong enough to excite most of the fluorophore molecules, additional increases in the illumination intensity will not give a linear increase in emission intensity (Fig. 5b). The nonlinearity is achieved near or above the saturation threshold for excitation. Using saturated illumination, a non-linear relationship between excitation and emission is obtained. This allows generation of moiré fringes with higher spatial frequencies than the initially applied illumination pattern and in turn results in a higher resolution [40].

SSIM theoretically offers unlimited resolution, as the resolution depends on the level of saturation [40]. However, in practice photobleaching of the dyes and decreasing signal-to-noise ratios are the limiting factors. For beads, resolution below 50 nm has been achieved [40]. An alternative to saturation (that employs high-laser power) is to use photoswitchable probes. Photoswitchable molecules can be reversibly switched between two different emission states by exposure to light of low intensities. Exploiting the photoswitching phenomenon, a non-linearity is created by saturation of the population of fluorophores in one or the other state. In an exemplar study, a photoswitchable protein, Dronpa, was used to get ~55 nm resolution in mammalian cells applying non-linear SIM with modest levels of excitation [85].

It is also possible to obtain high-resolution both in the lateral and axial direction using 3D SIM [86]. For example, multi-color 3D SIM has been applied to study the mammalian nucleus at 100 nm resolution in 3D and it was possible to resolve the chromatin, nuclear lamina and single nuclear pore complexes along the nuclear periphery [87]. Multicolor 3D SIM was also applied for live imaging of neurons (Fig. 5c) to visualize the complex actin structure in the neuronal growth cone [81]. Another 3D implementation, I⁵S [61], combines SIM with I⁵M to obtain ~100 nm resolution in three dimensions.

The most important advantage of SIM or SSIM is the ease of performing multi-color imaging (three or more) with any bright and photostable fluorophore. 3D high-resolution imaging is possible (although for SSIM this increases the photobleaching problem). Live imaging is generally restricted to slow-moving structures, because ideally the specimen should not move during the rotation of the grating. However, with the development of faster cameras or use of multifocal setups, it is possible to perform faster image acquisition for live imaging [81, 88]. In general, it is necessary to take ~10 raw images for SIM and ~100 for SSIM to computationally reconstruct a super-resolution image and heavy post-processing is required [82].

2.2.5 Pointillistic Techniques: Super-Resolution Imaging Using Single-Molecule Localization

To obtain high resolution, it is not always necessary to use sub-diffraction images. If a sufficient number of photons are collected, the position of a single fluorophore can be calculated with nanometer precision [89]. The issue is to evade the crowding that prevents getting signal from single emitters. In 2006, three different studies reported a new type of super-resolution microscopy strategy. Although the techniques were named differently as photoactivated localization microscopy (PALM [42]), STORM [41] and fluorescence photoactivation localization microscopy (FPALM [43]), they used essentially the same strategy: single-molecule switching to find the localization of single molecules with high precision. The strategy is based on photoactivation/photoswitching of fluorophores such that at one time point there would be only one fluorophore emitting per diffraction limited area, enabling calculation of the localization for each fluorophore. By repeating such imaging cycles many times to collect photons from all the fluorophores, an overall image can be reconstructed. Before imaging, the fluorophores are either in the default off-state (photoactivatable—PALM, FPALM) or a strong pulse is applied to switch them all off (photoswitchable—STORM). Then they are switched on/activated stochastically and imaged until they switch off or bleach. This is repeated until all the fluorophores are captured (Fig. 6a, b). The techniques based on this approach are collectively referred to as single-molecule localization microscopy or pointillistic microscopy [92].

The resolution, in this case, is independent of the wavelength and is dependent on the density and accuracy of the localization. In principle, the more photons detected from the molecule, the better is the localization precision. Essentially, the uncertainty in localization is inversely proportional to the square root of the number of photons detected [89, 93]. Approximately:

$$FWHM_{localization} \approx \frac{FWHM}{\sqrt{N}} \qquad (2)$$

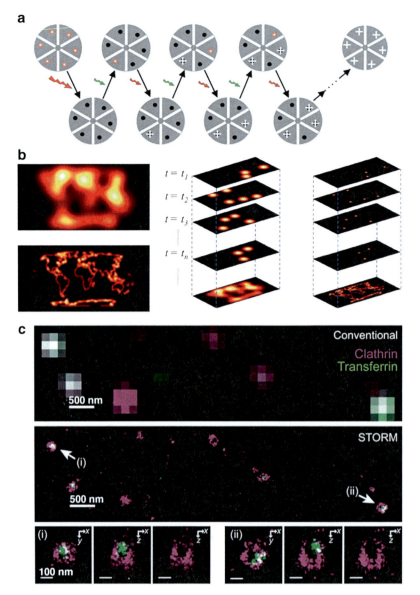

Fig. 6 Pointillistic imaging. (**a**) STORM with photo-switchable fluorophores. The scheme shows the imaging sequence for a hexameric object labeled with red fluorophores. The fluorophore can be switched between a fluorescent and a dark state by application of red and green laser, respectively. Initially, all fluorophores are switched off (the dark state) by a strong red laser pulse. Then, in each imaging cycle, a green laser pulse is applied to switch on just a fraction of the fluorophores so that an optically resolvable set of active fluorophores can be imaged. After being switched on, the emission is collected from the active fluorophores under red illumination. This allows their positions (*white crosses*) to be determined with high precision. The final image is generated by reconstruction of all the fluorophore positions collected through several imaging cycles. Reprinted with permission from [41]. (**b**) The principle of PALM. The sample is imaged by repeated cycles of activation, localization, and bleaching individual emitters to reconstruct a high-resolution picture. Reprinted with permission from [90]. (**c**) Live two-color 3D STORM with astigmatism-imaging of transferrin and clathrin-coated pits. SNAP-tagged clathrin was labeled with Alexa647 (*magenta*) and transferrin was directly labeled with Alexa568 (*green*). *Top panel* shows the conventional live image, and below is the *x–y* projection of the 3D STORM image (taken in 30 s). For the structures indicated by *arrows* (*i*) and (*ii*), *x–y* cross-sections near the plasma membrane (*left*), *x–z* cross-sections cut through the middle of the invaginating coated pit (*middle*) and corresponding *x–z* cross-section of the clathrin channel (*right*) are displayed at higher magnification in the *bottom panels*. Reprinted with permission from [91]

where FWHM$_{localization}$ is the localization precision, FWHM is obtained from Gaussian fitting of the PSF of the emitter and N is the number of photons captured from a single fluorescent molecule and varies from fluorophore to fluorophore. If, for example, 10,000 photons could be gathered in the active state of the fluorophore, it would be possible to reach 1–2 nm accuracy. One problem is the background and the residual emission of the photoactivatable/switchable fluorophores. Therefore, an important criterion for the fluorophore choice is the contrast ratio, which is the ratio of the emission after photoswitching/activation to the emission in the dark state. For this reason, in parallel to the technical developments in the pointillism imaging, many fluorescent probes are actively being characterized for their applicability for pointillism.

Initially, STORM has been demonstrated using a cyanine switch, which consists of a Cy3-Cy5 dye pair. Cy5, a commonly used cyanine dye, can be switched to a stable dark state by the same red laser light that is used for excitation of the dye. Green laser light is then used to convert Cy5 back to its fluorescent state. To increase the recovery rate, a Cy3 dye is placed in close proximity to Cy5 [41]. The fluorescent probes designed this way consist of a reporter that can be photo-switched on and off, and an activator that is required for the efficient photoactivation of the reporter. Using this pair it was possible to gather ~3,000 photons per switching cycle, predicting a theoretical localization accuracy of 4 nm. The primary cyanine switch has also been generalized to enable multi-color imaging. Variously paired switches can be combined for multi-color imaging. For example, three different activator molecules were paired with the same reporter, to visualize immobilized DNA molecules, microtubules and clathrin-coated pits in fixed mammalian cells with 25 nm resolution [94].

For PALM, photoswitchable/activatable variants of the fluorescent proteins, such as photoactivatable GFP (PAGFP [95]), have been the most commonly employed fluorophores. PALM has been used in probing protein clusters in the plasma membrane, by expressing PAGFP. For example, the nanoscale distribution and kinetics of influenza hemagglutinin clusters on fibroblast cell membranes were imaged at 40 nm resolution in living cells, using PAGFP [96]. In another study pointillism has been employed to uncover the clustered distribution of the chemotaxis receptors tagged with a different photoactivatable fluorescent protein, tandem-dimer Eos (tdEos), on the cell membrane of *E. coli* [97]. Using localization microscopy techniques, it is also possible to perform single-particle tracking with high spatial precision in living cells. For example, a highly spatially resolved maps of single-molecule motions have been created by imaging the membrane proteins Gag and VSVG, through labeling with EosFP [98].

Localization microscopy has also been modified for 3D high-resolution imaging. For instance, by insertion of a cylindrical lens in the imaging path, the information obtained from each fluorophore image can be increased. The centroid still gives the lateral positioning, whereas the shape or the ellipticity becomes a measure of the distance from the focal plane [99]. As a result, using the astigmatism of the image, the axial position can be determined. By this additional functionality, clathrin-coated pits in fixed cells have been imaged with ~25 nm lateral and ~55 nm axial resolution [100]. 3D STORM with astigmatism was also applied in live, two-color fashion yielding <25 nm lateral and ~60 nm axial resolution [91]. This way it was possible to resolve the morphology of the clathrin coat enclosing the transferrin cargo (Fig. 6c). Recently, dual-objective astigmatism-based 3D STORM has revealed the surprising actin ultrastructure in fixed neurons [101].

3D PALM has also been demonstrated through interferometric photoactivated localization microscopy (iPALM), which uses a dual lens system as in I^{5}M and collects the phase information through interference of photons coming from single fluorescent molecules and traveling different light paths. Employing iPALM, microtubules have been imaged with 25 nm 3D resolution [102].

Pointillistic approaches make the instrumentation relatively simple and can yield 10–40 nm resolution. Instead of scanning the sample, 100–100,000 frames are collected by a camera for one image. This means, on the other hand, that imaging is slower and post-processing is required as single emitters should be extracted from many frames and their localizations should be obtained computationally to reconstruct the final super-resolution image. Development of camera systems with single-photon sensitivity (electron multiplying charge-coupled devices, EMCCD) has been of great use for pointillistic methods, especially in terms of improving the time resolution.

The other requirement for pointillism is the necessity to use photoswitchable/activatable fluorophores. Theoretically, it is also possible to exploit the spontaneous blinking behavior of many fluorophores, including many of the Alexa and cyanine dyes. However, most of the time it is necessary to use oxygen-free redox buffers including reagents such as mercaptoethanol, glutathione or other oxygen scavengers to enhance blinking. Nevertheless, the repository for such fluorescent proteins and organic dyes is growing. Recently developed caged fluorophores (bearing a photocleavable cage moiety that can turn the fluorophore non-fluorescent) enable photoswitching without use of high concentrations of oxygen scavengers [103, 104].

3 A Comparison of the Super-Resolution Techniques

The super-resolution techniques presented above all come with specific advantages and disadvantages. It is hard to point out the best technique for super-resolution imaging, as the answer would be very much dependent on the particular application and the experimental requirements. Table 1 summarizes and compares the

Table 1
Comparison of far-field techniques

Technique		CLSM	STED	(S)SIM	PALM/STORM/FPALM
X, Y-resolution	Typical	200–250 nm	40–80 nm	100–130 nm	20–50 nm
	Best	–	10 nm	50 nm	10 nm
Z-resolution	Typical	500–700 nm	500–700 nm	250–300 nm	100 nm (TIRF implementation)
	Best	–	20–100 nm	100 nm	10–25 nm
Detection		Scanning	Scanning	Wide-field	Wide-field
Fluorophore choice		Broad	Restricted (dyes with appropriate STED characteristics)	Broad	Restricted (preferentially photoswitchable/activatable)
Live cell imaging		Yes	Restricted	Restricted	Restricted
Speed of imaging	Typical	Milliseconds to seconds	Seconds to minute	Seconds to minute	Seconds to minute
	Best	Milliseconds	Milliseconds	Milliseconds	Seconds
Multicolor imaging	Typical	2–3	1	1	1–2
	Best	>3	2	≥3	≥3
Data processing requirement		No	No	Yes	Yes
Complexity of the hardware		Medium	High	Medium to high	Low to medium
Bleaching		Medium	High	Medium to high	Medium to high
3D implementation		4Pi	isoSTED	3D SIM I^5S	3D STORM with cylindrical lenses iPALM

capabilities of the most commonly used far-field techniques. However, it should also be kept in mind that this table can change significantly over time, as these techniques are continuously being improved with new studies coming forward almost at a weekly basis. New results at the nanoscale also come with new challenges. The higher imaging precision gained by super-resolution is showing us not just prettier pictures, but also higher-complexity images that require a perspective change in terms of interpretation of new observations and re-thinking of the pre-existing ones [105, 106]. For example, "colocalization" experiments in super-resolution studies are presenting an improved precision and higher detail. Information obtained with some of the conventional imaging studies is becoming inadequate at the nanoscale, creating the necessity for more descriptive research at high-resolution. Previously unexpected observations, such as nanoscale domains of various kinds, are becoming apparent and new functional hypotheses have been pushed forward [106]. On the other hand, new schemes of data analysis and processing are being established to obtain more information from the newly arising and increasingly more complex data.

The improvements in imaging techniques also require methodological changes in sample preparation and labeling. As we reach higher resolution, the sizes of the fluorescent probes start to matter as much as their optical properties [107, 108]. Antibodies or fluorescent proteins are slowly becoming too big, and smaller probes and tags with minor target modifications are being developed.

Overall, with the new array of super-resolution microscopes and the methodologies developed in parallel, we are coming a bit closer to seeing and understanding the important cellular processes that have been waiting to be explored, just beneath our resolving power.

References

1. Synge EH (1928) XXXVIII. A suggested method for extending microscopic resolution into the ultra-microscopic region. Philos Mag 6:356–362
2. Ash EA, Nicholls G (1972) Super-resolution aperture scanning microscope. Nature 237:510–512
3. Lewis A, Isaacson M, Harootunian A, Muray A (1984) Development of a 500 Å spatial resolution light microscope. Ultramicroscopy 13:227–231. doi:10.1016/0304-3991(84)90201-8
4. Pohl DW (1994) Near-field optics: light for the world of NANO. J Vac Sci Technol B 12:1441. doi:10.1116/1.587313
5. Durig U, Pohl DW, Rohner F (1986) Near-field optical-scanning microscopy. J Appl Phys 59:3318. doi:10.1063/1.336848
6. Betzig E, Trautman JK (1992) Near-field optics: microscopy, spectroscopy, and surface modification beyond the diffraction limit. Science 257:189–195. doi:10.1126/science.257.5067.189
7. Gerton J, Wade L, Lessard G et al (2004) Tip-enhanced fluorescence microscopy at 10 nanometer resolution. Phys Rev Lett 93:180801. doi:10.1103/PhysRevLett.93.180801
8. Lessard G (2003) Apertureless near-field optical microscopy for fluorescence imaging. Dissertation, California Institute of Technology, Pasadena, CA
9. Hausmann M, Perner B, Rapp A et al (2006) Near-field scanning optical microscopy in cell biology and cytogenetics. Methods Mol Biol 319: 275–294. doi:10.1007/978-1-59259-993-6_14

10. Gheber LA, Hwang J, Edidin M (1998) Design and optimization of a near-field scanning optical microscope for imaging biological samples in liquid. Appl Optics 37:3574–3581
11. Hwang J, Gheber LA, Margolis L, Edidin M (1998) Domains in cell plasma membranes investigated by near-field scanning optical microscopy. Biophys J 74:2184–2190. doi:10.1016/S0006-3495(98)77927-5
12. Chen Y, Shao L, Ali Z et al (2008) NSOM/QD-based nanoscale immunofluorescence imaging of antigen-specific T-cell receptor responses during an in vivo clonal Vγ2Vδ2 T-cell expansion. Blood 111:4220–4232. doi:10.1182/blood-2007-07-101691
13. Binnig G, Quate C, Gerber C (1986) Atomic force microscope. Phys Rev Lett 56:930–933. doi:10.1103/PhysRevLett.56.930
14. Müller DJ, Dufrêne YF (2011) Atomic force microscopy: a nanoscopic window on the cell surface. Trends Cell Biol 21:1–9. doi:10.1016/j.tcb.2011.04.008
15. Rugar D, Hansma P (1990) Atomic force microscopy. Phys Today 43:23–30
16. Martin Y, Williams CC, Wickramasinghe HK (1987) Atomic force microscope—force mapping and profiling on a sub 100-Å scale. J Appl Phys 61:4723. doi:10.1063/1.338807
17. Zhong Q, Inniss D, Kjoller K, Elings VB (1993) Fractured polymer/silica fiber surface studied by tapping mode atomic force microscopy. Surf Sci Lett 290:L688–L692. doi:10.1016/0167-2584(93)90906-Y
18. Huang L, Su C (2004) A torsional resonance mode AFM for in-plane tip surface interactions. Ultramicroscopy 100:277–285. doi:10.1016/j.ultramic.2003.11.010
19. Schaap IAT, Carrasco C, de Pablo PJ, Schmidt CF (2011) Kinesin walks the line: single motors observed by atomic force microscopy. Biophys J 100:2450–2456. doi:10.1016/j.bpj.2011.04.015
20. Sharma S, Grintsevich EE, Phillips ML et al (2011) Atomic force microscopy reveals drebrin induced remodeling of f-actin with subnanometer resolution. Nano Lett 11:825–827. doi:10.1021/nl104159v
21. Dickenson NE, Moore D, Suprenant KA, Dunn RC (2007) Vault ribonucleoprotein particles and the central mass of the nuclear pore complex. Photochem Photobiol 83:686–691. doi:10.1111/j.1751-1097.2007.00050.x
22. Quist A, Doudevski I, Lin H et al (2005) Amyloid ion channels: a common structural link for protein-misfolding disease. Proc Natl Acad Sci U S A 102:10427–10432. doi:10.1073/pnas.0502066102
23. Hinterdorfer P, Dufrêne YF (2006) Detection and localization of single molecular recognition events using atomic force microscopy. Nat Methods 3:347–355. doi:10.1038/nmeth871
24. Bergkvist M, Cady NC (2011) Chemical functionalization and bioconjugation strategies for atomic force microscope cantilevers. Methods Mol Biol 751:381–400. doi:10.1007/978-1-61779-151-2_24
25. Roduit C, van der Goot FG, De Los Rios P et al (2008) Elastic membrane heterogeneity of living cells revealed by stiff nanoscale membrane domains. Biophys J 94:1521–1532. doi:10.1529/biophysj.107.112862
26. Rief M, Gautel M, Oesterhelt F et al (1997) Reversible unfolding of individual titin immunoglobulin domains by AFM. Science 276:1109–1112. doi:10.1126/science.276.5315.1109
27. Linke WA, Grützner A (2007) Pulling single molecules of titin by AFM—recent advances and physiological implications. Pflugers Arch 456:101–115. doi:10.1007/s00424-007-0389-x
28. Axelrod D (1981) Cell-substrate contacts illuminated by total internal reflection fluorescence. J Cell Biol 89:141–145. doi:10.1083/jcb.89.1.141
29. Toomre D, Bewersdorf J (2010) A new wave of cellular imaging. Annu Rev Cell Dev Biol 26:285–314. doi:10.1146/annurev-cellbio-100109-104048
30. Merrifield CJ, Feldman ME, Wan L, Almers W (2002) Imaging actin and dynamin recruitment during invagination of single clathrin-coated pits. Nat Cell Biol 4:691–698. doi:10.1038/ncb837
31. Yildiz A (2003) Myosin V walks hand-over-hand: single fluorophore imaging with 1.5-nm localization. Science 300:2061–2065. doi:10.1126/science.1084398
32. Saffarian S, Kirchhausen T (2008) Differential evanescence nanometry: live-cell fluorescence measurements with 10-nm axial resolution on the plasma membrane. Biophys J 94:2333–2342. doi:10.1529/biophysj.107.117234
33. Karatekin E, Tran VS, Huet S et al (2008) A 20-nm step toward the cell membrane preceding exocytosis may correspond to docking of tethered granules. Biophys J 94:2891–2905. doi:10.1529/biophysj.107.116756
34. Minsky M (1961) Microscopy apparatus. US Patent 3,013,467A.
35. Stelzer EHK (1998) Contrast, resolution, pixelation, dynamic range and signal-to-noise ratio: fundamental limits to resolution in fluorescence light microscopy. J Microsc 189:15–24. doi:10.1046/j.1365-2818.1998.00290.x
36. Webb RH (1996) Confocal optical microscopy. Rep Prog Phys 59:427–471. doi:10.1088/0034-4885/59/3/003

37. Cox IJ, Sheppard CJ, Wilson T (1982) Improvement in resolution by nearly confocal microscopy. Appl Optics 21:778–781. doi:10.1364/AO.21.000778
38. Hell SW, Wichmann J (1994) Breaking the diffraction resolution limit by stimulated emission: stimulated-emission-depletion fluorescence microscopy. Opt Lett 19:780–782. doi:10.1364/OL.19.000780
39. Heintzmann R, Cremer CG (1999) Laterally modulated excitation microscopy: improvement of resolution by using a diffraction grating. In: Altshuler GB, Benaron DA, Ehrenberg B et al (eds) SPIE proceedings. SPIE, pp 185–196. doi:10.1117/12.336833
40. Gustafsson MGL (2005) Nonlinear structured-illumination microscopy: widefield fluorescence imaging with theoretically unlimited resolution. Proc Natl Acad Sci U S A 102:13081–13086. doi:10.1073/pnas.0406877102
41. Rust MJ, Bates M, Zhuang X (2006) Sub-diffraction-limit imaging by stochastic optical reconstruction microscopy (STORM). Nat Methods 3:793–796. doi:10.1038/nmeth929
42. Betzig E, Patterson GH, Sougrat R et al (2006) Imaging intracellular fluorescent proteins at nanometer resolution. Science 313:1642–1645. doi:10.1126/science.1127344
43. Hess ST, Girirajan TPK, Mason MD (2006) Ultra-high resolution imaging by fluorescence photoactivation localization microscopy. Biophys J 91:4258–4272. doi:10.1529/biophysj.106.091116
44. Hell S, Stelzer EHK (1992) Properties of a 4Pi confocal fluorescence microscope. J Opt Soc Am A 9:2159. doi:10.1364/JOSAA.9.002159
45. Gustafsson MG, Agard DA, Sedat JW (1999) I5M: 3D widefield light microscopy with better than 100 nm axial resolution. J Microsc 195:10–16. doi:10.1046/j.1365-2818.1999.00576.x
46. Gugel H, Bewersdorf J, Jakobs S et al (2004) Cooperative 4Pi excitation and detection yields sevenfold sharper optical sections in live-cell microscopy. Biophys J 87:4146–4152. doi:10.1529/biophysj.104.045815
47. Nagorni M, Hell SW (1998) 4Pi-confocal microscopy provides three-dimensional images of the microtubule network with 100- to 150-nm resolution. J Struct Biol 123:236–247. doi:10.1006/jsbi.1998.4037
48. Lang M, Müller T, Engelhardt J, Hell SW (2007) 4Pi microscopy of type A with 1-photon excitation in biological fluorescence imaging. Opt Express 15:2459–2467. doi:10.1364/OE.15.002459
49. Hell S (1992) Fundamental improvement of resolution with a 4Pi-confocal fluorescence microscope using two-photon excitation. Opt Commun 93:277–282
50. Bewersdorf J, Egner A, Hell SW (2006) 4Pi Microscopy. In: Pawley JB (ed) Handbook of biological confocal microscopy. Springer, Boston, MA, pp 561–570
51. Denk W, Strickler JH, Webb WW (1990) Two-photon laser scanning fluorescence microscopy. Science 248:73–76
52. Hüve J, Wesselmann R, Kahms M, Peters R (2008) 4Pi microscopy of the nuclear pore complex. Biophys J 95:877–885. doi:10.1529/biophysj.107.127449
53. Egner A, Hell SW (2005) Fluorescence microscopy with super-resolved optical sections. Trends Cell Biol 15:207–215. doi:10.1016/j.tcb.2005.02.003
54. Bewersdorf J, Pick R, Hell SW (1998) Multifocal multiphoton microscopy. Opt Lett 23:655–657. doi:10.1364/OL.23.000655
55. Egner A, Andresen V, Hell SW (2002) Comparison of the axial resolution of practical Nipkow-disk confocal fluorescence microscopy with that of multifocal multiphoton microscopy: theory and experiment. J Microsc 206:24–32. doi:10.1046/j.1365-2818.2002.01001.x
56. Egner A, Jakobs S, Hell SW (2002) Fast 100-nm resolution three-dimensional microscope reveals structural plasticity of mitochondria in live yeast. Proc Natl Acad Sci U S A 99:3370–3375. doi:10.1073/pnas.052545099
57. Egner A, Verrier S, Goroshkov A et al (2004) 4Pi-microscopy of the Golgi apparatus in live mammalian cells. J Struct Biol 147:70–76. doi:10.1016/j.jsb.2003.10.006
58. Schmidt R, Wurm CA, Jakobs S et al (2008) Spherical nanosized focal spot unravels the interior of cells. Nat Methods 5:539–544. doi:10.1038/nmeth.1214
59. Gustafsson MGL, Agard DA, Sedat JW (1995) Sevenfold improvement of axial resolution in 3D wide-field microscopy using two objective lenses. In: Wilson T, Cogswell CJ (eds) SPIE proceedings. SPIE, pp 147–156. doi:10.1117/12.205334
60. Gustafsson MG (1999) Extended resolution fluorescence microscopy. Curr Opin Struct Biol 9:627–634. doi:10.1016/S0959-440X(99)00016-0
61. Shao L, Isaac B, Uzawa S et al (2008) I5S: wide-field light microscopy with 100-nm-scale resolution in three dimensions. Biophys J 94:4971–4983. doi:10.1529/biophysj.107.120352
62. Klar TA, Jakobs S, Dyba M et al (2000) Fluorescence microscopy with diffraction resolution barrier broken by stimulated emission. Proc Natl Acad Sci U S A 97:8206–8210. doi:10.1073/pnas.97.15.8206

63. Willig KI, Rizzoli SO, Westphal V et al (2006) STED microscopy reveals that synaptotagmin remains clustered after synaptic vesicle exocytosis. Nature 440:935–939. doi:10.1038/nature04592
64. Hoopmann P, Punge A, Barysch SV et al (2010) Endosomal sorting of readily releasable synaptic vesicles. Proc Natl Acad Sci U S A 107:19055–19060. doi:10.1073/pnas.1007037107
65. Hell SW (2003) Toward fluorescence nanoscopy. Nat Biotechnol 21:1347–1355. doi:10.1038/nbt895
66. Dyba M, Hell S (2002) Focal spots of size lambda/23 open up far-field florescence microscopy at 33 nm axial resolution. Phys Rev Lett 88:1–4. doi:10.1103/PhysRevLett.88.163901
67. Westphal V, Hell SW (2005) Nanoscale resolution in the focal plane of an optical microscope. Phys Rev Lett 94:143903. doi:10.1103/PhysRevLett.94.143903
68. Abbe E (1873) Beiträge zur Theorie des Mikroskops und der mikroskopischen Wahrnehmung. Arch Mikrosk Anat 9:413–418. doi:10.1007/BF02956173
69. Kittel RJ, Wichmann C, Rasse TM et al (2006) Bruchpilot promotes active zone assembly, Ca2+ channel clustering, and vesicle release. Science 312:1051–1054. doi:10.1126/science.1126308
70. Willig KI, Kellner RR, Medda R et al (2006) Nanoscale resolution in GFP-based microscopy. Nat Methods 3:721–723. doi:10.1038/nmeth922
71. Donnert G, Keller J, Wurm CA et al (2007) Two-color far-field fluorescence nanoscopy. Biophys J 92:L67–L69. doi:10.1529/biophysj.107.104497
72. Willig KI, Harke B, Medda R, Hell SW (2007) STED microscopy with continuous wave beams. Nat Methods 4:915–918. doi:10.1038/nmeth1108
73. Nägerl UV, Willig KI, Hein B et al (2008) Live-cell imaging of dendritic spines by STED microscopy. Proc Natl Acad Sci U S A 105:18982–18987. doi:10.1073/pnas.0810028105
74. Westphal V, Rizzoli SO, Lauterbach MA et al (2008) Video-rate far-field optical nanoscopy dissects synaptic vesicle movement. Science 320:246–249. doi:10.1126/science.1154228
75. Kamin D, Lauterbach MA, Westphal V et al (2010) High- and low-mobility stages in the synaptic vesicle cycle. Biophys J 99:675–684. doi:10.1016/j.bpj.2010.04.054
76. Watanabe S, Punge A, Hollopeter G et al (2011) Protein localization in electron micrographs using fluorescence nanoscopy. Nat Methods 8:80–84. doi:10.1038/nmeth.1537
77. Rankin BR, Moneron G, Wurm CA et al (2011) Nanoscopy in a living multicellular organism expressing GFP. Biophys J 100:L63–L65. doi:10.1016/j.bpj.2011.05.020
78. Berning S, Willig KI, Steffens H et al (2012) Nanoscopy in a living mouse brain. Science 335:551. doi:10.1126/science.1215369
79. Lichtman JW, Conchello J-A (2005) Fluorescence microscopy. Nat Methods 2:910–919. doi:10.1038/nmeth817
80. Huang B, Bates M, Zhuang X (2009) Super-resolution fluorescence microscopy. Annu Rev Biochem 78:993–1016. doi:10.1146/annurev.biochem.77.061906.092014
81. Fiolka R, Shao L, Rego EH et al (2012) Time-lapse two-color 3D imaging of live cells with doubled resolution using structured illumination. Proc Natl Acad Sci U S A 109:5311–5315. doi:10.1073/pnas.1119262109
82. Schermelleh L, Heintzmann R, Leonhardt H (2010) A guide to super-resolution fluorescence microscopy. J Cell Biol 190:165–175. doi:10.1083/jcb.201002018
83. Gustafsson MG (2000) Surpassing the lateral resolution limit by a factor of two using structured illumination microscopy. J Microsc 198:82–87
84. Heintzmann R, Jovin TM, Cremer C (2002) Saturated patterned excitation microscopy—a concept for optical resolution improvement. J Opt Soc Am A 19:1599–1609. doi:10.1364/JOSAA.19.001599
85. Rego EH, Shao L, Macklin JJ et al (2012) Nonlinear structured-illumination microscopy with a photoswitchable protein reveals cellular structures at 50-nm resolution. Proc Natl Acad Sci U S A 109:E135–E143. doi:10.1073/pnas.1107547108
86. Gustafsson MGL, Shao L, Carlton PM et al (2008) Three-dimensional resolution doubling in wide-field fluorescence microscopy by structured illumination. Biophys J 94:4957–4970. doi:10.1529/biophysj.107.120345
87. Schermelleh L, Carlton PM, Haase S et al (2008) Subdiffraction multicolor imaging of the nuclear periphery with 3D structured illumination microscopy. Science 320:1332–1336. doi:10.1126/science.1156947
88. York AG, Parekh SH, Dalle Nogare D et al (2012) Resolution doubling in live, multicellular organisms via multifocal structured illumination microscopy. Nat Methods 9:749–754. doi:10.1038/nmeth.2025
89. Thompson RE, Larson DR, Webb WW (2002) Precise nanometer localization analysis for individual fluorescent probes.

Biophys J 82:2775–2783. doi:10.1016/S0006-3495(02)75618-X
90. Dedecker P, Hofkens J, Hotta J-I (2008) Diffraction-unlimited optical microscopy. Mater Today 11:12–21. doi:10.1016/S1369-7021(09)70003-3
91. Jones SA, Shim S-H, He J, Zhuang X (2011) Fast, three-dimensional super-resolution imaging of live cells. Nat Methods 8:499–508. doi:10.1038/nmeth.1605
92. Lidke K, Rieger B, Jovin T, Heintzmann R (2005) Superresolution by localization of quantum dots using blinking statistics. Opt Express 13:7052–7062. doi:10.1364/OPEX.13.007052
93. Bobroff N (1986) Position measurement with a resolution and noise-limited instrument. Rev Sci Instrum 57:1152. doi:10.1063/1.1138619
94. Bates M, Huang B, Dempsey GT, Zhuang X (2007) Multicolor super-resolution imaging with photo-switchable fluorescent probes. Science 317:1749–1753. doi:10.1126/science.1146598
95. Patterson GH, Lippincott-Schwartz J (2002) A photoactivatable GFP for selective photolabeling of proteins and cells. Science 297:1873–1877. doi:10.1126/science.1074952
96. Hess ST, Gould TJ, Gudheti MV et al (2007) Dynamic clustered distribution of hemagglutinin resolved at 40 nm in living cell membranes discriminates between raft theories. Proc Natl Acad Sci U S A 104:17370–17375. doi:10.1073/pnas.0708066104
97. Greenfield D, McEvoy AL, Shroff H et al (2009) Self-organization of the Escherichia coli chemotaxis network imaged with super-resolution light microscopy. PLoS Biol 7:e1000137. doi:10.1371/journal.pbio.1000137
98. Manley S, Gillette JM, Patterson GH et al (2008) High-density mapping of single-molecule trajectories with photoactivated localization microscopy. Nat Methods 5:155–157. doi:10.1038/nmeth.1176
99. Kao HP, Verkman AS (1994) Tracking of single fluorescent particles in three dimensions: use of cylindrical optics to encode particle position. Biophys J 67:1291–1300. doi:10.1016/S0006-3495(94)80601-0
100. Huang B, Wang W, Bates M, Zhuang X (2008) Three-dimensional super-resolution imaging by stochastic optical reconstruction microscopy. Science 319:810–813. doi:10.1126/science.1153529
101. Xu K, Zhong G, Zhuang X (2013) Actin, spectrin, and associated proteins form a periodic cytoskeletal structure in axons. Science 339:452–456. doi:10.1126/science.1232251
102. Shtengel G, Galbraith JA, Galbraith CG et al (2009) Interferometric fluorescent super-resolution microscopy resolves 3D cellular ultrastructure. Proc Natl Acad Sci U S A 106:3125–3130. doi:10.1073/pnas.0813131106
103. Banala S, Maurel D, Manley S, Johnsson K (2012) A caged, localizable rhodamine derivative for superresolution microscopy. ACS Chem Biol 7:289–293. doi:10.1021/cb2002889
104. Grimm JB, Sung AJ, Legant WR et al (2013) Carbofluoresceins and carborhodamines as scaffolds for high-contrast fluorogenic probes. ACS Chem Biol 8(6):1303–1310. doi:10.1021/cb4000822
105. Lang T, Rizzoli SO (2010) Membrane protein clusters at nanoscale resolution: more than pretty pictures. Physiology (Bethesda) 25:116–124. doi:10.1152/physiol.00044.2009
106. Saka S, Rizzoli SO (2012) Super-resolution imaging prompts re-thinking of cell biology mechanisms: selected cases using stimulated emission depletion microscopy. Bioessays 34:386–395. doi:10.1002/bies.201100080
107. Opazo F, Levy M, Byrom M et al (2012) Aptamers as potential tools for super-resolution microscopy. Nat Methods 9:938–939. doi:10.1038/nmeth.2179
108. Ries J, Kaplan C, Platonova E et al (2012) A simple, versatile method for GFP-based super-resolution microscopy via nanobodies. Nat Methods 9:582–584. doi:10.1038/nmeth.1991

Chapter 3

Foundations of STED Microscopy

Marcel A. Lauterbach and Christian Eggeling

Abstract

This chapter presents the foundations of STED microscopy with a comparison to its generalization RESOLFT microscopy and to stochastic imaging methods (PALM, STORM, FPALM, and alike).

The first section reviews the advantages of optical microscopy, explains the diffraction limit, and shows how the classical resolution limit was finally broken. It also reviews some of the achievements in super-resolution imaging.

The second section explains in depth the principle of STED microscopy and highlights some special STED modalities like the use of continuous wave lasers, time-gating, fast imaging with up to 200 frames per second, and combination with fluorescence correlation spectroscopy.

The third section treats some aspects of resolution in the presence of noise, especially in the scope of high-resolution imaging.

Key words STED Microscopy, Resolution, Nanoscopy, Super-Resolution Microscopy

1 Introduction to Sub-Diffraction Far-Field Microscopy

The key figures of merit of an optical system are its spatial and temporal resolution. Spatial resolution indicates the ability to resolve two objects as distinct entities. For microscopists, the spatial resolving power of their microscope dictates how fine are the details they can distinguish, and accordingly how much knowledge they can gain about the static or variable structure of their interest. This may be a fixed or living cell, a colloidal system, an integrated circuit or anything else. Just as the limited spatial resolving power restricts the size of the smallest detail that can be analyzed, a limited temporal resolution impedes the study of dynamical processes.

The lateral resolution in a microscope seemed ultimately limited by diffraction to about half the wavelength used, as was stated saliently by Ernst Abbe in 1873 [1], but early-on also by other scientists [2]. It is given by:

$$d_{\min} \approx \frac{\lambda}{2\,\text{NA}}, \qquad (1)$$

where $d_{\min}$ is the minimal resolvable distance, λ the wavelength (in air) of the light that is used for imaging, and $\text{NA} = n\,\sin\alpha$ the numerical aperture with α being the half aperture angle of the microscope's objective, and n the refractive index of the immersion medium.

The following sections will give an insight into the different ways of pushing the limits of the spatial resolution in microscopy and to go finally beyond the diffraction limit as pursued since the 1990s.

1.1 Imaging in the Far-Field

Imaging living cells or even tissue under physiological conditions requires an imaging system that is minimally invasive and able to gather information from deep inside the sample. Optical microscopy with visible light provides such low invasiveness: The use of optical lenses that create a focus several micrometers away from any optical lens allows studying in the far-field, i.e., without the optical system getting too close to the system under study. Far-field microscopy has provided tremendous new insights into various scientific problems—especially in combination with fluorescent staining. It has become the workhorse of many scientific labs [3], although the far-field approach and the concomitant focusing of visible light comprises a limit to the spatial resolution of the microscope. As outlined before, this resolution limit implies that objects or details of a sample that are too close together cannot be resolved as distinct entities (Fig. 1).

The diffraction limit (Eq. 1) amounts to a lateral resolution $d_{\min}$ of 200 nm for visible light (λ = 500 nm) and typical numerical apertures (NA = 1.2–1.4). Smaller details cannot be resolved. The axial resolution is with $\approx$ 600 nm even worse.

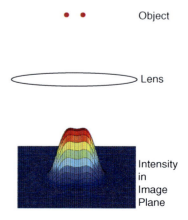

Fig. 1 The diffraction of light is inevitable in any (far-field) optical system and limits its resolution: Distinct objects cannot be separated if their diffraction patterns overlap too much

Many efforts are taken to acquire sharper images, i.e., to push the limits in spatial resolution imposed by diffraction. One approach is to abandon the advantageous visible light altogether because of its resolution limit. Reduction of the wavelength used for imaging is pursued in UV[1] [1] and X-ray microscopy [4, 5]. The higher resolution comes at the cost of complex optics and of difficulties of imaging living samples, which are harmed by the high-energy radiation.

Electron microscopy [6, 7] provides resolutions in the angstrom regime [8], using electrons with a very short de Broglie wavelength [9] for imaging. Only very thin samples (or surfaces) can be imaged and the specimen is placed in vacuum during imaging. Therefore imaging under physiological conditions is not straightforward.

Surface scanning techniques such as AFM[2] [10], STM[3] [11], SNOM[4] [12–14], TERS[5] [15–18], and SICM[6] [19] can only image surfaces. Some can reach molecular resolution on biological samples [20, 21], but not for observing the interior of cells.

It is far-field light microscopy which remains arguably the most important method for imaging in the biosciences due to the shortcomings of the alternative techniques. It is the only method that allows almost noninvasive imaging of thick samples under physiological conditions in three dimensions. Furthermore, there is not only a large array of specific fluorescence markers available for the study of countless cellular processes, but also many of them are compatible with imaging living samples. Immunostainings [22] are highly specific and regularly applied to fixed biological samples. Fluorescent proteins [23–25] like GFP[7] [26, 27] allow researchers to engineer organisms with self-staining properties. Labeling techniques such as FlAsH[8] [28], SnapTag[9] [29], and HaloTag[10] [30, 31] enable the use of organic dyes in living cells. Intracellular staining with quantum dots [32] is possible by microinjection [33]. However, its use as a cellular label has so far been mainly restricted to single isolated objects (such as used in single-particle tracking [34]), since dense labeling is problematic due to the relatively large size of the functionalized quantum dots. Moreover,

---

[1] Ultraviolet.
[2] Atomic Force Microscopy.
[3] Scanning Tunneling Microscopy.
[4] Scanning Near Field Optical Microscopy.
[5] Tip-Enhanced Raman Spectroscopy.
[6] Scanning Ion-Conductance Microscopy.
[7] Green Fluorescent Protein.
[8] Fluorescein Arsenical Helix Binder.
[9] Based on the use of the deoxyribonucleic acid repair protein alkyl guanine DNA alkyl transferase.
[10] Based on the use of a modified haloalkane dehalogenase.

functional studies are possible using dyes or proteins that change their fluorescence behavior depending on external parameters such as pH or calcium concentration [35, 36]. The option of studying dynamic processes provides an enormous field of applications in both biological [37] and synthetic samples (such as colloidal systems [38]).

A multitude of methods was therefore developed to push the limits in spatial resolution of far-field light microscopy as given by diffraction (for reviews, see, for example, [39–50]):

In a confocal microscope [51], the sample is scanned with a focused beam of light. The fluorescence is recorded with a point detector behind a pinhole. The pinhole rejects the out of focus light, thereby enabling axial sectioning. The lateral resolution may be increased by up to a factor of ≈ 1.4.

4Pi-microscopy [52, 53] and I^{2}M[11] [54, 55] increase the effective total aperture of the lens system by the use of two opposing objective lenses, thereby enhancing especially the axial resolution. TIRF microscopy[12] [56, 57] enables depth discrimination by illuminating only a very thin layer of the sample that is adjacent to the coverslip via an evanescent field of light. It remains limited to imaging structures next to the cover slip, though.

Hyperlenses [58], consisting of materials with a negative refractive index [59], not only promise to enhance the resolution by recovering the evanescent field [60]; but they also require close proximity to the sample, even if their image can be magnified in the far-field [61].

Various kinds of SIM[13] [50], sometimes combined with TIRF [62, 63] or with the two-lens approach, including I^{3}M[14] [54, 55], I^{5}M[15] [64], I^{5}S[16] [65], SWFM[17] [66], and HELM[18] [67] use nonuniform illumination of the specimen to extract high spatial frequencies. While the previous nonuniform (or structured-illumination) approaches usually work in the wide-field mode and are restricted to imaging optically rather thin samples, recent approaches for pushing the resolution limit based on a nonuniform illumination on a laser (point-) scanning device were proposed or even realized with a temporal modulation either of the excitation (SPIN[19] microscopy) or of the detection (SPADE[20] microscopy) [68] or via a camera-based

---

[11] Image Interference Microscopy.
[12] Total Internal Reflection Fluorescence Microscopy.
[13] Structured-Illumination Microscopy.
[14] Incoherent Interference Illumination Microscopy.
[15] The combination of I^{2}M and I^{3}M.
[16] A combination of I^{5}M with laterally structured illumination.
[17] Standing Wave Fluorescence Microscopy.
[18] Harmonic Excitation Light Microscopy.
[19] Scanning Patterned Illumination.
[20] Scanning Patterned Detection.

detection scheme (ISM[21]) [69]. These approaches would even allow the use of two-photon excitation or spontaneous Raman scattering.

These methods lead to extended resolution microscopy [50], but not to unlimited resolution microscopy in the far-field [40, 45]. The diffraction limit is not fundamentally broken by any of these methods. In all of the far-field methods, the attainable resolution is limited to a finite value. They reach a new limit, which is for the attainable lateral resolution up to a factor of two below Abbe's value, but they cannot provide a theoretically unlimited resolution.

1.2 Breaking the Diffraction Limit with STED Microscopy

More than a century after Abbe's seminal article was published, the diffraction limit in microscopy has been fundamentally broken by the invention of STED microscopy[22] [70], GSD[23] [71], and other RESOLFT[24] techniques [48, 72]. In theory, they provide a resolution without any optical limit.

In STED microscopy, inhibition of the fluorophore's ability to emit light is used to confine fluorescence emission to a small volume at a known position (see Sect. 2 for a detailed description of STED microscopy). After its theoretical description in 1994 [70], STED microscopy was first experimentally demonstrated in 1999 [73]. In the years since its first implementation, STED microscopy has seen many improvements, applications, and combinations with other techniques:

Theoretically, the attainable resolution is unlimited. The experimentally attained resolution was increased in several steps [74–78] to reach the current record of 2.4 nm [79].

STED microscopy was used for Fluorescence Fluctuation Spectroscopy [80–82], and combined with FLIM[25] [83]. Its applicability to immunostained samples [84], lithographic structures [85], fluorescent proteins [86], and organic dyes [87] in living cells, nitrogen-vacancy centers in diamond [78, 88], and quantum dots [89] was shown.

STED microscopy was realized with many different light sources, including Ti:Sapphire lasers [73], diode lasers [90], stimulated Raman scattering in optical fibers [91, 92], and cw-lasers[26] [93]. Super-continuum lasers [94] were suggested as light sources for STED microscopy [95] and used in STED microscopes for excitation [83] as well as for excitation and depletion [96]. The use of low-repetition-rate lasers reduces photobleaching through the

---

[21] Image Scanning Microscopy.
[22] Stimulated Emission Depletion Microscopy.
[23] Ground State Depletion.
[24] Reversible Saturable Optical Fluorescence Transitions.
[25] Fluorescence Lifetime Imaging.
[26] Continuous Wave Lasers.

T-Rex[27] and D-Rex[28] effects [77, 97]. The same photo-physical principle was used to improve live-cell STED imaging using fast scanners [98].

In many applications it is necessary not only to observe the spatial distribution of one molecular species/structure but also to analyze the relative position of two or more structures, at best with sub-diffraction resolution. This requires usually recordings in several channels. They are normally realized as different color channels, employing several types of fluorescent markers being excited and/or emitting at different wavelengths. In STED microscopy multi-channel recordings have been realized using either two excitation lasers in combination with two STED lasers [99, 100], corresponding wavelength ranges selected from a single super-continuum laser [101], or two excitation lasers and one STED laser by applying long Stokes-shift dyes [102–104]. Recently, separation of two color channels just by different detection windows was demonstrated [105]. Exploiting spectroscopic properties of the fluorescence labels could further simplify two-color STED microscopy and enable the simultaneous recording of more than two colors such as realized by the use of distinguished photoswitchable fluorescent proteins [106] or fluorophores with distinct fluorescence lifetimes [107]. It was furthermore demonstrated that the phase mask that shapes the STED beam can simultaneously serve for a phase contrast channel [108].

Applied in the far-field, microscopy with two-photon excitation is often used in thick samples such as tissue, since the excitation with (infra-)red photons reduces unwanted scattering and absorption and thus allows deeper penetration [109]. In addition, the two-photon excitation reduces photobleaching outside of the measurement volume. Therefore, two-photon excitation has been implemented in STED microscopy [110–112] allowing, for example, for the imaging of brain slices. Even two-photon-induced stimulated emission was demonstrated [113] but not yet applied in STED microscopy.

Whereas it is relatively simple to improve strongly just lateral [114] or axial [115] resolution by STED microscopy, a large improvement of resolution along all spatial directions requires more complex instrumentation. Two powerful concepts were shown so far. The first concept applies simultaneously two co-aligned STED beams, one improving the lateral and the other the axial resolution [116, 117]. The second concept uses a 4Pi microscope to simultaneously improve the resolution in all three dimensions (isoSTED) [39, 102]. In applications which require just optical sectioning in one layer without imaging along the optical

---

[27] Triplet Relaxation.
[28] Dark State Relaxation.

axis, and in which confinement of the measurement to a surface is acceptable, one can resort to TIRF excitation to reduce the axial extent of the observation volume [56, 57]. Combined with a conventional toroidal STED focus, TIRF-STED microscopy has recently been realized [118, 119] and combined with FCS[29] to reduce out-of-plane background [119].

Other than TIRF, SPIM[30] is not restricted to measuring close to a surface, but rather achieves a restricted axial extent of the observation volume by illuminating perpendicular to the detection axis [120]. This single plane illumination enables an axial resolution of down to around 200 nm for visible light and was recently combined with STED microscopy to feature an observation volume reduced along all spatial directions [121].

STED microscopy is well suited for biological imaging [122]. The enhanced resolution of STED microscopy has been applied in a wide variety of biological studies. To give a few examples: neurobiological questions were addressed by studying presynaptic active zones [123] and the fate of synaptic vesicles [124]. The distribution of the protein TRPM5[31] in olfactory neurons [125] was investigated. The distribution of Acetylcholine Receptors [126] and of Syntaxin 1 [127] on the plasma membrane, and of Synaptotagmin I and Synaptophysin on endosomes [77] was studied. Clusters of AMPA-receptors[32] and $Ca_V 1.3$ calcium channels were analyzed in the cochlea [128]. The cluster-size dependence of amyloid precursor protein on flotillin-2 was determined [129]. The structure of cristae in mitochondria [130], the endoplasmic reticulum [86], and the protein CD35 in nuclei of mammalian cells [77] was imaged as well as membranes of yeast [115] and bacteria [74]. High-resolution images of many parts of the cytoskeleton were obtained, including the microtubules [84, 96], neurofilaments [77, 93], and Vimentin filaments [92].

Just recently live-cell STED microscopy of non-static preparations emerged: the endoplasmic reticulum [131] and dendritic spines [110, 132] were imaged. The distributions of neuropeptide Y, Pil1 and Nce102 were visualized [133]. Dynamics of Actin in spines [134], dynamics of the protein Connexin-43 in membranes [87], and of lipids and proteins in the plasma membrane of living cells [81, 82, 135] was shown. Just the very last years brought high-resolution imaging with high time resolution using STED microscopy with frame rates up to 200 fps in colloidal samples [136, 137] and up to 28 fps in living neurons [138–141]. The endoplasmic reticulum was imaged with up to 5 fps [98].

---

[29] Fluorescence Correlation Spectroscopy.
[30] Single Plan Illumination Microscopy.
[31] Transient Receptor Potential Channel M5.
[32] α-amino-3-hydroxy-5-methyl-4-isoxazolepropionic acid receptor.

In polymer science, STED microscopy was applied to image block copolymer nanostructures [142], formation of colloidal crystals [137], and colloids in nanofibers [104]. The transfer of the STED concept to write and read out nanostructures was proposed [72] and realized [143–146].

1.3 RESOLFT Microscopy

The STED concept is generalized under the acronym RESOLFT. Essentially any photoswitchable reversible transition between a bright (fluorescent) and a dark (non-fluorescent) state can be exploited to achieve sub-diffraction spatial resolution. To this end the light-distribution driving such a transition in one direction has to contain one or more intensity zeros, while the light driving the respective opposite direction normally has intensity maxima at the positions of these zeros [70, 147]. Increasing the intensity of the light driving the first transition above a certain threshold, while keeping the intensity zeros, refines either the bright (e.g., fluorescence-emitting; for light driving the bright to dark direction) or the dark state population (e.g., fluorescence inhibition; for light driving the dark to bright direction) to sub-diffraction sized regions in the vicinity of the intensity zeros (for detailed explanations, refer also to the STED principle, Sect. 2.1 and Fig. 5). Scanning of the overlaid beams (and thus of the intensity zeros) over the sample consequently produces images with in principle unlimited spatial resolution. Whereas refinement of the bright state results in direct images, which do not need any post-processing, the refinement of the dark state produces negative or indirect images, which require further processing for obtaining the final image. STED microscopy is the most prominent direct imaging technique with the fluorophore's ground state being the dark and the excited state being the bright state. GSD[33] microscopy makes either use of metastable dark states populated via the fluorophore's excited (fluorescent) state for direct imaging or of the ground (dark) to excited (bright) state transition. It was proposed in 1995 [71] and 2002 [148] and later on successfully experimentally realized [149, 150] even on cellular samples [149], potentially providing unlimited spatial resolution, as indicated by measurements on nitrogen-vacancy centers in diamond [151, 152].

The intensities required for efficiently restricting either the bright or the dark state population depend on the lifetime of the two states, since the photo-induced transition has to outcompete spontaneous transitions. STED microscopy and indirect GSD microscopy use the first excited state of the fluorophore as the bright state. Since its lifetime is usually only a few nanoseconds, the laser intensities applied are rather high. With lifetimes of usually micro- to milliseconds of the metastable dark states, direct GSD

---

[33] Ground State Depletion.

microscopy can afford to use orders of magnitude lower laser intensities, potentially reducing phototoxic effects. However, these dark states are often efficient initiators of irreversible photobleaching reactions [77, 97, 153]. Therefore, reversible photoswitchable fluorophores with dark and bright states of different conformations are a better choice for further reducing the laser intensities, since these conformational states are usually much more stable with lifetimes of seconds to hours, especially in fluorescent proteins [72, 147]. Reversible photoswitchable proteins and organic dyes have consequently been successfully applied to generate sub-diffraction spatial image information with laser intensities of only a few kilo Watt per square centimeter [43, 154, 155]. Here, the most important constraint is that the fluorophores have to survive a sufficient number of photoswitching cycles. Recently, reversible photoswitchable proteins with a low switching fatigue have been generated [156] and successfully applied for sub-diffraction live-cell imaging with low laser intensities [157–159]. Faster switching variants will potentially pave the way for reduced image acquisition times, allowing the real-time study of cellular dynamics with a low level of light-induced perturbation. Owing to the lowered levels of light, some of the RESOLFT approaches have successfully been applied in wide-field illumination using a structured-illumination scheme (indirect GSD, SPEM,[34] SSIM[35] [150], photoswitchable proteins [159–161], and direct GSD [162]). In contrast to single-point scanning methods, these structured wide-field approaches require mathematical post-processing of the recorded data, which was recently minimized by scanning with thousands of donuts [161]. Finally, one may also make use of the temporal switching dynamics as proposed by DSOM [163] and later on realized in the GSD mode [164, 165].

1.4 Stochastic Switching

In 2006 a new approach to use the switching of dye molecules for sub-diffraction far-field imaging emerged: PALM[36] [166], STORM[37] [167, 168], and FPALM[38] [169]. The basic principle of all these concepts is to use photoswitching for keeping only a very small subpopulation of all dye molecules simultaneously in the bright state. This subpopulation is kept so low that only single isolated molecules are fluorescent at a time. Different subpopulations of molecules are then switched stochastically on and off and are imaged onto a camera. If the diffraction patterns of the molecules imaged in a single camera frame do not overlap, the molecules can be localized with high precision [170, 171]. An image with sub-diffraction spatial resolution can be reconstructed from these

---

[34] Saturated Patterned Excitation Microscopy.
[35] Saturated Structured-Illumination Microscopy.
[36] Photoactivation Localization Microscopy.
[37] Stochastic Optical Reconstruction Microscopy.
[38] Fluorescence Photoactivation Localization Microscopy.

molecular positions, gathered over hundreds to tens of thousands of camera frames.

Although the idea of using localization and on–off switching for enhancing spatial resolution has been presented before (see, for example, [172–174]), its applicability has only been boosted since the implementation of photoactivatable or switchable fluorophores.

The spatial resolution of these stochastic switching methods is mainly determined by the localization precision, which depends on the number of photons detected from a single molecule [171]. Since the emission of photons is Poisson distributed, the spatial resolution varies. This is in contrast to STED and RESOLFT microscopy, where the spatial resolution is given by the intensity of the switching light. Some single molecules may be missed or discarded due to a too low number of photons detected during a single on-period, introducing potential bias in the final image. Other effects, such as the dipole orientation of the molecules, may also bias the localization [175]. Compared to STED and RESOLFT microscopy, photo-bleaching of the fluorescent markers is less critical since every molecule has in principle to be switched only once.

The principle of stochastic switching was exploited in various facets and enhancements with many names, including PALMIRA[39] [176], SMACM[40] [177], SPDM[41] [178], iPALM[42] [179], DH-PALM[43] [180, 181], DSTORM[44] [182], GSDIM[45] [183], and SMS microscopy[46] [184]. Stochastic switching is realized with either photo-activatable [166, 169] or reversibly switchable [176, 185] fluorescent proteins and fluorophores [186] or Cyanine dye pairs [167, 168]. In addition, conventional organic dyes or fluorescent proteins whose blinking due to dark state transitions had been optimized by specific mounting media [182, 183, 187, 188], photochemical reactions of organic dyes [189, 190], or quantum-dot blueing [191] have been exploited. These stochastic single-molecule switching-based techniques have successfully been applied to living cells, multi-color, and 3D imaging [192].

Image acquisition is usually rather slow (seconds to minutes), since a rather large number of camera frames are needed to reconstruct the final image. Image acquisition time may, however, be reduced with the advent of brighter labels and more sophisticated acquisition and reconstruction techniques [193, 194].

---

[39] Photoactivation Microscopy with Independently Running Acquisition.
[40] Single-Molecule Active-Control Microscopy.
[41] Spectral Precision Distance Microscopy/Spectral Position Determination Microscopy.
[42] Interferometric Photoactivation Localization Microscopy.
[43] Double Helix Photoactivation Localization Microscopy.
[44] Direct Stochastic Optical Reconstruction Microscopy.
[45] Ground State Depletion with Individual Molecule Return.
[46] Single-Molecule Switching Microscopy/Single-Marker Switching Microscopy.

2 STED Microscopy

2.1 Principle

In STED microscopy, proposed in 1994 by Stefan Hell and Jan Wichmann [70] and for the first time experimentally implemented by Thomas Klar and Stefan Hell in 1999 [73], a focused laser beam excites fluorescent markers from their electronic ground state S_0 to their first excited state S_1. Inhibiting the ability of the fluorophores to spontaneously emit fluorescence light in the outer part of the excitation focus generates a smaller effective observation volume, i.e., a smaller volume from which fluorescence emission is still allowed. The inhibition of fluorescence emission is achieved by de-exciting the fluorophores via stimulated emission [195, 196] with a second, red-shifted, laser beam ("STED beam," "de-excitation beam," "depletion beam").

Figure 2 shows the involved energy levels and transitions with a Jabłoński diagram for a typical fluorophore [197, 198]: the fluorophores are excited from the ground state S_0 to higher vibronic states of their first excited electronic state S_1, which relaxes to the lowest vibronic state of S_1 within picoseconds. De-excitation of S_1 down to S_0 occurs within nanoseconds and may be accompanied by the spontaneous emission of a fluorescence photon. This spontaneous emission can be suppressed via stimulated emission which de-excites the state S_1 by optically forcing the molecule into S_0 immediately after excitation. This stimulated de-excitation process generates a stimulated photon, which is, however, not detected.

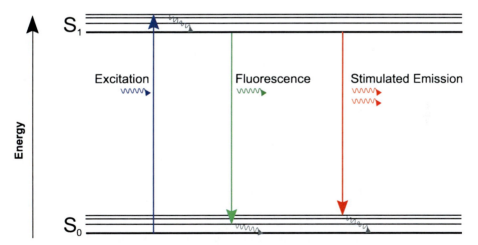

Fig. 2 Jabłoński diagram of the molecular states and transitions involved in STED microscopy. The fluorophore is excited from the lowest vibronic level of the electronic ground state S_0 to a higher vibronic level of the first excited electronic state S_1 by absorption of a photon of the excitation light. After relaxation to the lowest vibronic level of S_1, the fluorophore emits spontaneously a fluorescence photon and goes to a higher vibronic level of S_0. Stimulated emission forces the transition from S_1 to S_0. This transition can go to a different vibronic level. After spontaneous or stimulated emission, the fluorophore returns to the lowest vibronic level of S_0

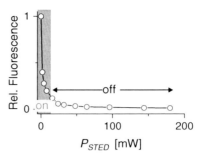

Fig. 3 Depletion curve: Dependence of the remaining spontaneous fluorescence emission on the applied STED power. For high STED powers, the probability of spontaneous fluorescence emission vanishes and the ability of the fluorophores to emit light spontaneously is essentially switched off

With increasing intensity of the de-excitation beam, the excited state S_1 is more and more likely depleted, i.e., the molecule is spending almost no time in this state. Hence the fluorophore is essentially confined to its ground state S_0, which is equivalent to switching the fluorescence ability of the molecule off, i.e., all spontaneous emission and thus potentially detected fluorescence signal is inhibited (Fig. 3) [40]. To stimulate emission, usually a wavelength at the red end of the emission spectrum is used, despite the fact that stimulated emission is usually most effective at the maximum of the fluorescence emission spectrum. This choice of wavelength results from two characteristics: First, the cross section of the S0–S1 excitation (i.e., for the absorption of light) must be small at the STED wavelength. Otherwise the STED light would also excite the dye to S_1 generating spontaneous fluorescence instead of stimulating the S_1 to S_0 transition. Second, for most dyes, the excitation spectrum overlaps with the emission spectrum and the excitation probability vanishes only towards the red end of the spectra. An alternative to choosing such a red-shifted wavelength may be the use of long Stokes-shift dyes or low temperatures [102, 199]. Second, stimulating the S_1 to S_0 transition at the red end of the emission spectrum allows a straightforward spectral separation of spontaneous and stimulated emission. To use the stimulated emission for increasing resolution in a STED microscope, an intensity distribution of the STED beam is used that has a local zero. An example is the toroidal ("donut") shape as shown in Fig. 4 [124, 200]; other intensity distributions are possible as well [115, 201]. By overlaying the engineered STED focus with a usually diffraction-limited Gaussian-shaped excitation spot, the fluorescence is inhibited everywhere but at the focal center. Using a high intensity of the STED beam, the S_1 state of the fluorophores is almost completely depleted also in those regions where the relative intensity of the beam is low, i.e., close to the midpoint. Only in the very center, where the de-excitation focus has zero intensity, the fluorophores remain in the fluorescing ("on") state (Fig. 5).

Foundations of STED Microscopy 53

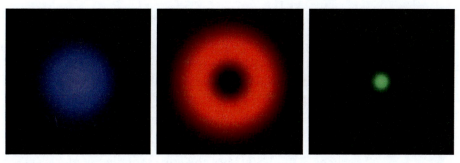

Fig. 4 Configuration of foci and observation spot in a STED microscope. Overlaying a Gaussian excitation focus (*left*) with a STED focus having one or several intensity zeros such as a torus (*middle*) results in a small region in which fluorescence emission is still allowed (observation spot, *right*). This small region is the effective PSF (point spread function) of a STED microscope

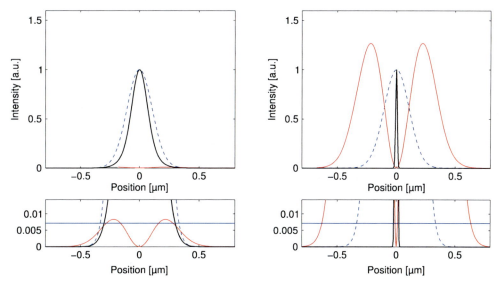

Fig. 5 Interaction and spatial intensity profiles of the excitation and depletion foci to generate a small region of sub-diffraction size where fluorescence emission is still allowed. The lower panels are enlargements of the upper panels. The *blue horizontal line* indicates the saturation intensity (the intensity at which half of the fluorescence is inhibited by stimulated emission). Intensity cross sections through the overlaid foci: *blue dotted*: excitation intensity distribution, *red*: STED intensity distribution, *black*: effective observation volume (describing the image of a point object). *Left*: low STED intensity. Only in the outer part of the focus the STED intensity is above the saturation intensity of the dye. The effective observation volume is slightly narrowed. *Right*: high STED intensity. Although the *spatial distribution* of the STED intensity is unchanged, the STED intensity is above the saturation intensity also close to the center. The effective observation volume is thus strongly narrowed because all molecules in the periphery are kept dark. Note that the different intensities are not drawn to scale: the STED intensity (*red line*) is usually much higher than the excitation intensity (*blue dotted line*)

The volume wherein spontaneous fluorescence emission is still allowed shrinks to zero and accordingly the spatial resolution of the STED microscope increases with increasing intensity of the STED beam [48]. The lateral resolution d, given as FWHM[47] of the volume in which fluorescence emission is still allowed, follows the relation [202]

$$d \approx d_c \Big/ \sqrt{1 + d_c^2 a^2 I / I_{\text{sat}}}, \qquad (2)$$

where d_c the FWHM of the corresponding confocal PSF[48] (Eq. 1), a a factor describing the shape of the STED intensity pattern, I the maximum intensity of the STED beam in the focus and I_{sat} the saturation intensity, a fluorophore-characteristic constant; at this intensity half of the fluorescence is inhibited by stimulated emission.

From Eq. (2) follows that with increasing STED intensity the size of the image of a point object (the "effective PSF") decreases: No lower limit exists for the size of d. With the depletion-beam turned off, the microscope operates as a standard confocal laser scanning microscope. For the formation of an image, the overlaid laser beams are scanned across the sample, as in conventional laser scanning microscopy. Parallelization, i.e., scanning with multiple laser spots is possible [203] and a confocal pinhole is in principle not necessary for the STED recordings.

2.2 Pulsed and Continuous Wave Operation

Usually pulsed lasers are used for evoking excitation and stimulated emission. Pulsed STED light provides high peak intensities and therefore high resolution (Eq. 2). This pulsed operation suppresses fluorescence most efficiently, since under optimized conditions the STED laser pulse immediately follows the excitation pulse, i.e., it reaches the fluorophore in the desired state, the excited S_1 state.

Because the pulse-to-pulse interval is usually much longer than the fluorescence lifetime of the dyes and the pulses usually much shorter than the lifetime, the dye is mostly in an idle state; it can emit maximally one photon within one pulse cycle. In contrast, with cw excitation,[49] photons can be emitted constantly with an average rate of about one per lifetime. This effect becomes important at intense excitation; more photons can then be captured per unit time with cw than with pulsed excitation [153, 204].

It is possible to use cw excitation and cw de-excitation for STED microscopy, (cw STED microscopy) [93]. To achieve the same resolution with cw STED microscopy as with pulsed STED microscopy in the case of low excitation intensities, the average intensity of the depletion beam must be increased by about a factor of $\Gamma = \delta / [\tau \ln(2)]$, where τ denotes the fluorescence lifetime and

---

[47] Full Width at Half Maximum.
[48] Point Spread Function, describing the image of a point object.
[49] Continuous Wave Excitation.

δ the pulse-to-pulse interval [205]. For dyes with $\tau = 4$ ns and a laser repetition rate of 80 MHz, this yields $\Gamma \approx 4$ [205].

Additionally, the shape of the effective PSF in cw STED microscopy is compromised by ongoing excitation. This leads to a less pronounced fluorescence on–off contrast. The poorer contrast manifests itself as a pedestal of the effective PSF compromising the separation of object details [206, 207], even if the FWHM of the PSF is the same as in pulsed operation.

2.3 Gated STED Microscopy

Albeit the inferior performance of the cw STED microscopy compared to the pulsed STED modality, cw STED microscopy bears some advantages: It allows, e.g., the use of much more compact and cheaper laser systems. The problem of the pedestal in cw STED microscopy (which compromises the separation of fine object details) can be surpassed by implementing pulsed-laser excitation in combination with a cw STED laser and time-gated detection. Time-gated detection is often used in conventional fluorescence microscopy for suppressing background, and in a pulsed STED scheme, it is known that photons should be detected after the STED pulse has left [114, 208], as shown in a recent experiment using time-correlated single-photon counting [83]. Time-gating can improve the spatial resolution provided by the cw STED implementation and finer details are resolved despite the use of low STED intensities [207, 209]. This follows from the fact that the temporal overlap between the excitation and the STED beam is restricted to the duration of the excitation pulse (usually less than 150 ps). Right afterwards, only the cw STED beam is acting and inhibiting fluorescence emission: The longer it lasts, the more likely it becomes that a fluorophore is switched off, i.e., the spatial resolution not only depends on the intensity of the STED light but also on the time span of the STED beam action [147, 210]. This can also be pictured in a way that STED reduces the fluorescence lifetime τ of the fluorescent state. Consequently, the collection of photons with a delay after the excitation pulse increases the fluorescence inhibition and ensures that fluorescence light is recorded mainly from fluorophores from the donut center, where the STED beam is inherently weak (and thus the fluorescence lifetime rather long). Such g-STED microscopy[50] has been realized for different fluorophores and in living cells, allowing the recording of live-cell images with sub-diffraction spatial resolution at moderate cw STED beam powers (<80 MW) [207]. This and the fact that cw beams are less prone to induce multi-photon phototoxic processes reduce light-induced stress on the sample. g-STED can be realized with simple cw lasers that are nowadays available at various wavelengths and by off-line processing of time-correlated single-photon counting recordings, or in real-time using a fast electronic gate. Of all current STED microscopy implementations with low peak powers reported so far, g-STED provides the sharpest images.

---

[50] Gated sted microscopy.

2.4 Fast STED Microscopy

Along with the increase in spatial resolution, the number of scanning points necessary and thus the time for the recording of an image increases. Nevertheless, very fast scanning enables high-resolution microscopy not only with high spatial resolution but also with high temporal resolution. Since the pixels are scanned sequentially in STED microscopy the time needed to record a frame scales approximately linear with the imaged area. Very high frame rates are possible, especially for small fields of view. This makes STED microscopy an ideal choice for imaging of live-cell dynamics.

A schematic setup of a beam-scanning STED microscope using a resonant beam scanner is shown in Fig. 6. A resonant scanner operating at 16 kHz allows acquiring a whole line of the image in 31 ms, using forward and backward scans (bidirectional scanning), but averaging might be necessary for a sufficient signal-to-noise ratio. This allows, e.g., recording movies of neurotransmitter vesicles in living cells at 28 fps [211] (Fig. 7), also with CW STED microscopy [139] and of nano-beads crystallizing

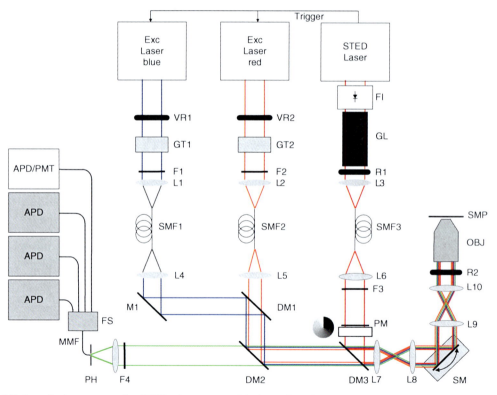

Fig. 6 Setup of a beam-scanning STED microscope with the option of two-color imaging. For a higher dynamic range of the detector, the fluorescence can be detected with several APDs or a PMT. Abbreviations: *APD*: avalanche photo diode; *DM*: dichroic mirror; *Exc*: excitation; *F*: filter; *FI*: Faraday isolator, *FS*: fiber splitter; *GL*: glass rod; *GT*: Glan Thompson polarizer; *L*: lens; *M*: mirror; *MMF*: multi mode fiber; *OBJ*: objective; *PM*: phase mask; *PMT*: photomultiplier tube; *R*: retarder wave plate; *SM*: scan mirror; *SMF*: single-mode fiber (polarization maintaining); *SMP*: sample; *VR*: variable retarder

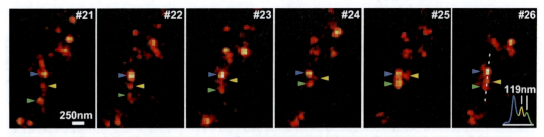

Fig. 7 Characteristics of synaptic vesicle movement. Successive frames of a movie showing the vesicle movement in an axon; recorded via Fast STED microscopy and digitally filtered. The *arrowheads* indicate three vesicles, which were tracked in all frames, localized in a sub-diffraction space. The *inset* in frame #26 shows an intensity profile along the *dotted white line*, demonstrating that vesicles with a spacing well below the diffraction limit are resolved. (Figure reproduced from [211])

into colloidal crystals at 200 fps [137] (Fig. 8). A detailed description of a beam-scanning STED microscope can be found in [103].

Fast scanning also reduces the illumination time per single spot and therefore the possibility of exciting and photobleaching the fluorophores from long-lived dark states. A similar reduction of photobleaching has been realized previously as T-Rex[51] and D-Rex[52] STED microscopy [77, 97] by lowering the repetition rates of the excitation and STED lasers. The low repetition rate leaves time for the fluorophores to relax from any vulnerable dark state back to their ground state, however (with repetition rates of 250 kHz) at the expense of increased recording time. Fast scanning achieves a similar effect and has allowed the recording of STED images in living cells even with CW laser sources [98, 139].

2.5 STED-FCS

Fluorescence correlation spectroscopy (FCS) is a noninvasive and very sensitive technique for the disclosure of complex dynamic processes such as diffusion and kinetics [212–214]. In FCS, the temporal fluctuation of a fluorescence signal is monitored at one point in the sample while fluorescent molecules diffuse in and out of the observation volume and the correlation function of these fluctuations is calculated. The decay time of this correlation function usually reflects the average transit time of the molecules through the observation volume. Hindrances of free diffusion therefore result in either a shift of the correlation curve towards larger times or an altered shape of it. The observation volume is often given by the micrometer large focal laser spot of a confocal microscope [215].

FCS is usually combined with far-field (confocal) microscopy and allows experiments in living cells. Often, however, prominent biological problems cannot be solved due to the limited resolution (larger 200 nm) of conventional (confocal) optical microscopy. For

---

[51] Triplet Relaxation.
[52] Dark State Relaxation.

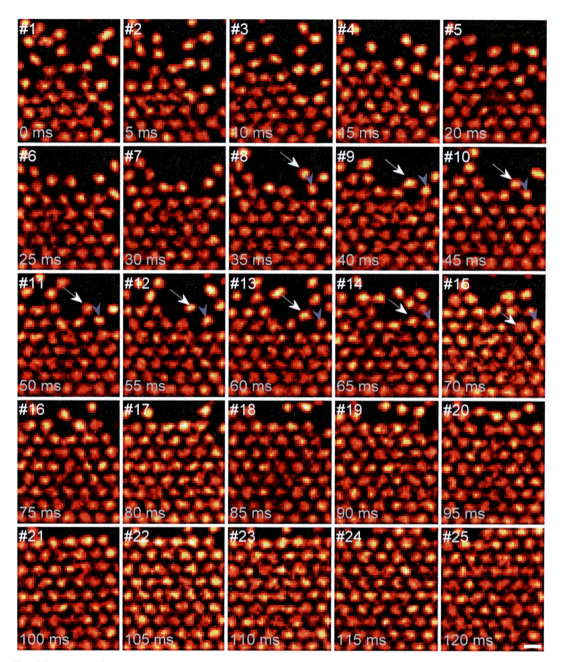

Fig. 8 Formation of a colloidal crystal. Successive movie frames of a colloidal crystal (monolayer), which is just forming. The movie is recorded at 200 fps. Two particles are highlighted in several frames by *white arrows* and *blue arrow heads*, respectively. Note how the bead marked with the *blue arrow heads* blocks in frames #10 – #12 the lattice site for the newly arriving bead marked with the *white arrows*. Data shown after linear deconvolution. Scale bar 250 nm. (Figure reproduced from [137])

example, on a diffraction-limited microscope, FCS requires rather low concentrations of fluorescently labeled molecules, because only a limited number of emitters must be in the observation volume at any given time. Sometimes, this can preclude its otherwise very promising application to (endogenous) systems where molecular concentrations of several micromol per liter are required. Further, conventional FCS with its rather large observation volume cannot fully characterize dynamic interactions on the nanoscale, since it cannot directly measure on the relevant spatial scales, although certain approaches extrapolate measurement results to the nanoscopic range [216]. Examples are lipid–lipid and lipid–protein interactions in the plasma membrane such as transient integration of proteins into nano-domains ("rafts"), which are considered to play a functional part in a whole range of membrane-associated processes [217]. The spatial dimensions of these domains are supposed to be on molecular scales and the direct, noninvasive observation of them in living cells is thus impeded by the resolution limit. Using the superior spatial resolution of STED microscopy with effective length scales of down to 20–30 nm in living cells, FCS can now be applied at rather large fluorophore concentrations [80, 82], or it can directly and noninvasively detect nanoscopic anomalous diffusion, e.g., of single lipid or protein molecules in the plasma membrane of living cells [81]. Using STED-FCS it was, for example, shown that sphingolipids or certain membrane proteins, unlike phosphoglycerolipids, were transiently (≈ 10 ms) trapped on the nanoscale in cholesterol-mediated molecular complexes [81, 82, 135]. Consequently STED-FCS is a sensitive and unique tool for studying nanoscale membrane organization, and determining the cellular functions and molecular interdependencies of membrane components, also extendable to intracellular nanoscale interactions.

3 Resolution

Most considerations about resolving power of a microscope are based on theoretical images. Many experimental details influence the resolution of the observed images as was emphasized by Ronchi [218]. Often, the effect of pixelation is not taken into account. But all images—independently of the (microscopy) method used—are finally displayed in pixels. Continuously moving resonant scanners can lead to further averaging over one pixel area. Whereas reconstruction filters can be used for displaying the final image, very often high-resolution images are published as raw data, where the gray level in one pixel represents the signal strength sampled at this position. With a finite pixel size, the recorded diffraction pattern does not necessarily represent well the true width of the PSF. In addition, it is unclear if two objects that are separated by one FWHM of the PSF are still seen with two distinct maxima.

Therefore, results on the influence of pixelation and noise on the separability of objects [103] are recapitulated here. In the following, two objects are regarded as resolved, if at least one pixel in between them is darker by a certain amount than its neighboring pixels. Of course, many different pixelations are possible. If a small pixel size is chosen, several pixels are between the intensity maxima of the diffraction patterns of neighboring objects and this definition becomes ambiguous. Therefore, the analysis is restricted to a pixel size of half the PSF width. This is a common choice in experiments, analogous to the Nyquist theorem [219–223], which refers to point sampling (and not pixelation), though. Gaussian diffraction patterns are assumed. The calculations are independent of the actual pixel size, only the ratio of the pixel size to the FWHM of the PSF matters.

3.1 Resolution in a Noiseless Image

First, we are considering the case of noiseless images. This corresponds to an infinite number of photons collected.

Two symmetric cases are analyzed: the midpoint between the diffraction patterns of two equally bright objects falling onto a pixel center (Fig. 9 left) or a pixel border (Fig. 9 right).

3.1.1 Midpoint on Pixel Center

The brightness of the center pixel between the two diffraction patterns (marked b in Fig. 9) was calculated in comparison with its two neighboring pixels (marked a and c in Fig. 9): integrating the intensities over the pixel areas shows that the center pixel is just 0.48% darker than its neighbors.

To have a dip of 19% (as for two lines separated according to the common Rayleigh criterion [224], without pixelation effects), two objects sitting symmetrically to a pixel center must have a distance of 1.18 times the FWHM of the Gaussian PSF. For a dip of 26% (as for two Airy patterns separated according to the Rayleigh criterion, without pixelation effects) their distance must even be 1.28 times the FWHM. No dip is seen below a distance of 0.995 times the FWHM.

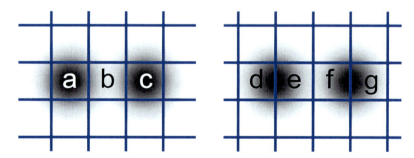

Fig. 9 Images of point objects with a symmetrical positioning relative to the pixel borders (*blue*). *Left*: the midpoint between the objects is centered on a pixel. *Right*: The midpoint is on a pixel border. The diffraction patterns are plotted with a FWHM of only one pixel size (half of what is used in the calculations) to show them clearly separated. Some pixels are labeled (a–g), as referenced in the text

3.1.2 Midpoint on Pixel Border

The chance to resolve the objects is worse, when the midpoint between the diffraction patterns falls onto a pixel border (Fig. 9 right): the central pixels (marked e and f) are brighter than the pixels d and g; there is no dip at all between the objects. The object distance must be larger than 1.16 times the FWHM to have a dip at all, 1.31 times the FWHM for an 19% dip and 1.39 times the FWHM for a 26% dip.

This means that for coarse pixelation, which is common in many experimental settings, two objects that are separated by their apparent FWHM are barely or not at all resolved (in the sense of an intensity dip between their two diffraction patterns), depending on their position relative to the pixels.

3.2 Resolution in the Presence of Noise

In contrast to the above analysis that presupposes a noiseless signal microscopy images are normally governed by shot noise [225]. This especially holds true for fast imaging, where relatively dark images are recorded due to the short exposure times. A similar situation is given for PALM, STORM, and alike microscopy techniques. Here, the final image maps the positions of all the molecules observed (in contrast to a conventional or STED image, which is a map of all the photons observed). The labeling density of the structures of interest must be kept low. Too dense staining leads to more than one simultaneously fluorescing molecule within each diffraction pattern (because the fraction of molecules in the dark state cannot be made arbitrarily high) and the single-molecule localization becomes difficult or even impossible [226]. Too densely stained samples are often even bleached on purpose before the image acquisition starts [183]. Therefore only few molecules are found in each pixel. This in turn means that the true structure may differ considerably from the observed dye distribution (just as a noisy photon distribution may not reflect faithfully the original dye distribution).

Some studies have examined how the quantum noise degrades resolution. But their scopes were rather special, for example regarding only the distance measurement of two point objects [227], discrete structures [228], low NA systems [229], or they did not consider the case of few photons [230, 231]. Fried [232] as well as Fannjiang and Sølna [233] analyzed resolution in the presence of noise, but considered only Gaussian noise. Terebizh [234] limited its analysis to relatively high signal-to-noise ratios and two incoherent point objects.

Here, we present simulation results on the influence of shot noise on the visibility of a dip between two point objects [103]. The total number of photons collected from each object is varied and the resulting probability to obtain a dip between the two objects is determined.

The case that the midpoint between the objects falls onto a pixel center (Fig. 9 left) is considered assuming an object distance

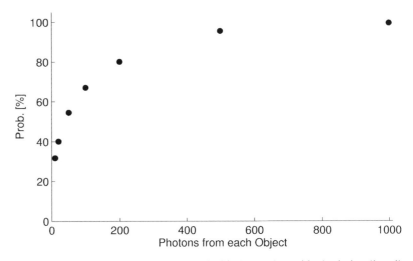

Fig. 10 Probability to have the center pixel between two objects darker than its neighbors for different numbers of photons observed from each object. The object distance is 1.28 times the FWHM of the PSF. Note that dark objects are often not resolved

of 1.28 times the FWHM of the PSF. This separation is according to the Rayleigh criterion and would give a 26 % dip in the noiseless case. The probability to find the middle pixel darker than its left and right neighbors is 67 % if 100 photons from each object are observed, i.e., in almost one third of the observations, two point objects would not be resolved. The probabilities for different photon numbers are shown in Fig. 10. As many as 500 photons detected per object are needed to resolve them in about 95 % of the observations.

3.3 Conclusions

On the one hand, the FWHM of the PSF as a measure for the resolution is advantageous for the good comparability of different PSF shapes. On the other hand, objects that are separated by a distance equal to the FWHM of the PSF are not in all cases resolved (in the sense of seeing an intensity dip between them) due to pixelation effects.

Considering shot noise, even objects separated by 1.28 times the FWHM of the PSF are only resolved with a 95 % probability if 500 photons are detected from each object.

To resolve objects with few counts in the image is not only problematic in conventional images when only few photons are counted; but it is also equally important in PALM, STORM and alike microscopy techniques, where normally only few molecules are observed. In the above example, 500 molecules need to be localized from each object to separate two point objects with 95 % probability. It is even worse for darker objects. Localization precision is not tantamount to resolution: for a high resolution, not only the localization precision but also the degree of sampling must be high as discussed by Shroff et al. in terms of the Nyquist criterion [226]; when employing the criterion, however, one must not forget that it strictly applies only to *band limited* signals that are *periodically sampled*.

4 Concluding Remarks

Optical microscopy remains arguably one of the most important imaging methods in the life sciences because it allows imaging with high specificity and at physiological conditions. Despite the use of highly developed microscopes with perfect objectives the resolution—and accordingly the useful magnification—remained, however, for a long time limited by diffraction. Developments of the last years allow now optical imaging in the far-field beyond this diffraction barrier.

References

1. Abbe E (1873) Beiträge zur Theorie des Mikroskops und der mikroskopischen Wahrnehmung. M. Schultze's Archiv für mikroskopische Anatomie 9:413–468
2. Lauterbach MA (2012) Finding, defining and breaking the diffraction barrier in microscopy – a historical perspective. Opt Nanoscopy 1:8
3. Pawley J (ed) (2006) Handbook of biological confocal microscopy. Springer, New York
4. Kirz J, Jacobsen C, Howells M (1995) Soft X-ray microscopes and their biological applications. Q Rev Biophys 28(1):33–130
5. Miao JW, Ishikawa T, Shen Q, Earnest T (2008) Extending X-ray crystallography to allow the imaging of noncrystalline materials, cells, and single protein complexes. Annu Rev Phys Chem 59:387–410
6. Ruska E (1934) Über Fortschritte im Bau und in der Leistung des magnetischen Elektronenmikroskops. Zeitschrift für Physik 87:580–602
7. von Borries B, Ruska E (1939) Magnetische Sammellinse kurzer Feldlänge. Patent Nr. 680284, Deutsches Reich, Reichspatentamt
8. Ruska E (1993) The development of the electron microscope and of electron microscopy. In: Frängsmyr T, Ekspong G (eds) Nobel lectures, physics 1981–1990, pp. 355–380. World Scientific Publishing, Singapore, New Jersey, London, Hong Kong
9. de Broglie L (1925) Recherches sur la théorie des quanta. Ann Phys 10e Série Tome 3:22–128
10. Binnig G, Quate CF, Gerber C (1986) Atomic force microscope. Phys Rev Lett 56(9):930–933
11. Binnig G, Rohrer H (1983) Scanning tunneling microscopy. Surf Sci 126(1–3):236–244
12. Synge EH (1928) A suggested method for extending microscopic resolution into the ultra-microscopic region. Philos Mag 6(35):356–362
13. Ash EA, Nicholls G (1972) Super-resolution aperture scanning microscope. Nature 237(5357):510–512
14. Pohl DW, Denk W, Lanz M (1984) Optical stethoscopy: Image recording with resolution $\lambda/20$. Appl Phys Lett 44(7):651–653
15. Anderson MS (2000) Locally enhanced Raman spectroscopy with an atomic force microscope. Appl Phys Lett 76(21):3130–3132
16. Kawata S, Inouye Y, Verma P (2009) Plasmonics for near-field nano-imaging and superlensing. Nat Photon 3(7):388–394
17. Hayazawa N, Inouye Y, Sekkat Z, Kawata S (2000) Metallized tip amplification of near-field Raman scattering. Opt Commun 183:333–336
18. Stöckle RM, Suh YD, Deckert V, Zenobi R (2000) Nanoscale chemical analysis by tip-enhanced Raman spectroscopy. Chem Phys Lett 318(1–3):131–136
19. Hansma PK, Drake B, Marti O, Gould SAC, Prater CB (1989) The scanning ion-conductance microscope. Science 243(4891):641–643
20. Müller DJ, Helenius J, Alsteens D, Dufrêne YF (2009) Force probing surfaces of living cells to molecular resolution. Nat Chem Biol 5(6):383–390
21. de Lange F, Cambi A, Huijbens R, de Bakker B, Rensen W, Garcia-Parajo M, van Hulst N,

Figdor CG (2001) Cell biology beyond the diffraction limit: near-field scanning optical microscopy. J Cell Sci 114(Pt 23):4153–4160
22. Lazarides E, Weber K (1974) Actin antibody: The specific visualization of actin filaments in non-muscle cells. Proc Natl Acad Sci USA 71(6):2268–2272
23. Tsien RY (1998) The green fluorescent protein. Annu Rev Biochem 67:509–544
24. Lippincott-Schwartz J, Patterson GH (2003) Development and use of fluorescent protein markers in living cells. Science 300(5616):87–91
25. Verkhusha VV, Lukyanov KA (2004) The molecular properties and applications of anthozoa fluorescent proteins and chromoproteins. Nat Biotechnol 22(3):289–296
26. Shimomura O, Johnson FH, Saiga Y (1962) Extraction, purification and properties of Aequorin, a bioluminescent protein from the Luminous Hydromedusan, Aequorea. J Cell Comp Physiol 59(3):223–239
27. Chalfie M, Tu Y, Euskirchen G, Ward WW, Prasher DC (1994) Green fluorescent protein as a marker for gene-expression. Science 263(5148):802–805
28. Griffin BA, Adams SR, Tsien RY (1998) Specific covalent labeling of recombinant protein molecules inside live cells. Science 281(5374):269–272
29. Keppler A, Gendreizig S, Gronemeyer T, Pick H, Vogel H, Johnsson K (2003) A general method for the covalent labeling of fusion proteins with small molecules in vivo. Nat Biotechnol 21(1):86–89
30. Los GV, Zimprich C, McDougall MG, Karassina N, Learish R, Klaubert DH, Darzins A, Bulleit RF, Wood K (2005) The HaloTag (TM): a novel technology for cellular analysis. J Neurochem 94:15
31. Los GV, Encell LP, McDougall MG, Hartzell DD, Karassina N, Zimprich C, Wood MG, Learish R, Ohana RF, Urh M, Simpson D, Mendez J, Zimmerman K, Otto P, Vidugiris G, Zhu J, Darzins A, Klaubert DH, Bulleit RF, Wood KV (2008) HaloTag: A novel protein labeling technology for cell imaging and protein analysis. ACS Chem Biol 3(6):373–382
32. Parak WJ, Pellegrino T, Plank C (2005) Labelling of cells with quantum dots. Nanotechnology 16(2):R9–R25
33. Dubertret B, Skourides P, Norris DJ, Noireaux V, Brivanlou AH, Libchaber A (2002) In vivo imaging of quantum dots encapsulated in phospholipid micelles. Science 298(5599):1759–1762
34. Pinaud F, Michalet X, Iyer G, Margeat E, Moore H-P, Weiss S (2009) Dynamic partitioning of a Glycosyl-Phosphatidylinositol-anchored protein in Glycosphingolipid-rich microdomains imaged by single-quantum dot tracking. Traffic 10(6):691–712
35. Shimomura O, Johnson FH, Saiga Y (1963) Microdetermination of calcium by aequorin luminescence. Science 140(357):1339–1340
36. Tsien RY (1980) New calcium indicators and buffers with high selectivity against magnesium and protons: Design, synthesis, and properties of prototype structures. Biochemistry 19(11):2396–2404
37. Stephens DJ, Allan VJ (2003) Light microscopy techniques for live cell imaging. Science 300(5616):82–86
38. Crocker JC, Grier DG (1996) Methods of digital video microscopy for colloidal studies. J Colloid Interface Sci 179(1):298–310
39. Hell S, Schmidt R, Egner A (2009) Diffraction-unlimited three-dimensional optical nanoscopy with opposing lenses. Nat Photon 3:381–387
40. Hell SW (2009) Microscopy and its focal switch. Nat Methods 6(1):24–32
41. Huang B, Bates M, Zhuang X (2009) Super-resolution fluorescence microscopy. Annu Rev Biochem 78:993–1016
42. Chi KR (2009) Super-resolution microscopy: breaking the limits. Nat Methods 6(1):15–18
43. Dedecker P, Hofkens J, Hotta JI (2008) Diffraction-unlimited optical microscopy. Mater Today 11:12–21
44. Lippincott-Schwartz J, Manley S (2009) Putting super-resolution fluorescence microscopy to work. Nat Methods 6(1):21–23
45. Hell SW (2007) Far-field optical nanoscopy. Science 316(5828):1153–1158
46. Rice JH (2007) Beyond the diffraction limit: far-field fluorescence imaging with ultrahigh resolution. Mol Biosyst 3(11):781–793
47. Hell SW, Dyba M, Jakobs S (2004) Concepts for nanoscale resolution in fluorescence microscopy. Curr Opin Neurobiol 14(5):599–609
48. Hell SW (2003) Toward fluorescence nanoscopy. Nat Biotechnol 21(11):1347–1355
49. Egner A, Hell SW (2005) Fluorescence microscopy with super-resolved optical sections. Trends Cell Biol 15(4):207–215
50. Gustafsson MGL (1999) Extended resolution fluorescence microscopy. Curr Opin Struct Biol 9(5):627–634
51. Minsky M (1961) Microscopy apparatus. U.S. Patent 3013467

52. Hell S, Stelzer EHK (1992) Properties of a 4Pi confocal fluorescence microscope. J Opt Soc Am A Opt Image Sci Vis 9(12):2159–2166
53. Hell SW, Stelzer EHK, Lindek S, Cremer C (1994) Confocal microscopy with an increased detection aperture: Type-B 4Pi confocal microscopy. Opt Lett 19(3):222–224
54. Gustafsson MGL, Agard DA, Sedat JW (1995) Sevenfold improvement of axial resolution in 3D widefield microscopy using two objective lenses. In: Proc SPIE – Three-Dimensional Microscopy: Image Acquisition and Processing II, vol 2412, pp 147–156
55. Gustafsson MGL, Agard DA, Sedat JW (1996) 3D widefield microscopy with two objective lenses: Experimental verification of improved axial resolution. In: Proc SPIE – Three-Dimensional Microscopy: Image Acquisition and Processing III, vol 2655, pp 62–66
56. Axelrod D (1981) Cell-substrate contacts illuminated by total internal-reflection fluorescence. J Cell Biol 89(1):141–145
57. Temple PA (1981) Total internal-reflection microscopy: A surface inspection technique. Appl Opt 20(15):2656–2664
58. Veselago VG (1968) The electrodynamics of substances with simultaneously negative values of ε and μ. Sov Phys Uspekhi 10(4):509–514
59. Veselago VG (1967) Properties of materials having simultaneously negative values of dielectric (ε) and magnetic (μ) susceptibilities. Sov Phys Solid State 8(12):2854–2856
60. Pendry JB (2000) Negative refraction makes a perfect lens. Phys Rev Lett 85(18):3966–3969
61. Liu ZW, Lee H, Xiong Y, Sun C, Zhang X (2007) Far-field optical hyperlens magnifying sub-diffraction-limited objects. Science 315(5819):1686–1686
62. Cragg GE, So PTC (2000) Lateral resolution enhancement with standing evanescent waves. Opt Lett 25(1):46–48
63. Kner P, Chhun BB, Griffis ER, Winoto L, Gustafsson MGL (2009) Super-resolution video microscopy of live cells by structured illumination. Nat Methods 6(5):339–U36
64. Gustafsson MGL, Agard DA, Sedat JW (1999) I^{5}M: 3D widefield light microscopy with better than 100 nm axial resolution. J Microsc 195:10–16
65. Shao L, Isaac B, Uzawa S, Agard DA, Sedat JW, Gustafsson MGL (2008) I^{5}S: Widefield light microscopy with 100-nm-scale resolution in three dimensions. Biophys J 94(12):4971–4983
66. Bailey B, Farkas DL, Taylor DL, Lanni F (1993) Enhancement of axial resolution in fluorescence microscopy by standing-wave excitation. Nature 366(6450):44–48
67. Frohn JT, Knapp HF, Stemmer A (2000) True optical resolution beyond the Rayleigh limit achieved by standing wave illumination. Proc Natl Acad Sci USA 97(13):7232–7236
68. Lu J, Min W, Conchello J-A, Xie XS, Lichtman JW (2009) Super-resolution laser scanning microscopy through spatiotemporal modulation. Nano Lett 9(11):3883–3889
69. Müller CB, Enderlein J (2010) Image scanning microscopy. Phys Rev Lett 104(19):198101
70. Hell SW, Wichmann J (1994) Breaking the diffraction resolution limit by stimulated emission: stimulated-emission-depletion fluorescence microscopy. Opt Lett 19(11):780–782
71. Hell SW, Kroug M (1995) Ground-state-depletion fluorescence microscopy: a concept for breaking the diffraction resolution limit. Appl Phys B Lasers Opt 60(5):495–497
72. Hell SW (2004) Strategy for far-field optical imaging and writing without diffraction limit. Phys Lett A 326(1–2):140–145
73. Klar TA, Hell SW (1999) Subdiffraction resolution in far-field fluorescence microscopy. Opt Lett 24(14):954–956
74. Dyba M, Hell SW (2002) Focal spots of size $\lambda/23$ open up far-field florescence microscopy at 33 nm axial resolution. Phys Rev Lett 88(16):163901
75. Westphal V, Kastrup L, Hell SW (2003a) Lateral resolution of 28 nm ($\lambda/25$) in far-field fluorescence microscopy. Appl Phys B Lasers Opt 77(4):377–380
76. Westphal V Hell SW (2005a) Nanoscale resolution in the focal plane of an optical microscope. Phys Rev Lett 94:143903
77. Donnert G, Keller J, Medda R, Andrei MA, Rizzoli SO, Lührmann R, Jahn R, Eggeling C, Hell SW (2006) Macromolecular-scale resolution in biological fluorescence microscopy. Proc Natl Acad Sci USA 103(31):11440–11445
78. Rittweger E, Han KY, Irvine SE, Eggeling C, Hell SW (2009a) STED microscopy reveals crystal colour centres with nanometric resolution. Nat Photon 3(3):144–147
79. Wildanger D, Patton BR, Schill H, Marseglia L, Hadden JP, Knauer S, Schönle A, Rarity JG, O'Brien JL, Hell SW, Smith JM (2012) Solid immersion facilitates fluorescence microscopy with nanometer resolution and sub-Ångström emitter localization. Adv Mater 24(44):OP309–OP313
80. Kastrup L, Blom H, Eggeling C, Hell SW (2005) Fluorescence fluctuation spectroscopy

in subdiffraction focal volumes. Phys Rev Lett 94(17):178104
81. Eggeling C, Ringemann C, Medda R, Schwarzmann G, Sandhoff K, Polyakova S, Belov VN, Hein B, von Middendorff C, Schönle A, Hell SW (2009) Direct observation of the nanoscale dynamics of membrane lipids in a living cell. Nature 457(7233):1159–1162
82. Ringemann C, Harke B, von Middendorff C, Medda R, Honigmann A, Wagner R, Leutenegger M, Schönle A, Hell SW, Eggeling C (2009) Exploring single-molecule dynamics with fluorescence nanoscopy. New J Phys 11(10):103054
83. Auksorius E, Boruah BR, Dunsby C, Lanigan PMP, Kennedy G, Neil MAA, French PMW (2008) Stimulated emission depletion microscopy with a supercontinuum source and fluorescence lifetime imaging. Opt Lett 33(2):113–115
84. Dyba M, Jakobs S, Hell SW (2003) Immunofluorescence stimulated emission depletion microscopy. Nat Biotechnol 21(11):1303–1304
85. Westphal V, Seeger J, Salditt T, Hell SW (2005) Stimulated emission depletion microscopy on lithographic nanostructures. J Phys B At Mol Opt Phys 38(9):S695–S705
86. Willig KI, Kellner RR, Medda R, Hein B, Jakobs S, Hell SW (2006a) Nanoscale resolution in GFP-based microscopy. Nat Methods 3(9):721–723
87. Hein B, Willig KI, Wurm CA, Westphal V, Jakobs S, Hell SW (2010) Stimulated emission depletion nanoscopy of living cells using SNAP-Tag fusion proteins. Biophys J 98(1):158–163
88. Han KY, Willig KI, Rittweger E, Jelezko F, Eggeling C, Hell SW (2009) Three-dimensional stimulated emission depletion microscopy of nitrogen-vacancy centers in diamond using continuous-wave light. Nano Lett 9(9):3323–3329
89. Irvine SE, Staudt T, Rittweger E, Engelhardt J, Hell SW (2008) Direct light-driven modulation of luminescence from Mn-doped ZnSe quantum dots. Angew Chem Int Ed 47(14):2685–2688
90. Westphal V, Blanca CM, Dyba M, Kastrup L, Hell SW (2003b) Laser-diode-stimulated emission depletion microscopy. Appl Phys Lett 82(18):3125–3127
91. Rankin BR, Kellner RR, Hell SW (2008) Stimulated-emission-depletion microscopy with a multicolor stimulated-Raman-scattering light source. Opt Lett 33(21):2491–2493
92. Rankin BR, SW Hell (2009) STED microscopy with a MHz pulsed stimulated-Raman-scattering source. Opt Express 17(18):15679–15684
93. Willig KI, Harke B, Medda R, Hell SW (2007) STED microscopy with continuous wave beams. Nat Methods 4(11):915–918
94. Ranka JK, Windeler RS, Stentz AJ (2000) Visible continuum generation in air-silica microstructure optical fibers with anomalous dispersion at 800 nm. Opt Lett 25(1):25–27
95. Courvoisier C, Giust R (2006) Using a continuum of light in STED confocal microscopy. In: Proc. SPIE – biophotonics and new therapy frontiers, vol 6191, pp 619108
96. Wildanger D, Rittweger E, Kastrup L, Hell SW (2008) STED microscopy with a supercontinuum laser source. Opt Express 16(13):9614–9621
97. Donnert G, Eggeling C, Hell SW (2007a) Major signal increase in fluorescence microscopy through dark-state relaxation. Nat Methods 4(1):81–86
98. Moneron G, Medda R, Hein B, Giske A, Westphal V, Hell SW (2010) Fast STED microscopy with continuous wave fiber lasers. Opt Express 18(2):1302–1309
99. Donnert G, Keller J, Wurm CA, Rizzoli SO, Westphal V, Schönle A, Jahn R, Jakobs S, Eggeling C, Hell SW (2007b) Two-color far-field fluorescence nanoscopy. Biophys J 92(8):L67–L69
100. Meyer L, Wildanger D, Medda R, Punge A, Rizzoli SO, Donnert G, Hell SW (2008) Dual-color STED microscopy at 30-nm focal-plane resolution. Small 4(8):1095–1100
101. Neumann D, Bückers J, Kastrup L, Hell SW, Jakobs S (2010) Two-color STED microscopy reveals different degrees of colocalization between hexokinase-I and the three human VDAC isoforms. PMC Biophys 3(1):4
102. Schmidt R, Wurm CA, Jakobs S, Engelhardt J, Egner A, Hell SW (2008) Spherical nanosized focal spot unravels the interior of cells. Nat Methods 5(6):539–544
103. Lauterbach MA (2009) Fast STED Microscopy. PhD thesis, Georg-August-Universität zu Göttingen, Göttingen, Germany
104. Friedemann K, Turshatov A, Landfester K, Crespy D (2011) Characterization via two-color STED microscopy of nanostructured materials synthesized by colloid electrospinning. Langmuir 27(11):7132–7139
105. Tønnesen J, Nadrigny F, Willig KI, Wedlich-Söldner R, Nägerl UV (2011) Two-color STED microscopy of living synapses using a single laser-beam pair. Biophys J 101:2545–2552
106. Willig KI, Stiel AC, Brakemann T, Jakobs S, Hell SW (2011) Dual-label sted nanoscopy of

living cells using photochromism. Nano Lett 11(9):3970–3973
107. Bückers J, Wildanger D, Vicidomini G, Kastrup L, Hell SW (2011) Simultaneous multi-lifetime multi-color STED imaging for colocalization analyses. Opt Express 19(4):3130–3143
108. Lauterbach MA, Guillon M, Soltani A, Emiliani V (2013) STED microscope with spiral phase contrast. Sci Rep 3:2050
109. Denk W, Strickler JH, Webb WW (1990) Two-photon laser scanning fluorescence microscopy. Science 248(4951):73–76
110. Ding JB, Takasaki KT, Sabatini BL (2009) Supraresolution imaging in brain slices using stimulated-emission depletion two-photon laser scanning microscopy. Neuron 63(4):429–437
111. Moneron G, Hell SW (2009) Two-photon excitation STED microscopy. Opt Express 17(17):14567–14573
112. Li Q, Wu SSH, Chou KC (2009a) Subdiffraction-limit two-photon fluorescence microscopy for GFP-tagged cell imaging. Biophys J 97:3224–3228
113. Belfield KD, Bondar MV, Yanez CO, Hernandez FE, Przhonska OV (2009) One- and two-photon stimulated emission depletion of a sulfonyl-containing fluorene derivative. J Phys Chem B 113(20):7101–7106
114. Westphal V, Hell SW (2005b) Nanoscale resolution in the focal plane of an optical microscope. Phys Rev Lett 94(14):143903
115. Klar TA, Jakobs S, Dyba M, Egner A, Hell SW (2000) Fluorescence microscopy with diffraction resolution barrier broken by stimulated emission. Proc Natl Acad Sci USA 97(15):8206–8210
116. Harke B, Ullal CK, Keller J, Hell SW (2008a) Three-dimensional nanoscopy of colloidal crystals. Nano Lett 8(5):1309–1313
117. Wildanger D, Medda R, Kastrup L, Hell SW (2009) A compact STED microscope providing 3D nanoscale resolution. J Microsc 236:35–43
118. Gould TJ, Myers JR, Bewersdorf J (2011) Total internal reflection STED microscopy. Opt Express 19(14):13351–13357
119. Leutenegger M, Ringemann C, Lasser T, Hell SW, Eggeling C (2012) Fluorescence correlation spectroscopy with a total internal reflection fluorescence STED microscope (TIRF-STED-FCS). Opt Express 20(5):5243–5263
120. Huisken J, Swoger J, Del Bene F, Wittbrodt J, Stelzer EHK (2004) Optical sectioning deep inside live embryos by selective plane illumination microscopy. Science 305:1007–1009
121. Friedrich M, Gan Q, Ermolayev V, Harms GS (2011) STED-SPIM: Stimulated emission depletion improves sheet illumination microscopy resolution. Biophys J 100:L43–L45
122. Simpson GJ (2006) Biological imaging – the diffraction barrier broken. Nature 440(7086):879–880
123. Kittel RJ, Wichmann C, Rasse TM, Fouquet W, Schmidt M, Schmid A, Wagh DA, Pawlu C, Kellner RR, Willig KI, Hell SW, Buchner E, Heckmann M, Sigrist SJ (2006) Bruchpilot promotes active zone assembly, Ca^{2+} channel clustering, and vesicle release. Science 312 (5776):1051–1054
124. Willig KI, Rizzoli SO, Westphal V, Jahn R, Hell SW (2006b) STED microscopy reveals that synaptotagmin remains clustered after synaptic vesicle exocytosis. Nature 440(7086):935–939
125. Lin W, Margolskee R, Donnert G, Hell SW, Restrepo D (2007) Olfactory neurons expressing transient receptor potential channel M5 (TRPM5) are involved in sensing semiochemicals. Proc Natl Acad Sci USA 104(7):2471–2476
126. Kellner RR, Baier CJ, Willig KI, Hell SW, Barrantes FJ (2007) Nanoscale organization of nicotinic acetylcholine receptors revealed by stimulated emission depletion microscopy. Neuroscience 144(1):135–143
127. Sieber JJ, Willig KI, Heintzmann R, Hell SW, Lang T (2006) The SNARE motif is essential for the formation of syntaxin clusters in the plasma membrane. Biophys J 90(8):2843–2851
128. Meyer AC, Frank T, Khimich D, Hoch G, Riedel D, Chapochnikov NM, Yarin YM, Harke B, Hell SW, Egner A, Moser T (2009) Tuning of synapse number, structure and function in the cochlea. Nat Neurosci 12(4):444–453
129. Schneider A, Rajendran L, Honsho M, Gralle M, Donnert G, Wouters F, Hell SW, Simons M (2008) Flotillin-dependent clustering of the amyloid precursor protein regulates its endocytosis and amyloidogenic processing in neurons. J Neurosci 28(11):2874–2882
130. Schmidt R, Wurm CA, Punge A, Egner A, Jakobs S, Hell SW (2009) Mitochondrial cristae revealed with focused light. Nano Lett 9(6):2508–2510
131. Hein B, Willig KI, Hell SW (2008) Stimulated emission depletion (STED) nanoscopy of a fluorescent protein-labeled organelle inside a living cell. Proc Natl Acad Sci USA 105(38):14271–14276
132. Nägerl UV, Willig KI, Hein B, Hell SW, Bonhoeffer T (2008) Live-cell imaging of dendritic spines by STED microscopy. Proc Natl Acad Sci USA 105(48):18982–18987

133. Hein B (2009) Live Cell STED Microscopy Using Genetically Encoded Markers. PhD thesis, Georg-August-Universität zu Göttingen
134. Urban NT, Willig KI, Hell SW, Nägerl UV (2011) STED nanoscopy of actin dynamics in synapses deep inside living brain slices. Biophys J 101(5):1277–1284
135. Mueller V, Ringemann C, Honigmann A, Schwarzmann G, Medda R, Leutenegger M, Polyakova S, Belov VN, Hell SW, Eggeling C (2011) STED nanoscopy reveals molecular details of cholesterol- and cytoskeleton-modulated lipid interactions in living cells. Biophys J 101(7):1651–1660
136. Westphal V, Lauterbach MA, Di Nicola A, Hell SW (2007) Dynamic far-field fluorescence nanoscopy. New J Phys 9:435
137. Lauterbach MA, Ullal C, Westphal V, Hell SW (2010a) Dynamic imaging of colloidal-crystal nanostructurs at 200 frames per second. Langmuir 26(18):14400–14404
138. Westphal V, Rizzoli SO, Lauterbach MA, Kamin D, Jahn R, Hell SW (2008a) Video-rate far-field optical nanoscopy dissects synaptic vesicle movement. Science 320(5873): 246–249
139. Lauterbach MA, Keller J, Schönle A, Kamin D, Westphal V, Rizzoli SO, Hell SW (2010b) Comparing video-rate STED nanoscopy and confocal microscopy of living neurons. J Biophotonics 3(7):417–424
140. Hoopmann P, Punge A, Barysch SV, Westphal V, Bückers J, Opazo F, Bethani I, Lauterbach MA, Hell SW, Rizzoli SO (2010) Endosomal sorting of readily releasable synaptic vesicles. Proc Natl Acad Sci USA 107(44): 19055–19060
141. Kamin D, Lauterbach MA, Westphal V, Keller J, Schönle A, Hell SW, Rizzoli SO (2010) High- and low-mobility stages in the synaptic vesicle cycle. Biophys J 99:675–684
142. Ullal CK, Schmidt R, Hell SW, Egner A (2009) Block copolymer nanostructures mapped by far-field optics. Nano Lett 9(6): 2497–2500
143. Li LJ, Gattass RR, Gershgoren E, Hwang H, Fourkas JT (2009b) Achieving $\lambda/20$ resolution by one-color initiation and deactivation of polymerization. Science 324(5929): 910–913
144. Scott TF, Kowalski BA, Sullivan AC, Bowman CN, McLeod RR (2009) Two-color single-photon photoinitiation and photoinhibition for subdiffraction photolithography. Science 324(5929):913–917
145. Andrew TL, Tsai HY, Menon R (2009) Confining light to deep subwavelength dimensions to enable optical nanopatterning. Science 324(5929):917–921
146. Fischer J, Freymann G, Wegener M (2010) The materials challenge in diffractionunlimited direct-laser-writing optical lithography. Adv Mater 22:3578–3582
147. Hell SW, Jakobs S, Kastrup L (2003) Imaging and writing at the nanoscale with focused visible light through saturable optical transitions. Appl Phys A Mater Sci Process 77: 859–860
148. Heintzmann R, Jovin TM, Cremer C (2002) Saturated patterned excitation microscopy—a concept for optical resolution improvement. J Opt Soc Am A Opt Image Sci Vis 19(8): 1599–1609
149. Bretschneider S, Eggeling C, Hell SW (2007) Breaking the diffraction barrier in fluorescence microscopy by optical shelving. Phys Rev Lett 98:218103
150. Gustafsson MGL (2005) Nonlinear structured-illumination microscopy: Widefield fluorescence imaging with theoretically unlimited resolution. Proc Natl Acad Sci USA 102(37):13081–13086
151. Rittweger E, Wildanger D, Hell SW (2009b) Far-field fluorescence nanoscopy of diamond color centers by ground state depletion. Europhys Lett 86(1):14001
152. Han KY, Kim SK, Eggeling C, Hell SW (2010) Metastable dark states enable ground state depletion microscopy of nitrogen vacancy centers in diamond with diffraction-unlimited resolution. Nano Lett 10(8): 3199–3203
153. Eggeling C, Volkmer A, Seidel CAM (2005) Molecular photobleaching kinetics of Rhodamine 6G by one- and two-photon induced confocal fluorescence microscopy. ChemPhysChem 6(5):791–804
154. Hofmann M, Eggeling C, Jakobs S, Hell SW (2005) Breaking the diffraction barrier in fluorescence microscopy at low light intensities by using reversibly photoswitchable proteins. Proc Natl Acad Sci USA 102(49):17565–17569
155. Bossi M, Fölling J, Dyba M, Westphal V, Hell SW (2006) Breaking the diffraction resolution barrier in far-field microscopy by molecular optical bistability. New J Phys 8:275
156. Stiel AC, Andresen M, Bock H, Hilbert M, Schilde J, Schoenle A, Eggeling C, Egner A, Hell SW, Jakobs S (2008) Generation of monomeric reversibly switchable red fluorescent proteins for far-field fluorescence nanoscopy. Biophys J 95(6):2989–2997
157. Grotjohann T, Testa I, Leutenegger M, Bock H, Urban NT, Lavoie-Cardinal F, Willig KI,

157. Eggeling C, Jakobs S, Hell SW (2011) Diffraction-unlimited all-optical imaging and writing with a photochromic GFP. Nature 478(7368):204–208
158. Brakemann T, Stiel AC, Weber G, Andresen M, Testa I, Grotjohann T, Leutenegger M, Plessmann U, Urlaub H, Eggeling C, Wahl MC, Hell SW, Jakobs S (2011) A reversibly photoswitchable GFP-like protein with fluorescence excitation decoupled from switching. Nat Biotechnol 29(10):942–947
159. Rego EH, Shao L, Macklin JJ, Winoto L, Johansson GA, Kamps-Hughes N, Davidson MW, Gustafsson MGL (2012) Nonlinear structured-illumination microscopy with a photoswitchable protein reveals cellular structures at 50-nm resolution. Proc Natl Acad Sci USA 109(3):E135–E143
160. Schwentker MA, Bock H, Hofmann M, Jakobs S, Bewersdorf J, Eggeling C, Hell SW (2007) Wide-field subdiffraction RESOLFT microscopy using fluorescent protein photoswitching. Microsc Res Tech 70(3):269–280
161. Chmyrov A, Keller J, Grotjohann T, Ratz M, d'Este E, Jakobs S, Eggeling C, Hell SW (2013) Nanoscopy with more than 100,000 'doughnuts'. Nat Methods 10(8):737–740
162. Schwentker M. Parallelized ground state depletion. PhD thesis, Ruperto-Carola University of Heidelberg, 2007
163. Enderlein J (2005) Breaking the diffraction limit with dynamic saturation optical microscopy. Appl Phys Lett 87(9):094105
164. Sýkora J, Dertinger T, Enderlein J (2006) Dynamic optical saturation microscopy. In: Enderlein J, Gryczynski ZK (eds) Ultrasensitive and single-molecule detection technologies, vol 6092. Proc SPIE
165. Humpolíčková J, Benda A, Enderlein J (2009) Optical saturation as a versatile tool to enhance resolution in confocal microscopy. Biophys J 97(9):2623–2629
166. Betzig E, Patterson GH, Sougrat R, Lindwasser OW, Olenych S, Bonifacino JS, Davidson MW, Lippincott-Schwartz J, Hess HF (2006) Imaging intracellular fluorescent proteins at nanometer resolution. Science 313(5793):1642–1645
167. Rust MJ, Bates M, Zhuang XW (2006) Sub-diffraction-limit imaging by stochastic optical reconstruction microscopy (STORM). Nat Methods 3(10):793–795
168. Zhuang XW (2009) Nano-imaging with STORM. Nat Photon 3(7):365–367
169. Hess ST, Girirajan TPK, Mason MD (2006) Ultra-high resolution imaging by fluorescence photoactivation localization microscopy. Biophys J 91(11):4258–4272
170. Heisenberg W (1930) Die physikalischen Prinzipien der Quantentheorie. Hirzel, Leipzig
171. Thompson RE, Larson DR, Webb WW (2002) Precise nanometer localization analysis for individual fluorescent probes. Biophys J 82(5):2775–2783
172. Gordon MP, Ha T, Selvin PR (2004) Single-molecule high-resolution imaging with photobleaching. Proc Natl Acad Sci USA 101:6462–6465
173. Qu X, Wu D, Mets L, Scherer NF (2004) Nanometer-localized multiple single-molecule fluorescence microscopy. Proc Natl Acad Sci USA 101(31):11298–11303
174. Lidke KA, Rieger B, Jovin TM, Heintzmann R (2005) Superresolution by localization of quantum dots using blinking statistics. Opt Express 13(18):7052–7062
175. Engelhardt J, Keller J, Hoyer P, Reuss M, Staudt T, Hell SW (2011) Molecular orientation affects localization accuracy in superresolution far-field fluorescence microscopy. Nano Lett 11:209–213
176. Egner A, Geisler C, von Middendorff C, Bock H, Wenzel D, Medda R, Andresen M, Stiel AC, Jakobs S, Eggeling C, Schönle A, Hell SW (2007) Fluorescence nanoscopy in whole cells by asynchronous localization of photoswitching emitters. Biophys J 93(9):3285–3290
177. Biteen JS, Thompson MA, Tselentis NK, Shapiro L, Moerner WE (2009) Superresolution imaging in live caulobacter crescentus cells using photoswitchable enhanced yellow fluorescent protein. In: Proc SPIE – Single Molecule Spectroscopy and Imaging II, vol 7185, pp 71850I
178. Lemmer P, Gunkel M, Baddeley D, Kaufmann R, Urich A, Weiland Y, Reymann J, Müller P, Hausmann M, Cremer C (2008) SPDM: light microscopy with single-molecule resolution at the nanoscale. Appl Phys B Lasers Opt 93(1):1–12
179. Shtengel G, Galbraith JA, Galbraith CG, Lippincott-Schwartz J, Gillette JM, Manley S, Sougrat R, Waterman CM, Kanchanawong P, Davidson MW, Fetter RD, Hess HF (2009) Interferometric fluorescent super-resolution microscopy resolves 3D cellular ultrastructure. Proc Natl Acad Sci USA 106(9):3125–3130
180. Peng W (2009) PALM reading. Nat Methods 6(4):243–243
181. Pavani SRP, Thompson MA, Biteen JS, Lord SJ, Liu N, Twieg RJ, Piestun R, Moerner WE (2009) Three-dimensional, single-molecule fluorescence imaging beyond the diffraction limit by using a double-helix point spread function. Proc Natl Acad Sci USA 106(9):2995–2999

182. Heilemann M, van de Linde S, Schuttpelz M, Kasper R, Seefeldt B, Mukherjee A, Tinnefeld P, Sauer M (2008) Subdiffraction-resolution fluorescence imaging with conventional fluorescent probes. Angew Chem Int Ed 47(33):6172–6176

183. Fölling J, Bossi M, Bock H, Medda R, Wurm CA, Hein B, Jakobs S, Eggeling C, Hell SW (2008) Fluorescence nanoscopy by ground-state depletion and single-molecule return. Nat Methods 5(11):943–945

184. Fölling J (2008) High-resolution microscopy with photoswitchable organic markers. PhD thesis, Georg-August-Universität zu Göttingen, Göttingen, Germany

185. Andresen M, Stiel AC, Fölling J, Wenzel D, Schonle A, Egner A, Eggeling C, Hell SW, Jakobs S (2008) Photoswitchable fluorescent proteins enable monochromatic multilabel imaging and dual color fluorescence nanoscopy. Nat Biotechnol 26(9):1035–1040

186. Fölling J, Belov V, Kunetsky R, Medda R, Schonle A, Egner A, Eggeling C, Bossi M, Hell SW (2007) Photochromic rhodamines provide nanoscopy with optical sectioning. Angew Chem Int Ed 46(33):6266–6270

187. Bock H, Geisler C, Wurm CA, von Middendorff C, Jakobs S, Schönle A, Egner A, Hell SW, Eggeling C (2007) Two-color far-field fluorescence nanoscopy based on photoswitchable emitters. Appl Phys B Lasers Opt 88(2):161–165

188. Steinhauer C, Forthmann C, Vogelsang J, Tinnefeld P (2008) Superresolution microscopy on the basis of engineered dark states. J Am Chem Soc 130(50):16840–16841

189. Sharonov A, Hochstrasser RM (2006) Widefield subdiffraction imaging by accumulated binding of diffusing probes. Proc Natl Acad Sci USA 103(50):18911–18916

190. Schwering M, Kiel A, Kurz A, Lymperopoulos K, Sprödefeld A, Krämer R, Herten D-P (2011) Far-field nanoscopy with reversible chemical reactions. Angew Chem Int Ed 50(13):2940–2945

191. Hoyer P, Staudt T, Engelhardt J, Hell SW (2011) Quantum dot blueing and blinking enables fluorescence nanoscopy. Nano Lett 11(1):245–250

192. Patterson G, Davidson M, Manley S, Lippincott-Schwartz J (2010) Superresolution imaging using single-molecule localization. Annu Rev Phys Chem 61:345–367

193. Endesfelder U, van de Linde S, Wolter S, Sauer M, Heilemann M (2010) Subdiffraction-resolution fluorescence microscopy of myosin-actin motility. Chem Phys Chem 11(4):836–840

194. Jones SA, Shim S-H, He J, Zhuang X (2011) Fast, three-dimensional super-resolution imaging of live cells. Nat Methods 8(6):499–508

195. Einstein A (1916) Zur Quantentheorie der Strahlung. Mitteilungen der Physikalischen Gesellschaft Zürich 18:47–62

196. McCumber DE (1964) Einstein relations connecting broadband emission and absorption spectra. Phys Rev A Gen Phys 136(4A):A954–A957

197. Jabłoński A (1933) Efficiency of Anti-Stokes fluorescence in dyes. Nature 131:839–840

198. Smentek L (2009) Different sides of the Jabłoński diagram on its 75th anniversary. Newsletter of the Forum on International Physics, American Physical Society, http://www.aps.org/units/fip/newsletters/200906/upload/jablonski.pdf

199. Giske A (2007) CryoSTED microscopy A new spectroscopic approach for improving the resolution of STED microscopy using low temperature. PhD thesis, Ruperto-Carola University of Heidelberg, Germany

200. Willig KI (2006) STED microscopy in the visible range. PhD thesis, Ruperto-Carola University of Heidelberg, Germany

201. Klar TA, Engel E, Hell SW (2001) Breaking Abbe's diffraction resolution limit in fluorescence microscopy with stimulated emission depletion beams of various shapes. Phys Rev E 64(6):066613

202. Harke B, Keller J, Ullal CK, Westphal V, Schönle A, Hell SW (2008b) Resolution scaling in STED microscopy. Opt Express 16(6):4154–4162

203. Bingen P, Reuss M, Engelhardt J, Hell SW (2011) Parallelized STED fluorescence nanoscopy. Opt Express 19(24):23716–23726

204. Gregor I, Patra D, Enderlein J (2004) Optical saturation in fluorescence correlation spectroscopy under continuous-wave and pulsed excitation. ChemPhysChem 5:1–7

205. Harke B (2008) 3D STED Microscopy with Pulsed and Continuous Wave Lasers. PhD thesis, Georg-August-Universität zu Göttingen, Göttingen, Germany

206. Leutenegger M, Eggeling C, Hell SW (2010) Analytical description of STED microscopy performance. Opt Express 18(25):26417–26429

207. Vicidomini G, Moneron G, Han KY, Westphal V, Ta H, Reuss M, Engelhardt J, Eggeling C, Hell SW (2011) Sharper low-power STED nanoscopy by time gating. Nat Methods 8(7):571–573

208. Schrader M, Meinecke F, Bahlmann L, Kroug M, Cremer C, Soini E, Hell SW (1995) Monitoring the excited state of a fluorophore

208. in a microscope by stimulated emission. Bioimaging 3(4):147–153
209. Moffitt JR, Osseforth C, Michaelis J (2011) Time-gating improves the spatial resolution of sted microscopy. Opt Express 19(5):4242–4254
210. Vicidomini G, Schönle A, Ta H, Han KY, Moneron G, Eggeling C, Hell SW (2013) STED nanoscopy with time-gated detection: Theoretical and experimental aspects. PLoS ONE 8(1):e54421
211. Westphal V, Rizzoli SO, Lauterbach MA, Kamin D, Jahn R, Hell SW (2008b) Video-rate far-field optical nanoscopy dissects synaptic vesicle movement. Science 320(5873):246–249
212. Magde D, Elson E, Webb WW (1972) Thermodynamic fluctuations in a reacting system—measurement by fluorescence correlation spectroscopy. Phys Rev Lett 29:705–708
213. Ehrenberg M, Rigler R (1974) Rotational brownian motion and fluorescence intensify fluctuations. Chem Phys 4(3):390–401
214. Haustein E, Schwille P (2003) Ultrasensitive investigations of biological systems by fluorescence correlation spectroscopy. Methods 29(2):153–166
215. Widengren J, Rigler R (1990) Ultrasensitive detection of single molecules using fluorescence correlation spectroscopy. In: Klinge B, Owman C (eds) Bioscience. Lund University Press, Lund, pp 180–183
216. Wawrezinieck L, Rigneault H, Marguet D, Lenne P-F (2005) Fluorescence correlation spectroscopy diffusion laws to probe the submicron cell membrane organization. Biophys J 89(6):4029–4042
217. Lingwood D, Simons K (2010) Lipid rafts as a membrane-organizing principle. Science 327(5961):46–50
218. Ronchi V (1961) Resolving power of calculated and detected images. J Opt Soc Am 51(4):458–460
219. Whittaker ET (1915) On the functions which are represented by the expansions of the interpolation-theory. Proc R Soc 35:181–194
220. Lüke HD (1999) The origins of the sampling theorem. IEEE Commun Mag 37(4):106–108
221. Shannon CE (1948) A mathematical theory of communication. Bell Syst Tech J 27(3):379–423 and 623–656
222. Shannon CE (1949) Communication in the presence of noise. Proc Inst Radio Eng 37(1):10–21
223. Nyquist H (1928) Certain topics in telegraph transmission theory. J Am Inst Electr Eng 47:214–216
224. Rayleigh JW (1874) On the manufacture and theory of diffraction-gratings. Philos Mag Ser 4 47(310):81–93
225. Carlsson K, Wallen P, Brodin L (1989) 3-dimensional imaging of neurons by confocal fluorescence microscopy. J Microsc 155:15–26
226. Shroff H, Galbraith CG, Galbraith JA, Betzig E (2008) Live-cell photoactivated localization microscopy of nanoscale adhesion dynamics. Nat Methods 5(5):417–423
227. Ram S, Ward ES, Ober RJ (2006) Beyond rayleigh's criterion: A resolution measure with application to single-molecule microscopy. Proc Natl Acad Sci USA 103(12):4457–4462
228. Kolobov MI (2008) Quantum limits of super-resolution for imaging discrete subwavelength structures. Opt Express 16(1):58–66
229. Beskrovny VN, Kolobov MI (2008) Quantum-statistical analysis of superresolution for optical systems with circular symmetry. Phys Rev A 78(4):043824
230. Falconi O (1967) Limits to which double lines, double stars, and disks can be resolved and measured. J Opt Soc Am 57(8):987–993
231. Lucy LB (1992) Statistical limits to superresolution. Astron Astrophys 261(2):706–710
232. Fried DL (1979) Resolution, signal-to-noise ratio, and measurement precision. J Opt Soc Am 69(3):399–406
233. Fannjiang A, Sølna K (2007) Broadband resolution analysis for imaging with measurement noise. J Opt Soc Am A Opt Image Sci Vis 24(6):1623–1632
234. Terebizh VYu (1999) Using a priori information in image restoration: Natural resolution limit. Astron Rep 43(1):42–58

Chapter 4

Application of Real-Time STED Imaging to Synaptic Vesicle Motion

Benjamin G. Wilhelm and Dirk Kamin

Abstract

In the emerging field of super-resolution microscopy, the branch of live-cell imaging is still in its infancy. Regardless of its importance for addressing relevant biological questions, live super-resolution imaging has to face several obstacles when compared to conventional imaging: (1) speeding up the naturally slow image acquisition process, (2) choosing appropriate fluorophores (both in terms of photostability and spectral properties), and (3) handling increased illumination intensities (as usually higher intensities are needed for live imaging at adequate frequencies compared to fixed-cell imaging). In this chapter, we review recent progress made with stimulated emission depletion (STED) microscopy in imaging single synaptic vesicles at video-rate and discuss the technical difficulties that were to overcome. We give a brief overview on the use of conventional fusion proteins as live-cell markers for STED microscopy (such as the green fluorescent protein, GFP, and its derivates) as well as on new labeling techniques for live-cell STED imaging (SNAP-, CLIP-, and HaLo-tags). Besides STED microscopy also other super-resolution approaches have performed imaging of living cells in the past couple of years. Here, the progress of structured illumination microscopy (SIM) and single-molecule detection approaches, such as STORM and (F)PALM, are also discussed.

Key words Live super-resolution imaging, STED microscopy, Synaptic vesicles, Fluorescent proteins and labeling techniques

1 Synaptic Vesicles as a Paradigm for Live Super-Resolution Microscopy

Communication between neurons is crucial for essentially any body function. In this respect, the majority of our neurons communicate via chemical synapses, which employs the fusion of the so-called synaptic vesicles (SVs) with the plasma membrane of the presynaptic cell. The SVs accumulate in clusters in the presynaptic terminal (Fig. 1a–I). Upon arrival of an electrical signal at the synapse, the vesicles fuse with the plasma membrane and release their cargo—the neurotransmitters—into the synaptic cleft (exocytosis; Fig. 1a–II). Here, the neurotransmitters diffuse and dock to receptors in the plasma membrane of the postsynaptic cell, which causes a change in the electric potential therein, meaning that the signal has propagated from one cell to the next [1, 2].

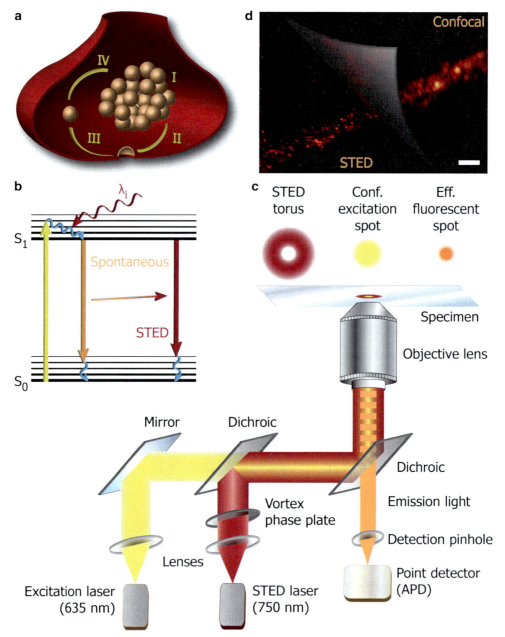

Fig. 1 (**a**) Cartoon of the mobility in the synaptic vesicle cycle. SVs are immobilized in a vesicle cluster (I) inside the small synapse. Stimulation leads to their movement towards the active zone (II) where they fuse with the plasma membrane to release neurotransmitter into the synaptic cleft. The vesicle material gets retrieved via endocytosis to form a new vesicle (III), which is integrated into the vesicle cluster (IV). (**b**) Jablonski diagram depicting energetic states of fluorophores during STED imaging. Excitation of a fluorophore results in the elevation (*yellow arrow*) of an electron from its ground state (S_0) to a higher energetic level (S_1). In conventional fluorescence, the electron then returns to the ground state by spontaneous emission of light (spontaneous, *orange arrow*). In STED microscopy, the return to the ground state is forced by stimulated emission via illumination of the sample by photons of a higher wavelength (λ_l). The fluorescence, which is released in this case (STED), is not in the spectral range of the spontaneous emission but red-shifted and is therefore not detected.

Subsequently, SV components are retrieved from the plasma membrane (endocytosis), vesicles are refilled with neurotransmitter (Fig. 1a–III) and eventually returned to the vesicle cluster (Fig. 1a–IV). This process of exo- and endocytosis is commonly described as the synaptic vesicle cycle [1]. According to this classical model of SV recycling, the vesicles need to be mobile to serve their function (i.e., fusion and release of the neurotransmitter); hence, it is important to characterize the movement of individual vesicles in order to completely understand synaptic function.

Unfortunately, the live investigation of single vesicle movements using conventional imaging techniques is impeded by two parameters: (1) SVs are rather small with a diameter of approximately 42 nm [3] and (2) they are densely packed in the small-sized space of the synapse, the vesicle cluster (see Fig. 1a). As the spatial resolution of a conventional fluorescence microscope is restricted by the diffraction-limit to approximately 200–300 nm, it is impossible to resolve individual vesicles that are in such close proximity to each other. Hence, it is difficult to determine the motion behavior of individual SVs, and past studies have either addressed the motion of single vesicles one at a time [4] or the motion of the entire cluster [5, 6], but never differentiated a substantial amount of vesicles simultaneously. Since this is especially relevant in the context of different SV populations which are expected to play distinctive roles during synaptic transmission (for a review on such populations of pools see [7]), it is desirable to investigate the movement of individual SVs.

2 Super-Resolution STED Microscopy

The first of several far-field microscopy techniques that allows imaging below the diffraction-limit was *stimulated emission depletion* (STED) microscopy, invented by Stefan W. Hell [8]. The principle of this method is based on the process of *stimulated emission*, which was already postulated by Albert Einstein in 1917 [9].

Fig. 1 (continued) (**c**) Principle of a STED microscope. The fluorophore is excited by the excitation beam [usually ATTO 647N with a 635 nm laser (*yellow beam*)]. The corresponding STED (depletion) laser has a higher wavelength of 750 nm (*red beam*) to allow depletion of excited molecules (see (**b**)). A vortex phase plate in the beam path shapes the torus-like STED beam, which is superimposed with the excitation beam through a dichroic mirror. Both beams are focused onto the specimen via the objective lens. The molecules in the outer region of the diffraction-limited excitation spot (*yellow*) are hindered in spontaneous emission by the STED torus (*red*), resulting in a decreased effective excitation spot (i.e., increased resolution). The fluorescence passes a dichroic mirror and is detected by avalanche photo diodes (APD). (**d**) Comparison of confocal and STED resolution. Primary hippocampal neurons were labeled for the synaptic vesicle protein synaptotagmin and imaged with a commercially available Leica TCS STED system. Please note that in the confocal image only large spots are detected while the STED equivalent allows to discern separate vesicles within these spots. Scale bar 1 μm

Stimulated emission describes a process (illustrated in Fig. 1b) in which an excited fluorophore (transition from S_0 to S_1) is hit [while residing in the short-lived S_1 state (the life-time has a few nanoseconds)] by a photon of a higher wavelength (λi—STED depletion beam) and instead of emitting fluorescence at its typical wavelength, spontaneous emission, the fluorophore returns to the ground state (transition from S_1 to S_0) by emitting light of the wavelength of the depleting beam (STED). This process can be used in microscopy to restrict the size of the effective excitation beam thus increasing the net resolution.

In STED microscopy (Fig. 1c), the excitation beam (excitation laser, yellow) is superimposed with a STED beam (STED laser, red) modified to feature a local intensity zero in its center (i.e., doughnut-shaped) that coincides with the focal spot of the excitation beam. The STED beam has a longer wavelength (red-shifted) and is thus able to deplete the fluorescence of exited fluorophores in the specimen via stimulated emission. Hence, fluorescence is only allowed in the small area of the intensity zero of the STED beam (basically the center/hole in the doughnut), i.e., the effective fluorescence spot is significantly reduced [10]. In the past, STED was able to achieve resolutions below 50 nm in multicolor imaging of biological samples [11] and even down to 6 nm imaging nitrogen vacancy centers in crystal diamonds [12]. The increase in resolution in STED microscopy allows discerning small structures within a synapse where a confocal microscope is only able to resolve a blurred large spot (see Fig. 1d).

3 Real-Time Super-Resolution Imaging Using STED Microscopy

In order to apply super-resolution STED microscopy to living samples, specific imaging requirements have to be met. Our group has performed several studies employing live super-resolution STED microscopy in the past [13–15]. In this respect, we found the following aspects the most important for applying STED microscopy to living samples:

1. Scanning speed: we restricted the field of view to a relatively small area of 2.5 × 1.8 μm. This automatically increases the scanning speed as the total imaging area is decreased. Additionally, the excitation and the depletion beam were scanned across the specimen using a 16-kHz resonant mirror, which allowed fast scanning of the specimen in one axis, whereas the perpendicular axis was accessed using a piezo-controlled specimen stage.

2. Signal detection: one important parameter for live-STED microscopy is the intensity of the excitation as well as of the STED laser. Both need to be chosen to allow sufficient signal

intensities without bleaching the fluorophores for following rounds of imaging. In our study on the mobility of SVs, we used an effective laser power of 3.5–6 and 400 MW/cm^2 for the excitation and the STED beam, respectively. For detection of the signal, we used several highly sensitive avalanche photodiodes (APD) in parallel, which ensured a high quantum efficiency and signal-to-noise ratio.

3. Fluorophore properties: in STED microscopy, the fluorophore is exposed to substantially more light during the imaging process compared to conventional microscopy (which is inherent to the technique as at least two lasers are used: one for excitation and one for depletion). In this respect, a well-established marker for studying SV cycling—the FM dyes [16]—cannot be used due to its rather low photostability. Also, the most prominent fluorescence markers, the green fluorescent protein (GFP) and its derivates, are not the ideal candidates for live-STED microscopy due to their photochemical properties [17] (also see discussion below on using fluorescent proteins [FPs] for live super-resolution imaging). Therefore, we employed the chemical dye ATTO 647N (Atto-Tec, Siegen, Germany) in our study on SV mobility. Its spectral properties (excitation 625–660 nm and optimal depletion with a 750-nm STED laser) as well as its superior photostability and quantum yield render it an ideal dye for STED microscopy. Further, we made use of highly specific antibodies for labeling the proteins of interest in our studies [18]. Using specific antibodies effectively decreases the amount of unspecific fluorescence, thus enhancing the signal-to-noise ratio.

Combining the above-outlined features allowed us to perform real-time STED imaging of SVs in synapses [13–15]. Also, as our approach is not dependent on sparse labeling of the vesicles (see [4]), we were able to characterize the motility of several individual vesicles simultaneously.

4 Real-Time STED Imaging of Synaptic Vesicle Motion

The "classical" model of SV recycling proposed by Heuser and Reese in 1973 [19] suggests that SVs present a rather constricted motion behavior. Further, this notion was confirmed by several studies, which investigated the averaged motion behavior of SV ensembles (see above). In this respect, we studied the mobility of SVs using super-resolution STED microscopy in real-time. Our findings allow a more differentiated description of the SV mobility in particular regarding the dependence of the vesicle mobility on synaptic activity.

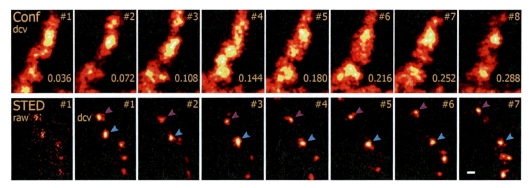

Fig. 2 Real-time STED imaging of SVs. Primary hippocampal neurons were live-labeled against the synaptic vesicle protein synaptotagmin and imaged with video-rate (28 frames per second) using a custom-built STED microscope. While live imaging in confocal mode does not resolve single vesicles (*upper row*), live-STED imaging allows to discern movement of individual vesicles in consecutive frames (frames #1 to #7). To increase signal-to-noise ratio, the raw frames (see first STED frame) were deconvoluted (dcv) with a Gaussian algorithm (80 nm at full width at half maximum; compare first frames in *lower row*, raw vs. dcv), making it possible to track single vesicles automatically. The *numbers* in the confocal frames represent recording times in seconds (which is the same for the respective STED frames). Scale bar 250 nm

In our study [14], we first investigated the mobility of the recently endocytosed vesicles. We found that the majority of vesicles were in a low mobility stage (see purple arrowhead in Fig. 2), whereas only a few were highly mobile at rest (see blue arrowhead in Fig. 2). In Fig. 2, two different time series of moving SVs imaged with conventional confocal microscopy (upper panel) and super-resolution STED microscopy (lower panel) are depicted. While STED microscopy allows capturing the movement of individual vesicles along a neuronal process (see for instance the individual vesicles indicated by the arrowheads), the confocal images only show an almost homogeneous signal with slight shifts in intensity. Postprocessing of the data (Gaussian deconvolution) did in fact lower the resolution but did significantly enhance the signal-to-noise ratio, which allowed unambiguously tracking single vesicles throughout consecutive frames. Despite the distinct mobility stages of the vesicles, the median trace speed we observed was 2 nm/ms.

In summary, we found that (1) individual vesicles undergo high- and low-mobility stages which (2) are not influenced by moderate synaptic activity. Furthermore, imaging experiments of entire boutons confirmed that (3) SVs are exchanged intersynaptically with both resting and recycling populations being similarly involved [14]. Altogether, these studies demonstrate that super-resolution STED microscopy can reliably be applied to living specimens and provide valuable insights into vesicle trafficking, which would not have been possible with conventional microscopy.

5 Live-STED Imaging with Genetically Encoded Fluorescent Probes

The major advantage of genetically encoded FPs is their ability to be expressed as fusion-tags on native proteins, which makes it possible to address structural or functional questions in living cells. Here, GFP and its derivates are the most commonly applied markers in biological sciences [20] and their discovery and development was recognized with the 2008 Nobel Prize in Chemistry. However, their use in super-resolution microscopy has thus far been limited. While pointillistic approaches (such as photoactivation localization microscopy [PALM] and stochastic optical reconstruction microscopy [STORM] [21]) are largely dependent on photoactivatable fluorophores, the limiting factor in STED microscopy is usually their relatively low brightness and photostability compared to conventional dyes used for STED imaging.

Whereas the first study to report the use of GFP in STED microscopy still used fixed cells [17], several more recent studies demonstrated live-STED on biological samples with FPs [22, 23]. Here, the hurdle of the low chromophore brightness was overcome by increasing the recording time up to 10–40 s per frame. Photobleaching was at least partially compensated by the fact that the FPs were used as volume markers (i.e., proteins diffuse freely in the cytosol, which allows rapid replenishment of bleached molecules). In the study of Nägerl et al. [22] STED microscopy was used to visualize activity-dependent structural changes of dendritic spines in brain slices from transgenic mice. Their data shows that STED microscopy is able to reveal structural changes in spine morphology beyond the diffraction barrier (~130 nm resolution) within 10 µm of the slice (with up to 20 s per frame, see Fig. 3). The maximum depths ever reached with STED microscopy so far was achieved using multiphoton excitation in the near infrared [24]. In this study, STED microscopy—when compared to conventional multiphoton microscopy—yields a threefold better resolution (~280 nm) at approximately 100 µm below the surface of the brain slice. However, for this study a pure chemical dye (Alexa Fluor 594) was used. Nevertheless, the results are a good illustration of the potential use of STED microscopy for deep tissue imaging.

These studies demonstrate the potential power of STED microscopy in combination with FPs to address biologically relevant questions. However, regardless of the progress made in the past couple of years (see also *in vivo* work by Rankin et al. [25]), engineering better genetically encoded probes for super-resolution imaging remains critical (i.e., brighter and more photostable). Changing their photophysical properties is difficult, as it usually requires mutating several amino acids of the chromophore, which in turn might alter the overall protein structure and function. One recent successful attempt towards this direction is the development of a novel monomeric far-red fluorescent protein, which has higher

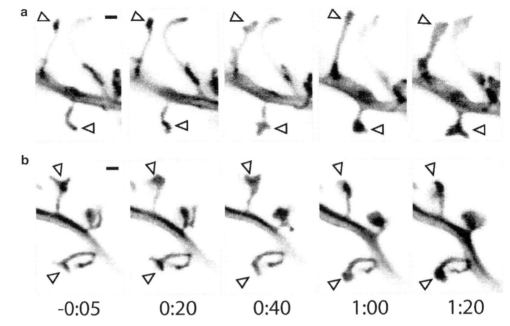

Fig. 3 STED imaging of dendritic spines after plasticity induced stimulation shows structural changes in spine morphology ((**a**) and (**b**)). The *arrows* indicate spines that show structural changes while the times (h:mm) are all in respect to the plasticity inducing stimulation which occurred at 0:00 (time point not shown)

excitation efficiency with red lasers as well as higher photo- and pH-stability [26]. Thus far, it has been used for STED imaging in mammalian cells but not yet live.

6 Potential Alternative Labeling Techniques for Live-STED Microscopy

To benefit from the superior spectral properties of chemical dyes without losing the advantages of genetically encoded fusion proteins (i.e., targeting specific proteins *in vivo*), novel labeling techniques have been developed and successfully used in STED microscopy. One approach uses genetically encoded fluorogen-activating proteins (FAPs). The FAPs are fusion proteins, which can bind and activate certain molecules (fluorogens) which then switch from a dark into a fluorescent state [27]. The advantage of this technique is the combination of a bright chemical dye with a genetic labeling system and as the fluorogen is dark until activation (binding) by the FAP the background fluorescence is negligible. The first STED study using the FAP labeling technique even used the commercially available Leica TCS STED system to investigate living Yeast and HeLa cells with resolutions up to 110 nm (compared to 327 nm in confocal mode) [28]. Besides the FAP tags, there are several other tagging systems available which have been employed for STED microscopy of living cells: Halo-tags [29], CLIP-tags [30], and

SNAP-tags [23, 30]. In all of these technologies, the tag is expressed as a fusion protein. Upon application of the corresponding ligand coupled to a fluorophore (the different technologies have different ligands which allows multicolor experiments), the ligand is covalently bound to its tag, thus marking the protein of interest. In contrast to the FAP technology, the fluorophore is "per se" fluorescent and does not need to be activated. Therefore, it is indispensable to wash off the remaining unbound ligands prior to imaging in order to increase the signal-to-noise ratio.

By combining CLIP and SNAP technologies, Pellett and colleagues were even able to perform dual-color live-STED microscopy using the commercially available Leica TCS STED system [30]. Using HEK293 cells they could record uptake and internal resorting of the epidermal growth factor (EGF-CLIP-Atto 647N) via the EGF receptor (EGFR-SNAP-Chromeo494) in movies with up to 22 frames before photobleaching (12 s per frame). The resolution they obtained was also well below the diffraction barrier with 78 (Atto 647N) and 82 nm (Chromeo494).

In conclusion, alternative labeling techniques for fusion proteins offer a powerful tool to address biological questions with live-STED microscopy. The outlined studies concerning different tagging systems and the use of GFP variants further emphasize two different aspects that resolution itself is not the limiting factor anymore but rather the availability of bright and stable fluorophores.

7 Alternative Techniques for Super-Resolution Live Imaging

Besides STED microscopy, several other super-resolution imaging techniques have been used for live-cell imaging in the past. In the following section, we will introduce two commonly used techniques, which are also commercially available.

7.1 Structured Illumination Microscopy

In structured illumination microscopy (SIM), a striped pattern (grid) is superimposed on the sample while imaging. If a series of images is recorded while rotating the grid, a characteristic variation in the fluorescent signal can be observed, which delivers high-resolution information that is otherwise inaccessible. The final super-resolution image is afterwards reconstructed by the signal variation between all single images [31]. SIM is a wide-field microscopy approach that is particularly useful for imaging large fields of view. The resolution achieved so far is at least 100 nm in the lateral and 200 nm in the axial dimension. As SIM is a non-scanning technique, it is naturally faster than many other super-resolution microscopy techniques (see single-molecule localization techniques below). However, image acquisition and processing (as every single frame is composed of typically 9–15 images) slows down the process substantially compared to imaging small regions of interest with STED microscopy [32]. A recent study reports high-speed SIM of tubulin and kinesin dynamics in living *Drosophila melanogaster* S2 cells with acquisition rates of up to 11 images per second [33] and thus elegantly demonstrates the capability of live imaging with SIM.

Unlike STED microscopy, the limiting factor for live SIM does not rely on the photochemical properties of the dye but rather on the image acquisition and processing system. Although the maximum resolution in SIM is generally lower than the one of STED (<60 nm compared to ~100 nm), its acquisition speed is substantially faster for large fields of view. Further, SIM offers increased lateral and axial resolution while the increased resolution in live-STED microscopy is so far limited to the x–y axis. In terms of fluorophore selection, SIM is not limited to specific dyes and in principle can be performed with any fluorophore available. This renders SIM also more suitable for multicolor imaging when two or more colors are needed. Since SIM is based on conventional wide-field microscopy with non-laser illumination sources, it also features reduced bleaching and phototoxicity compared to STED microscopy [21].

7.2 Single-Molecule Localization Methods

All biological structures are composed of many individual molecules. In this respect, the precise localization of a representative population of these molecules would allow to reconstruct a super-resolution image of the particular structure. This concept is used in single-molecule localization technologies such as PALM [34], FPALM (fluorescence photoactivation localization microscopy [35]), and STORM [36]. A sample containing photoswitchable markers is illuminated homogeneously with a defined low-power light source. Using minimal amounts of light (i.e., adjusted to only excite a small fraction of all fluorophores in the sample) ensures the activation of only a few molecules. If these molecules are at least separated by the length of one diffraction-limit, it is possible to calculate the position of the individual molecules with <30 nm precision by determining the center of its point spread function. By running multiple consecutive imaging cycles (100–100,000), the position of many molecules is recorded and further used to reconstruct the super-resolution image. Hence, the major difficulty of applying single-molecule localization techniques to live-cell imaging is already inherent to the technique—i.e., the rather long image acquisition time limits live imaging to approximately 0.1 Hz [21]. Nevertheless, several single-molecule localization studies have recently demonstrated live-cell imaging of biological samples; for example, by imaging focal adhesion molecules [37] or the MreB protein to visualize the structure of actin filaments in bacteria [38]. Furthermore, the technique showed high potential in single-particle tracking measurements as in visualizing the behavior of hemagglutinin in the plasma membrane [39] and in recording the diffusion properties of membrane proteins in COS7 cells [40].

In summary, single-molecule localization microscopy approaches are still far from video-rate image acquisition during live-cell imaging but offer a great advancement in both lateral and axial resolution. Compared to STED microscopy, the number of suitable dyes is substantially larger and growing rapidly (see for example [41, 42]). The increasing availability of fluorophores further indicates a higher potential for multicolor imaging. However, this will only be beneficial if image acquisition time is reduced significantly. In terms of toxicity for the preparation, it is comparable to STED microscopy [21].

8 Potential of Super-Resolution Microscopy for Live-Cell Imaging of Biological Samples

Although super-resolution live imaging is still a rather young discipline, its importance for biological research is emerging rapidly. So far, several relevant biological questions have been tackled and it becomes clear that super-resolution microscopy can provide valuable information concerning cellular physiology. For imaging large areas, SIM is most likely the method of choice as it is based on wide-field microscopy. Single-molecule localization microscopy approaches are still limited in image acquisition speed and are not able to capture fast intracellular processes. However, once this hurdle is overcome, it will be an extremely powerful tool for investigating cellular movements. For now, from all the different techniques introduced in this chapter, STED microscopy has probably the biggest potential for real-time imaging. As outlined in this chapter, it has successfully been used in the past to characterize the dynamics of single SVs during different activity states of the cell [13–15]. It is to be expected that development of advanced fluorescent proteins and dyes (increased brightness and photostability) as well as imaging systems (increased speed and sensitivity) will further extend the applicability of live-STED microscopy. Once this is achieved, it will be possible to investigate dynamic processes in a wide variety of different biological systems—possibly in multicolor and over longer time periods—substantially improving our knowledge of cellular functions.

References

1. Südhof TC (2004) The synaptic vesicle cycle. Annu Rev Neurosci 27:509–547
2. Katz B (2003) Neural transmitter release: from quantal secretion to exocytosis and beyond. J Neurocytol 32:437–446
3. Takamori S et al (2006) Molecular anatomy of a trafficking organelle. Cell 127:831–846
4. Lemke EA, Klingauf J (2005) Single synaptic vesicle tracking in individual hippocampal boutons at rest and during synaptic activity. J Neurosci 25:11034–11044
5. Gaffield MA, Rizzoli SO, Betz WJ (2006) Mobility of synaptic vesicles in different pools in resting and stimulated frog motor nerve terminals. Neuron 51:317–325
6. Kraszewski K et al (1995) Synaptic vesicle dynamics in living cultured hippocampal neurons visualized with CY3-conjugated antibodies

directed against the lumenal domain of synaptotagmin. J Neurosci 15:4328–4342
7. Denker A, Rizzoli SO (2010) Synaptic vesicle pools: an update. Front Synaptic Neurosci 2:135
8. Hell SW, Wichmann J (1994) Breaking the diffraction resolution limit by stimulated emission: stimulated-emission-depletion fluorescence microscopy. Opt Lett 19:780–782
9. Einstein A (1917) Zur Quantentheorie der Strahlung. Phys Z 18:121–128
10. Hell SW (2009) Microscopy and its focal switch. Nat Methods 6:24–32
11. Wilhelm BG, Groemer TW, Rizzoli SO (2010) The same synaptic vesicles drive active and spontaneous release. Nat Neurosci 13:1454–1456
12. Rittweger E, Han KY, Irvine SE, Eggeling C, Hell SW (2009) STED microscopy reveals crystal colour centres with nanometric resolution. Nat Photonics 3:144–147
13. Westphal V et al (2008) Video-rate far-field optical nanoscopy dissects synaptic vesicle movement. Science 320:246–249
14. Kamin D et al (2010) High- and low-mobility stages in the synaptic vesicle cycle. Biophys J 99:675–684
15. Lauterbach MA et al (2010) Comparing video-rate STED nanoscopy and confocal microscopy of living neurons. J Biophotonics 3:417–424
16. Gaffield MA, Betz WJ (2006) Imaging synaptic vesicle exocytosis and endocytosis with FM dyes. Nat Protoc 1:2916–2921
17. Willig KI et al (2006) Nanoscale resolution in GFP-based microscopy. Nat Methods 3:721–723
18. Matteoli M, Takei K, Perin MS, Südhof TC, De Camilli P (1992) Exo-endocytotic recycling of synaptic vesicles in developing processes of cultured hippocampal neurons. J Cell Biol 117:849–861
19. Heuser JE, Reese TS (1973) Evidence for recycling of synaptic vesicle membrane during transmitter release at the frog neuromuscular junction. J Cell Biol 57:315–344
20. Tsien RY (1998) The green fluorescent protein. Annu Rev Biochem 67:509–544
21. Toomre D, Bewersdorf J (2010) A new wave of cellular imaging. Annu Rev Cell Dev Biol 26:285–314
22. Nägerl UV, Willig KI, Hein B, Hell SW, Bonhoeffer T (2008) Live-cell imaging of dendritic spines by STED microscopy. Proc Natl Acad Sci U S A 105:18982–18987
23. Hein B et al (2010) Stimulated emission depletion nanoscopy of living cells using SNAP-tag fusion proteins. Biophys J 98:158–163
24. Ding JB, Takasaki KT, Sabatini BL (2009) Supraresolution imaging in brain slices using stimulated-emission depletion two-photon laser scanning microscopy. Neuron 63:429–437
25. Rankin BR et al (2011) Nanoscopy in a living multicellular organism expressing GFP. Biophys J 100:L63–L65
26. Morozova KS et al (2010) Far-red fluorescent protein excitable with red lasers for flow cytometry and superresolution STED nanoscopy. Biophys J 99:L13–L15
27. Szent-Gyorgyi C et al (2007) Fluorogen-activating single-chain antibodies for imaging cell surface proteins. Nat Biotechnol 26:235–240
28. Fitzpatrick JAJ et al (2009) STED nanoscopy in living cells using fluorogen activating proteins. Bioconjug Chem 20:1843–1847
29. Schröder J, Benink H, Dyba M, Los GV (2009) In vivo labeling method using a genetic construct for nanoscale resolution microscopy. Biophys J 96:L01–L03
30. Pellett PA et al (2011) Two-color STED microscopy in living cells. Biomed Opt Express 2:2364–2371
31. Gustafsson MG (2000) Surpassing the lateral resolution limit by a factor of two using structured illumination microscopy. J Microsc 198:82–87
32. Gustafsson MGL et al (2008) Three-dimensional resolution doubling in wide-field fluorescence microscopy by structured illumination. Biophys J 94:4957–4970
33. Kner P, Chhun BB, Griffis ER, Winoto L, Gustafsson MGL (2009) Super-resolution video microscopy of live cells by structured illumination. Nat Methods 6:339–342
34. Betzig E et al (2006) Imaging intracellular fluorescent proteins at nanometer resolution. Science 313:1642–1645
35. Hess ST, Girirajan TPK, Mason MD (2006) Ultra-high resolution imaging by fluorescence photoactivation localization microscopy. Biophys J 91:4258–4272
36. Rust MJ, Bates M, Zhuang X (2006) Sub-diffraction-limit imaging by stochastic optical reconstruction microscopy (STORM). Nat Methods 3:793–795
37. Shroff H, Galbraith CG, Galbraith JA, Betzig E (2008) Live-cell photoactivated localization microscopy of nanoscale adhesion dynamics. Nat Methods 5:417–423
38. Biteen JS et al (2008) Super-resolution imaging in live Caulobacter crescentus cells using photoswitchable EYFP. Nat Methods 5:947–949
39. Hess ST et al (2007) Dynamic clustered distribution of hemagglutinin resolved at 40 nm in

living cell membranes discriminates between raft theories. Proc Natl Acad Sci U S A 104:17370–17375

40. Manley S et al (2008) High-density mapping of single-molecule trajectories with photoactivated localization microscopy. Nat Methods 5:155–157

41. Vaughan JC, Jia S, Zhuang X (2012) Ultrabright photoactivatable fluorophores created by reductive caging. Nat Methods 9:1181–1184

42. Shim S-H et al (2012) Super-resolution fluorescence imaging of organelles in live cells with photoswitchable membrane probes. Proc Natl Acad Sci U S A 109:13978–13983

Chapter 5

Photoactivated Localization Microscopy for Cellular Imaging

Paulina Achurra, Seamus Holden, Thomas Pengo, and Suliana Manley

Abstract

In a method termed photoactivated localization microscopy (PALM), super-resolution fluorescence imaging can be achieved through the localization of single molecules. This allows the resolution of specific proteins fused to the appropriate fluorescent protein label. Here, we summarize fluorescent proteins suitable for PALM, the technical aspects of multicolor and three-dimensional imaging, and the software packages that are available. Additionally, we highlight several biological applications with an emphasis on neuroscience.

Key words Super-resolution microscopy, Fluorescence proteins, Live imaging, Data analysis for PALM

1 Introduction

Over the past 30 years, light microscopy has become an increasingly powerful tool in the life sciences. The ability to fluorescently label specific proteins of interest and image them in fixed and live specimens has transformed our understanding of the mechanisms and dynamics underlying cellular processes. However, one limitation of every classical method based on light microscopy is the diffraction limit, which restricts image resolution to around 250 nm laterally and 600–800 nm axially. Since a single protein is just a few nanometers in size, information on the scale of a single molecule or even a single molecular complex is beyond the reach of classical imaging. Furthermore, this resolution limit severely restricts the information available on the scale of cellular features important to neuroscience, such as neuronal spines and synapses. Several key methods have emerged in recent years to address this limitation.

In a standard fluorescence microscope, individual proteins appear nearly a factor of 100 larger than their actual size because their fluorescence is spread by diffraction. This spreading of the light from individual molecules obscures all information on protein organization below the scale of a few hundred nanometers. Before the advent of fluorescence near-field scanning optical microscopy (NSOM) [1, 2], nanoscopic structural information was readily available only through immunogold labeling electron microscopy (EM) [3], which provides relatively limited information about protein identity in fixed samples. However, near-field microscopy accesses only molecules within a few nanometers of the surface.

Super-resolution imaging was named one of the top ten breakthroughs of 2006 by Science Magazine and method of the year by Nature Methods in 2008. Far-field super-resolution imaging is now coming into its full potential, as evidenced by the recent commercialization of three of the main methods: stimulated emission-depletion (STED) [4, 5], structured illumination microscopy (SIM) [6], and (fluorescence) photoactivated localization microscopy ((f)PALM) [7, 8]/(direct) stochastic optical reconstruction microscopy ((d)STORM) [9, 10]. We note that (f)PALM and (d)STORM use the same imaging and reconstruction methodologies, and differ only in that PALM was developed using fluorescent protein (FP) labels, whereas STORM/dSTORM were developed using antibody-targeted inorganic dyes. Briefly, STED relies on the nonlinear optical transition of a molecule from an excited state to its ground state, known as stimulated emission. In addition to a scanning excitation beam (as for confocal microscopy), a second, toroidal beam that de-excites the dye is scanned simultaneously, creating an effectively smaller fluorescence emission volume. PALM and SIM are both widefield methods and will be described in greater detail below. Each of these methods has unique strengths and weaknesses, but all are converging on similar capabilities in terms of flexibility and multicolor imaging. STED has an edge currently in speed of imaging, with multibeam approaches reducing imaging time by orders of magnitude, with large fields of view imaged in less than a second [11]. However, PALM has the best resolution demonstrated thus far in biological samples, down to ~20 nm laterally and 10 nm axially [12, 13].

In PALM (Fig. 1), photoactivatable fluorescent proteins (FP) [14] are continuously activated, imaged, and bleached to temporally separate molecules that would otherwise be spatially indistinguishable. With low background noise and efficient photon collection, molecular locations can be determined with high precision [15]. This is because the center of a molecular image or point spread function (PSF) can be localized, typically by fitting to a Gaussian function, even down to the nanometer scale [16]. Thus, by combining point localization with the stochastic switching of thousands of single molecules, resolving molecular distributions at the nanoscale is possible. Molecular positions from multiple images are mapped to create a single, composite PALM image [8].

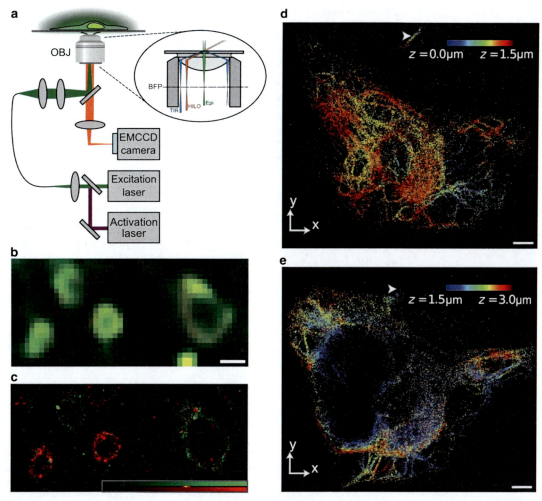

Fig. 1 PALM instrumentation and measurement capabilities. (**a**) Instrumentation and illumination strategies for 2D single-color PALM. Sample is excited and imaged using the same objective lens (*OBJ*); emitted fluorescence is directed onto an EMCCD camera. Common widefield imaging schemes are illustrated: *Epi* epifluorescence excitation, *HILO* highly inclined thin illumination excitation, *TIR* total internal reflection excitation. BFP indicates back focal plane. (**b, c**) Two-color spectral unmixing PALM of cryosections of *E. coli* cells expressing membrane-anchored bsDronpa or Dronpa. Reconstructed widefield image (**b**) and PALM image (**c**). Scale bar, 500 nm. (**d, e**) 3D PALM imaging of a 3 μm section of vimentin network. PA-mCherry1–vimentin imaged using two-photon activation and astigmatic 3D PALM. *z* location indicated as a color map. For clarity, localizations corresponding to 0–1.5 μm (**d**) and 1.5–3 μm (**e**) are shown separately. *Arrowheads* in (**d**) and (**e**) indicate a fibril that persists over >2 μm axially. (**b, c**) Reproduced from [37]. (**d, e**) Reproduced from [50]

2 Sample Preparation and Hardware: Dyes and Microscopes for PALM Imaging

2.1 Fluorescent Proteins for PALM

Several considerations determine the suitability of a particular fluorescent protein (FP) for PALM; the key requirement is that they can be photoactivated (irreversible photoconversion of an FP from a dark state to a fluorescent state), photoswitched (*reversible* photoconversion of an FP from a dark state to a fluorescent state), or photoconverted (irreversible photoconversion of an FP from one fluorescent bandwidth to another). These properties allow the "optical highlighting" [17] of a sparse subset of FPs within a diffraction-limited ensemble, which is the process at the core of PALM imaging.

In addition to optical highlighting, there are several further considerations that determine the suitability of an FP for PALM:

- *Photostability* [18] should be high to maximize the number of photons collected from each molecule before photobleaching, and thus maximize localization precision.

- *Spontaneous photoactivation rate* (due to thermal activation or photoactivation at the wavelength of fluorescent excitation) should be low, to ensure a high ratio of dark to fluorescent molecules during data acquisition (analogous to the rate ratio for STORM [19]). This allows the target structure to be labeled at high density, to give a well-sampled image (as defined by Nyquist) [20].

- The *contrast ratio* between the fluorescent and the dark state, defined as the emission intensity ratio between the fluorescent and the dark state [21, 22], should be high to facilitate high density labeling without generating problematic background fluorescence.

- *FP aggregation* should be minimized where possible to avoid perturbation of the target protein. This requirement is particularly important in studies where quantification of protein spatial distribution is required (such as in studies of clustering and cluster size), which is a primary use of PALM.

A growing number of FPs are being created specifically tailored for the requirements of PALM imaging, continually increasing the capabilities of the technique [18, 23, 24]. A number of recent reports provide detailed tables comparing the properties of various FPs for PALM to assist in choosing the best fluorophore for a particular application [18, 21, 25–27]. Most PALM FPs are "non-standard," in the sense that most researchers must generate and characterize new mutants to utilize these FPs. We note that an already widely used fluorescent protein, EYFP, displays photoswitching characteristics suitable for PALM [28]; this can prove useful in facilitating initial studies if existing EYFP mutants are available. However, EYFP must initially be bleached before single molecules can be detected, and in the process, the density of molecules is compromised. Thus, a critical step in preparing any biological sample for PALM imaging at its best resolution includes optimizing an FP fusion that maintains protein function.

2.2 The Basics: Single-Color 2D PALM Instrumentation

A single-color PALM setup (Fig. 1a) can be straightforwardly assembled out of commercially available components and is essentially identical to a single-molecule total internal reflection fluorescence (TIRF) microscope. The basic requirements are a laser illumination system, a TIRF excitation module, an inverted microscope with a high numerical aperture TIRF objective, and a single photon-sensitive (EMCCD) camera.

PALM is based on widefield imaging detection of single fluorescent proteins. Epifluorescence illumination, where a parallel laser beam illuminates the sample above the objective (Fig. 1a) [29] in some circumstances gives sufficient signal-to-noise for efficient single molecule detection, notably for PALM studies in bacteria [28, 30]. However, in thicker eukaryotic samples, the large background signal due to out of focus fluorescence may overwhelm the signal from in-focus molecules.

An alternative scheme, TIRF illumination [31], uses a laser beam directed onto the microscope coverslide at an angle greater than the critical angle for total internal reflection (Fig. 1a) [29]. This generates a rapidly decaying evanescent illumination field in the sample medium, which illuminates only about 100 nm near the coverslip surface. TIRF illumination can thus be used to enhance signal-to-noise to a sufficient level for PALM, even in thicker eukaryotic samples, with the limitation that only structures close to the surface of the cell may be studied. TIRF illumination is thus especially suited for studies of membrane proteins.

A compromise illumination scheme, sometimes called highly inclined and laminated optical sheet (HILO) microscopy (Fig. 1a) [32], uses laser beams incident upon the microscope coverslide an angle near, but less than, the critical angle for total internal reflection. This significantly reduces the excitation depth of field about the focal plane (to about 7 μm) [32], reducing background fluorescence, and enabling imaging of single FPs within a few microns from the surface.

Single-color PALM requires an excitation laser of appropriate wavelength, together with a photoactivation laser or photoconversion laser (usually at 405 nm). It is often convenient to couple the multiple lasers into one single mode fiber (as is the case for commercial TIRF systems), since this allows straightforward beam alignment and coupling of the multiple beams into the microscope objective. Excitation laser powers on the order of 50–100 MW should be sufficient for PALM; however, if the research also intends to carry out (d)STORM on the same microscope, higher laser powers may be useful.

The researcher intending to obtain their own PALM setup has several options. The first option is construction of a custom PALM setup. Here the researcher can choose to what degree they wish to customize or construct their own apparatus. Because the microscope requirements are essentially the same as a regular TIRF microscope, most of the components are available as quite user-friendly systems. Complete TIRF illumination modules for standard inverted microscopes are available, as are pre-aligned laser excitation modules containing several lasers coupled into a single optical fiber for use with TIRF illumination systems. Together with an inverted microscope, a TIRF objective, an EMCCD camera, and freely available analysis software (see below) this represents the minimal equipment for PALM. This home-built type of system is quite an attractive option in terms of cost and flexibility. However, the degree of specialist expertise required to put together even quite a "user-friendly" home-built PALM system, consisting mainly of commercial TIRF modules, should not be underestimated. A thorough grasp of practical single molecule fluorescence imaging is necessary in order to achieve the optical alignment and optimization of signal-to-noise required for high quality PALM images.

The second option is to purchase a commercial PALM/STORM microscope; several manufacturers now offer such systems including Nikon, Ziess, Leica, and Vutara. Although such microscopes come with a higher price tag and reduced flexibility, for more biologically focused laboratories, and especially for multiuser facilities, an integrated system with full vendor support, and (advertised) ease of use, is clearly an attractive option.

2.3 Multicolor PALM

Multicolor PALM (Fig. 1b, c) imaging initially proved challenging, due to low photon counts of early red photoactivatable/photoswitchable FPs, or due to spectral overlap of bright green-to-red photoconvertible FPs with available green photoswitchable FPs. Initial multicolor PALM experiments [22] combining green-to-red photoconvertible tdEos [33] and green photoswitchable Dronpa [34] suffered from photobleaching of a large fraction of the Dronpa molecules before they could be imaged, causing a reduction in the quality of the green channel image compared to the red channel.

The subsequent development of improved photoactivatable FPs such as PAmCherry [23] and PATagRFP [24] largely resolved this issue. By pairing these FPs with green-emitting PA-GFP [35], PS-CFP [36], or Dronpa [34], high quality two-color PALM images may be obtained [23, 24]. The PA-GFP-PATagRFP pair has also been used to demonstrate live cell two-color PALM, with simultaneous acquisition in both emission channels. A frequently reported drawback to these green-emitting FPs is their inadvertent photoconversion with the excitation wavelength. This can lead to high background levels that will compromise single molecule localization precision.

An intriguing approach for multicolor PALM is based on ratiometric imaging of two closely spaced spectral windows spanning about 100 nm [37]. The principle is that even though two FPs may have emission maxima separated by as little as 15 nm [37], the ratio of the intensities in each emission channel will differ sufficiently between the two FPs to allow them to be distinguished. This concept is a special case of the more general theoretical approach of "spectral unmixing" [38–40], originally developed for remote sensing applications [39], and now finding applications in microscopy [38, 40].

"Spectral unmixing" PALM (Fig. 1b, c) was initially demonstrated for two FPs, bsDronpa and Dronpa, in *E. coli* cryosections [37]. Recently, the approach was used for the first demonstration of three-color PALM [41, 42], for Dendra2 [43], PAmCherry1 [23], and PAmKate [42]. Spectral unmixing PALM was demonstrated in live fibroblast cells and could in principle be extensible to four or more different FPs.

2.4 PALM in Three Dimensions

A number of approaches have been developed to extend PALM to three dimensions (Fig. 1d, e), all of which are based upon determining the position of each individual FP not only in the x- and y-axes, but also the z-axis.

The first method developed for 3D PALM, astigmatic imaging (3D-STORM) [44] introduces a shift in the relative focal positions between the x- and y-axes of the microscope, causing the FP's PSF to appear elliptical in most image planes. The FP's z-position can then be determined from the degree and orientation of the ellipticity. Several alternative methods for 3D PALM have subsequently been developed. In *biplane* PALM, the same FP is observed in two closely spaced image planes [45]. In *double-helical PSF imaging*, manipulation of the phase of emitted fluorescence in the Fourier plane is used to modify the FP's PSF into a two-lobed distribution, with the z-position of the FP encoded in angle between the two lobes [46]. For total photon counts on the order of 100–1,000 per FP, z-resolution on the order of 50 nm (around half the xy-resolution) can be achieved.

Two additional methods have been developed which give isotropic resolution in all three dimensions. *Virtual volume super-resolution microscopy* (VVSRM) [47] utilizes a right-angled mirror placed just above the microscope cover glass to rotate the collected z-axis fluorescence onto the x–y plane, giving z-resolution identical to xy-resolution (around 20 nm). However, a drawback of this method is that samples must be mounted directly on the mirror substrate, and then mounted at 45–30° to the cover glass; this gives a small useable sample area and is experimentally inconvenient. *Interferometric photoactivated localization microscopy* (iPALM) [11] uses single-photon interferometry between two objectives placed on either side of the sample. This approach yields a z-resolution of about 10 nm for typical FP samples, actually twice the xy-resolution, 20 nm. However, in addition to the requirement that the sample be sandwiched between two microscope objectives, the technical complexity of the iPALM experimental apparatus is high; this currently limits widespread adoption of the method.

2.5 3D PALM in "Thick" Samples

The 3D PALM methods above can yield sub-diffraction limited images in single eukaryotic cells and in bacteria, within only a few microns from the coverslip for "standard" illumination strategies (epifluorescence) [29] or HILO [32], because a large number of FPs outside of the plane of focus are activated and fluorescent during imaging. At further distances from the coverslip, the signal from single FPs in the image plane is overwhelmed by background from out of focus FPs. In order to study whole cells (typically up to 15 μm thick), alternative methods are required.

One solution is to use two-photon illumination to activate the FPs. Two-photon photoactivation (2PA) reduces the axial thickness in which FPs are activated from at least 7 μm (for one-photon photoactivation under HILO illumination [32]; even larger for epifluorescence), to 1–2 μm depending on the 2PA implementation [48–50]. Because the depth of field for 3D PALM is about 1 μm, 2PA allows sequential 3D PALM across multiple z-planes separated by about 1 μm (Fig. 1d, e), without significant photobleaching outside of the image plane. In practice this method currently allows 3D PALM imaging over an axial range of around 10 μm [49, 50].

2.6 Live Cell PALM

Because PALM does not require special buffers or imaging conditions, it translates quite readily to live cell imaging. However, since several thousand diffraction-limited single molecule images are required to obtain a single PALM image, the frame rate of PALM is only around 0.01–0.04 Hz [20, 28], limiting the range of cellular processes which can be studied by live cell PALM.

In order to study fast biological processes, PALM can be combined with single particle tracking (sptPALM) [51]. Traditional SPT has millisecond time resolution and single molecule sensitivity, but is limited to very low concentrations of labeled molecules by the requirement that molecular separations be greater than the diffraction limit. By utilizing stochastic photoactivation of diffusing molecules, sptPALM can be performed at high densities, allowing several orders of magnitude more trajectories per cell to be obtained than traditional SPT, extending the utility of the method.

3 Software: Data Analysis for PALM Imaging

All forms of localization microscopy require a series of computational steps to translate the raw data coming from the camera, into a super-resolution image. To better compare the different possibilities, we can divide the process into the following steps. Preprocessing reduces the effect of background and noise on the raw data. Localization isolates and characterizes each single molecule with sub-pixel resolution. After localization, post-processing selects those peaks that actually belong to photon emissions from a single molecule, discarding spurious localizations ("sieving"). Post-processing includes operations such as drift correction, to compensate the movement of the sample during the image acquisition, and chromatic aberration correction. Finally, once a collection of localized points has been formed, rendering translates it into a super-resolution image for visualization. An overview of the choices for each step present in various experiments and software is shown in Table 1.

3.1 Preprocessing

The first step is to preprocess the data to remove the influence of background fluorescence and noise. Several different approaches have been proposed. In the first PALM publication [8], the difference between successive frames was calculated, which highlights temporal changes in intensity and thus effectively removes the relatively constant background. Alternatively, background subtraction can be performed by using the mean [15], the median, or a local background estimation [52]. Noise reduction is in most cases performed by using a smoothing filter, which can take the form of a uniform filter [53], a Gaussian filter with different criteria for the size [54–56], or a spot enhancing filter such as a Laplacian-of-Gaussian [52], a difference-of-Gaussians [57], or more sophisticated solutions such as the patch-based adaptive filter combined with deconvolution [58].

All of these preprocessing steps affect the ability to distinguish real signal from noise in different ways. The only quantitative analysis comparing different preprocessing steps and the effect on peak detection has been performed by Krizek et al., using Monte Carlo simulations [54–56]. Their recommendation is to use a Gaussian filter with a sigma twice the size of the PSF.

Preprocessed data should in general only be used for the peak detection, as the optimization algorithms used in localization usually account for noisy data and thus can work directly with the unprocessed data. However, if the software used for localization does not provide control over preprocessing it already performs (such is the case for most of the freely available software), then care has to be taken not to choose a large filter size for additional preprocessing, as it would smooth out the peaks and result in localization precision loss. Nonetheless, if most of the molecules are in the focus plane, as when in TIRF mode, then a reasonable choice is to choose a smoothing size comparable to the PSF (matched filtering). Data filtered with this smaller filter can be fed directly into the localization software, even it performs additional preprocessing, improving peak detection without greatly affecting the localization precision.

Generally, a first attempt can be made to use the raw data directly and only perform additional preprocessing if either the peak count is low or many spurious peaks appear. In such cases though, the sample preparation and the data acquisition should be optimized first.

3.2 Localization

The preprocessed images then undergo the localization process. This is a step that requires finding spot-like objects in the image (maxima extraction) and finding the center with sub-pixel precision

Table 1
Summary of different software packages and their features

	Betzig et al. [8]	Wolter et al. [53]	Hedde et al. [52]	Henriques et al. [57]	Krizek et al. [54]	Matsuda et al. [58]	Thompson et al. [15]	Smith et al. [56]	Niu and Yu [59]	Rogers et al. [55]
Alias for algorithm										
Software	IDL and MATLAB code	RapidSTORM	LivePALM	QuickPALM					Octane	PolyParticleTracker Octane; PolyParticleTracker
3D		Astigmatism, biplane		Astigmatism				Astigmatism		
Preprocessing	Differential in time	Averaging	Background subtraction; Laplace-like filter	Band-pass (DoG)	Gaussian, 2*FWHM	"Patch-based adaptive filter"; deconvolution	Background subtraction	Gaussian, sigma of the PSF		Gaussian, 3*sigma
Localization										
Model	Gaussian	Gaussian (optional covariance)	Gaussian	N/A	Gaussian	Gaussian with a third power bkg term	Gaussian	Multiple Gaussians	Gaussian	Weighted polynomial
Maxima extraction	Threshold	Non-maxima suppression	Threshold	CLEAN	4-n. threshold	Threshold	Local threshold	Local maxima (5 sigmas)	8-n. threshold	8-n. threshold
Refinement	Levenberg-Marquardt, double pass	Levenberg-Marquardt	Linear approximation (fluoroBancroft)	Center of mass	Levenberg-Marquardt	Least-squares [15]	Least-squares	Iterative MLE (Newton-Raphson)	Powell's method	QR-based direct LS solution
Extracted region	2.5 Sigma	4 Sigma+1 square	5×5 pixels	4×FWHM+1 square	7×7			6 Sigma+1 (=19 px) square		4×Estimated radius+1 square

Postprocessing

Sieving	Threshold on total photon count above shot noise (e.g., 5×); grouping	Distance from initial position; amplitude threshold; covariance deviation (if covariance model used)	Distance from initial position; max. amplitude threshold	Size threshold; symmetry threshold	Nonlinear threshold on localization precision and SNR	Distribution analysis of peak width	$N > 25$; grouping	Duplicate removal	Eccentricity brightness; radius; skewness
Correction	Drift			Drift, chromatic	Chromatic	Drift			
Rendering	Gaussians, sigma dependent on localization precision	Betzig/van der Linde rendering + equalization	2D histogram	2D histogram	Average of jittered 2D histogram	Gaussians, constant width, sampled at twice the FWHM of the fitted Gaussian		Gaussians, covariance calculated from the uncertainty in the localization process	Gaussian, sigma of the PSF

(refinement, but also referred to as the proper localization). If the images after preprocessing have good SNR and uniform intensities, a global threshold on the preprocessed image suffices [8, 52, 54, 58], but in case local variations of intensities are present, a local threshold is preferred, either in the form of a 4- [54] or 8-neighbor local maxima criterion [55, 59], or with larger windows [15, 53, 56]. More sophisticated algorithms have also been used, borrowing from the analysis of astronomical data, such as the CLEAN algorithm [60] used in [26].

Once the candidate positions have been extracted, the center is estimated with sub-pixel precision. The simplest approach is to calculate the center-of-mass of the pixel data [26], but most algorithms perform a more accurate model fit on a small region around each candidate. The most common model is a single bivariate Gaussian function [8, 15, 52–54, 59], but other options have also been studied, such as a third exponent on the background term [58] or a weighted polynomial approximation [55]. Usually, in the presence of multiple spots active in the same region at the same time, the localization is discarded, but recent work demonstrated that fitting these overlapping spots with a multiple Gaussian model allows localization microscopy to be performed at significantly higher density of fluorescently active molecules [61].

Maximum localization accuracy is given by the best match of the fitted function to the PSF shape. A Gaussian model is a good compromise between accuracy and performance [15], but if also estimating the axial position of the molecules using an engineered PSF or a biplane setup, then specialized algorithms have been proposed to deal with the non-Gaussian nature of the PSF.

The fastest algorithms use an approximation of the Gaussian, like a weighted polynomial (Octane), or substitute the full least-squares fit for a simpler estimation algorithm (QuickPALM). An alternative approach uses an iterative weighted centroid, with weights calculated from a Gaussian integrated over each pixel [15]. As the signal-to-noise ratio (related mostly to the number of photons per molecule) of the data increases the improvement given by a more accurate algorithm is less appreciable so preference can be given to faster algorithms.

3.3 Post-processing

Once the localization has been performed, a decision has to be made for each localized peak to keep it or discard it as a false positive. The most common criteria, shared by most of the current software packages [8, 15, 52, 53, 55] is the amplitude of the fitted function, also called brightness, or the photon count. The magnitude of the estimated refinement is also used to discard erroneous localizations [52, 53]. In some software, one can set criteria to threshold peaks based on their size [55, 57], symmetry [55, 57], skewness [55], or eccentricity [55]. In others, more sophisticated criteria are available such as a nonlinear threshold on the estimated localization precision and the SNR [54], or automated outlier rejection based on the peak width distribution [58].

An important correction in any time-lapse experiment is to compensate stage or sample drift using a de-drifting procedure [8, 57, 58, 62]. Another important correction for multicolor experiments is to compensate the effect of chromatic aberration [54, 57]. Both can be achieved by including external fiduciary points, such as gold or fluorescent beads, or in densely labeled samples by tracking a known fixed structure within the sample.

3.4 Rendering

For visualization purposes, the set of localized peaks needs to be transformed into an image. The easiest technique is to bin the positions of the molecules in a 2D histogram, using a bin size comparable to a fraction of the PSF width [52, 57]. An improvement is to perform jittering: each point is displaced by an amount of the order of the localization precision and the resulting histograms are averaged [54]. However, a better representation estimates the density by using kernel density estimation: the image is formed by summing a series of bivariate functions of a certain family, centered at each localized molecule position. The most common model function is the Gaussian and the width can either be constant, which is equivalent to assuming a uniform localization precision [58, 59], or can be variable with a width that depends on the localization precision [8, 53, 56]. The latter is the preferred method, as it takes into account the localization precision and reduces the visual contribution of poor localizations.

3.5 Software

There are many options as far as the existing software, but some have drawbacks to researchers with applications that may require modifications to adapt the analysis. For example, not all software is available. Among those that are, some do not come with the source code, which makes it difficult to understand the underlying algorithms that can only be used as black boxes.

RapidSTORM [53] provides the source code and the ability to perform analysis on-the-fly and on 3D data. LivePALM [52] and PolyParticleTracker [55] are written MATLAB but only PolyParticleTracker provides the source code. The most popular algorithms, QuickPALM [57] and Octane [59], are implemented as a plugin for ImageJ [63], a nearly ubiquitous biological image processing platform. Both also provide the source code. DAOSTORM [61] also provides the source code.

With the wide adoption of localization microscopy, the software is going to play a key role in making it accessible to the widest possible scientific audience. Which software is finally going to survive the test of time is going to be a question of its ease-of-use, features, and flexibility, but also, because of the rapidly changing and specialized needs of researchers, the availability of the original code.

4 Biological Applications: Examples of PALM Imaging in Neuroscience

Advances in super-resolution imaging methods have brought remarkable improvements in recent years, enabling the collection of images with nanoscale resolution. This has allowed researchers to explore the subcellular and molecular organization of neurons with unprecedented detail. Here, we will review some seminal applications of PALM to the study of neurons and discuss its strengths in giving previously inaccessible insights. We will also consider current challenges and finally discuss future perspectives.

4.1 Synaptic Protein Distribution

Synapses are highly specialized connections between neurons, usually occurring between a presynaptic axonal terminal and a postsynaptic dendritic spine, typically a few hundreds of nanometers in size. Within these small features, hundreds of proteins and receptors are clustered together. The molecular organization of these different components and their dynamic behavior is fundamental for effective signal transmission [64, 65]. Conversely, alterations in synaptic composition and/or organization have been linked to many neuropsychiatric disorders such as Alzheimer's disease, schizophrenia, and autism among others [66, 67]. Because of its fundamental role in neurological function, great efforts have been directed to understand the molecular topography of synaptic components both in normal and pathological situations.

The high protein density present in synapses poses an enormous challenge to understanding their molecular organization. Conventional imaging techniques lack either the resolution or the specificity required to map the positions of individual proteins. Fluorescence microscopy and advances in genetically engineered fluorescent proteins allow the visualization of highly specific protein labels; however, its diffraction-limited nature means that it cannot resolve proteins closer than about 200 nm apart. Electron microscopy (EM), on the other hand, provides molecular resolution, but requires fixing and staining of samples, a process that can disrupt labile structures. Furthermore, staining is not generally targeted to specific proteins. While immunolabeling with gold offers some specificity, its low labeling density offers only low-resolution information on protein locations [25, 68, 69]. There have been several labeling technologies that use genetically encoded tags to locally generate EM contrast, but these are still in their early days and allow only identification of a single protein species at best [70, 71].

PALM is exceptionally well suited to address this problem, since by employing photoactivatable proteins one can target the desired protein with a high degree of specificity, and by imaging a subset of the labeled molecules at a time one can localize them with nanometer resolution [8, 22].

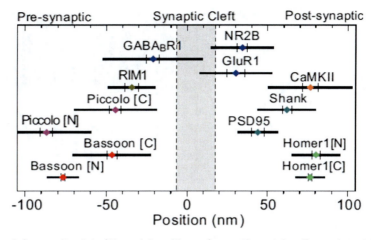

Fig. 2 Composite plot of the axial positions of synaptic proteins. For each protein, the *colored dot* specifies the mean axial position, the two *vertical lines* represent the associated standard error of the mean, and the half length of the horizontal bar denotes the standard deviation, derived from multiple synapses. Antibodies against either the N- or the C-terminal region of the protein of interest are specified. Reproduced from [68]

The organization of several pre- and postsynaptic proteins was resolved in a study using STORM (Fig. 2), where photoswitchable synthetic dyes are used instead of photoactivatable proteins. Measurements of the axial positions of more than ten proteins localized to synapses revealed the highly ordered structure of the presynaptic active zone and the postsynaptic density [68]. They also showed significant variations in synaptic neurotransmitter composition and distribution, revealing the presence of previously undetected immature synapses.

Another advance in understanding the organization of synaptic proteins came from correlative fluorescence electron microscopy, which in addition to near-molecular resolution obtained with PALM gives independent information on cellular context from electron microscopy. The presynaptic protein α-liprin fused to tdEos was expressed in *Caenorhabditis elegans* neurons. Imaging ultrathin sections of fixed tissue with PALM to localize α-liprin-tdEos and correlating them to EM images of the same regions showed α-liprin localized to the presynaptic dense projection, as was already known [72]. Importantly, in the absence of ultrastructural information the α-liprin signal would have been uninterpretable.

4.2 Receptor Trafficking

A recent study demonstrates the potential of super-resolution techniques when studying a phenomenon that is heterogeneous in nature, and occurs in very small compartments, such as neuronal spines. The authors study AMPA receptor trafficking in neurons applying sptPALM [73]. Cells were transfected with fluorescent protein fusions HA-mEos2-GluA1 to label AMPAR subunit

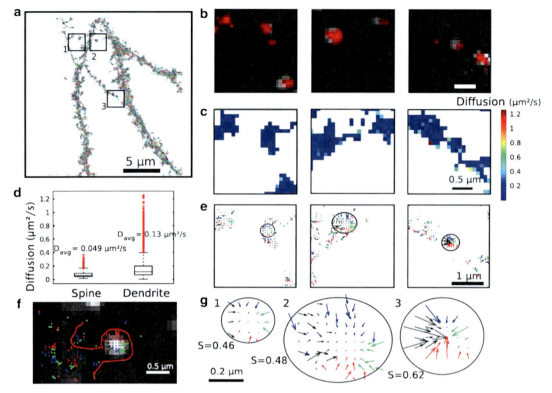

Fig. 3 Stochastic analysis of super-resolution trajectories for AMPARs. (**a**) 30,000 trajectories of GluA1-containing AMPARs located in the plasma membrane; three regions of interest are marked. (**b**) The overlay of the highest receptor densities (*red*) with the hippocampal confocal neuronal image reveals that these regions colocalize with the synaptic marker Homer (*white spots*). (**c**) Map of diffusion coefficients. (**d**) Median diffusion coefficients in dendritic spines (sampled over 3,341 points) and in the dendritic shaft (9,385 points). D_{avg} is the average diffusion coefficient. (**e**) Three disjoint interaction potentials marked in (**a**–**c**). (**f**) Field of forces in the neuronal membrane. (**g**) Three potential well patterns, characterized by a converging force, (extracted from (**e**)) quantified using the index S, confirming that the wells are due to direct interactions. Reproduced from [73]

GluR1 and Homer Cerulean. By tracking thousands (~30k) of mEos2-GluA1 trajectories on the neuronal surface they identified regions of high AMPAR densities that localized at synapses, consistent with previous studies [74]. Estimates of the underlying interaction potentials for AMPAR in these high density regions indicate that they are formed by molecular interactions with an ensemble of proteins, presumably postsynaptic scaffolding proteins, and not directed by molecular crowding or aggregation. Furthermore, it was possible to extract diffusion coefficients of single AMPAR (Fig. 3) and by studying the velocity map of this trajectories, two families of spines where observed, ones with a net inward drift and ones with a net outward drift.

4.3 Dendritic Spine Morphology

Dendritic spines are small, actin-rich protrusions found along dendrites of neurons. They are of fundamental importance for neuronal communication since it is on the spine head that most excitatory synapses are located [75]. Recent studies have shown spines to be remarkably dynamic, and capable of undergoing morphological changes in response to long-term potentiation (LTP) [76, 77] as well as long-term depression (LTD) [77, 78], cellular correlates of learning and memory. These forms of activity-dependent plasticity are essential for neuronal circuitry remodeling. What morphological changes a spine undergoes in response to activity-induced stimulus, and how these morphological changes affect synaptic strength are central questions in understanding synaptic plasticity.

In the past, indirect methods have been used to probe the question of activity-induced spine remodeling. Electron microcopy sections of neurons after LTP induction indicated an enlargement of the spine head when compared to untreated neurons [79]. Fluorescence recovery after photobleaching (FRAP) has shown an increase in spine volume after LTP [80, 81], but it was not until the advent of super-resolution techniques that it became possible to directly observe spines in living cells with nanometric precision [82–84].

Live mature rat hippocampal neurons in culture (20–30 days in vitro) were imaged with PALM [84]. Because the actin cytoskeleton is responsible for spine shape, neurons were transfected to express a photoactivatable probe that reversibly binds to actin with low affinity (actin binding peptide, ABP-tdEosFP). They found spine dimensions of 140 ± 45 nm in diameter for the neck, 550 ± 240 nm for spine length, 600 ± 180 nm width of the spine head, and 340 ± 85 nm diameter of actin-free regions (holes) and cup-shaped structures (Fig. 4). These results are in agreement with previous electron microscopy studies [85, 86] as well as STED measurements on living neurons [82].

Spine dynamics were also measured before and after chemically induced LTP to quantify changes in spine morphological parameters (Fig. 5). Because spine morphological plasticity occurs on a timescale of tens of minutes [87], long-term imaging is needed to study the dynamic behavior of spines. This brings new challenges to PALM imaging because typically the fluorescent probe will be depleted, given that a fraction of the labels are photobleached during each localization cycle. By using ABP-tdEosFP, this problem was circumvented since there is a constant replenishing of unbleached probe from the cytosolic pool. This allowed long-term PALM imaging, and in this manner, dendritic spines were imaged for several minutes.

To investigate the effects of synaptic activity on spine morphology, AMPA receptors were stimulated by incubating neurons with AMPA receptor agonist. PALM imaging revealed a drastic reduction on spine head diameter suggesting an underlying depolymerization and redistribution of the actin cytoskeleton. These results are in agreement with previous work [88].

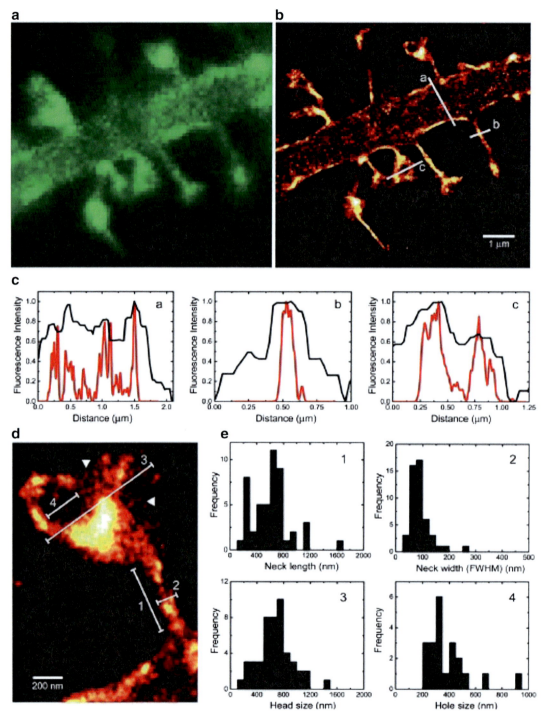

Fig. 4 PALM super-resolution microscopy of dendritic spines. (**a**) Conventional fluorescence microscopy of a rat hippocampal neuron expressing ABP-tdEosFP and fixed at DIV 25 (unconverted form of the fluorophore). (**b**) Super-resolution PALM image of the same dendritic segment, acquired with a frame rate of 50 ms (8,000 frames). (**c**) Profiles across the dendritic shaft (**a**), the spine neck (**b**), and the spine (**c**). (**d**) Detail of a single dendritic spine. (**e**) Quantification of the spine parameters indicated in (**d**): neck length (panel *1*; $N = 48$ spines from four cells and three independent experiments), neck width (*2*; $N = 48$, four cells, three experiments), spine head diameter (*3*; $N = 48$, four cells, three experiments), and diameter of actin-free regions ("hole," *4*; $N = 21$, four cells, three experiments). Reproduced from [84]

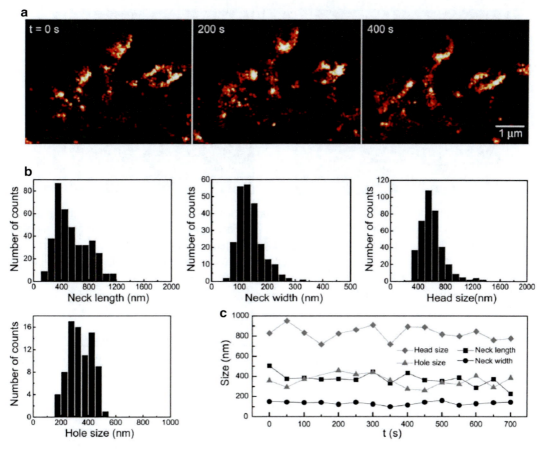

Fig. 5 Spine dynamics in mature hippocampal neurons. (**a**) Super-resolution time-lapse imaging of dendritic spines from a neuron expressing ABP-tdEosFP at DIV 27. Each PALM image was reconstructed from 2,000 frames recorded at 25 ms. (**b**) Quantification of morphological spine parameters (as in Fig. 2e) in living neurons (neck width: 140,645 nm mean 6 standard deviation, $N = 236$ spines; neck length: 550 ± 240 nm, $N = 385$; head diameter: 600 ± 180 nm, $N = 385$; diameter of holes and cup-shapes: 340 ± 85 nm, $N = 81$; from 12 cells and 3 independent experiments). (**c**) Baseline dynamics of the morphological parameters of an individual spine under control conditions. Reproduced from [84]

4.4 Actin Cytoskeleton Remodeling

Actin is continually undergoing rapid turnover in spines on time scales of less than a minute [39], while polymerizing at the fast-growing (barbed) end and depolymerizing at the slow-growing (pointed) end of actin filaments. Changes observed in spine morphology and synaptic plasticity are believed to be mainly the result of actin cytoskeleton remodeling [76, 78]. Hence, understanding the dynamics of actin in spines is fundamental to understanding spine structure and function.

Two recent studies, one using two-photon uncaging of photoactivatable actin, and another using FRAP showed that actin can polymerize both at the spine head and base [89, 90]. However, a detailed map of actin flow inside spines has not been possible to obtain due to the small length scales involved.

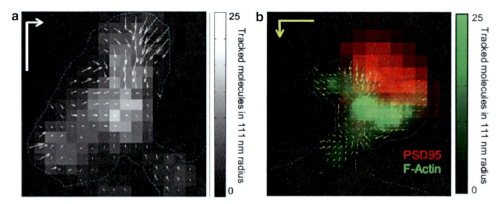

Fig. 6 Heterogeneous actin dynamics within individual spines. (**a**) Representative spine showing inward orientation of actin flow. *Arrow* length represents relative velocity. *Gray scale* represents tracked molecular density. Scale bar, 200 nm; vector, 100 nm/s. (**b**) Locally averaged molecular movement vector plotted superimposed on local tracked molecule density (*green*) and the deconvolved widefield image of PSD-95-cerulean (*red*). Some but not all foci of high-velocity motion are closely associated with the synapse. Scale bar, 250 nm; vector, 200 nm/s. Reproduced from [92]

Using single particle tracking and PALM (sptPALM) [91], actin flow dynamics was resolved in hippocampal neurons transfected with actin-tdEos [92]. Because of the wide separation of timescales between the diffusion of monomeric actin-tdEos (G-actin), and motion of actin-tdEos in the polymerized form (F-actin), only actin filaments were visualized with this method. Average velocity density maps of actin inside spines revealed highly heterogeneous flow, in accordance with previous observations [93, 94]. Not only was an inward flow present from the postsynaptic region, but surprisingly, actin flow occurred throughout the spine, sometimes as discrete foci (Fig. 6). This result shows a much more complex scenario than previously imagined. The authors hypothesized that this apparent subdomain organization and heterogeneity of actin within the spine might help regulate many diverse functions of actin despite the confined location inside the spine.

4.5 Microtubules and Axonal Transport

Intracellular transport is fundamental for proper neuronal function. In neurons, transport is polarized with some cargoes such as proteins, lipids, and organelles transported anterogradely from the cell body down the axon, and others such as multivesicular bodies and endosomes transported retrogradely from the axon towards the soma. Yet other cargoes such as mitochondria are transported in both directions. What control these transport processes are motor proteins moving along microtubules [95]. Alterations in axonal transport have been observed in multiple neurodegenerative diseases, such as Alzheimer's disease, Huntington's disease, and amyotrophic lateral sclerosis [96].

In [97], polarized transport was studied using a truncated version of the motor protein Kinesin I known to accumulate at the tip of the axon. PALM was used to localize KIF5 (a member of the kinesin superfamily motor proteins) fused to Dronpa and found to colocalize with the antibody hMB11, an antibody that recognizes GTP-tubulin. These results implicate GTP-tubulin as a possible cue responsible for polarized transport in axons.

4.6 Challenges and Perspectives

Neurons, because of their small features and highly dense compartments, are natural candidates to be imaged with PALM. The fact that there are such few published works thus far probably reflects the recent development of the technique since its beginnings in 2006, and we should expect to see many more applications in the near future.

The works described here show the extraordinary ability PALM has to resolve protein localization, even in highly dense structures such as synapse. However, to dissect synapse composition, and more interestingly, interactions occurring between proteins within the synapse, it is necessary to approach molecular resolution, which for PALM translates into better probes that exhibit higher photon counts. Yet, higher photon counts generally mean higher excitation fluxes, which can become a concern when doing live imaging in cells that suffer from phototoxicity as neurons are known to.

Besides spatial resolution, temporal resolution is relevant when studying dynamic events. Imaging rates of ~10 s per frame are currently attainable but in order to study faster events such as synaptic vesicle release rates, which happen in the order of milliseconds, imaging rates need to increase by at least a factor of 10–100. When using irreversible photoconvertible fluorophores, temporal resolution is limited by the photobleaching rate, hence there is a trade-off between faster photobleaching and adequate photon count.

Finally, most of the PALM works are performed in a TIRF configuration, yet neurons are three-dimensional objects, and as such three-dimensional imaging is required. Recent advances of PALM combining two-photon microscopy [49], use of astigmatic lenses [44], and the advent of iPALM [12] hold great promise to allow the three-dimensional imaging of neurons.

5 Summary

In this chapter, we have highlighted key aspects of PALM imaging, focusing on hardware and software requirements with concrete examples of how these methodologies have been applied in the domain of neuroscience. Super-resolution imaging is a rapidly developing field, with new dyes, imaging modalities, and software packages constantly improving the flexibility and power of the approach. As demonstrated by the first wave of publications applying PALM imaging, the ability to visualize smaller and smaller protein structures is sure to have a powerful impact on the future of neuroscience.

References

1. Lewis A, Lieberman K (1991) Near-field optical imaging with a non-evanescently excited high-brightness light source of sub-wavelength dimensions. Nature 354(6350):214–216
2. Betzig E, Trautman JK (1992) Near-field optics: microscopy, spectroscopy, and surface modification beyond the diffraction limit. Science 257(5067):189–195
3. Page Faulk W, Malcolm Taylor G (1971) Communication to the editors. An immuno-colloid method for the electron microscope. Immunochemistry 8(11):1081–1083
4. Hell SW, Wichmann J (1994) Breaking the diffraction resolution limit by stimulated emission: stimulated-emission-depletion fluorescence microscopy. Opt Lett 19(11):780–782
5. Hell SW (2003) Toward fluorescence nanoscopy. Nat Biotechnol 21(11):1347–1355
6. Gustafsson MGL (2005) Nonlinear structured-illumination microscopy: wide-field fluorescence imaging with theoretically unlimited resolution. Proc Natl Acad Sci U S A 102(37):13081–13086
7. Hess et al (2006) Biophysical Journal 91(11):4258–4272
8. Betzig E et al (2006) Imaging intracellular fluorescent proteins at nanometer resolution. Science 313(5793):1642–1645
9. Rust MJ, Bates M, Zhuang X (2006) Sub-diffraction-limit imaging by stochastic optical reconstruction microscopy (STORM). Nat Methods 3(10):793–795
10. Heilemann M et al (2008) Subdiffraction-resolution fluorescence imaging with conventional fluorescent probes. Angew Chem Int Ed 47(33):6172–6176
11. Chmyrov et al (2013) Nature Methods 10:737–740
12. Shtengel G et al (2009) Interferometric fluorescent super-resolution microscopy resolves 3D cellular ultrastructure. Proc Natl Acad Sci U S A 106(9):3125–3130
13. Kanchanawong P et al (2010) Nanoscale architecture of integrin-based cell adhesions. Nature 468(7323):580–584
14. Lippincott-Schwartz J, Patterson GH (2009) Photoactivatable fluorescent proteins for diffraction-limited and super-resolution imaging. Trends Cell Biol 19(11):555–565
15. Thompson RE, Larson DR, Webb WW (2002) Precise nanometer localization analysis for individual fluorescent probes. Biophys J 82(5):2775–2783
16. Yildiz A et al (2003) Myosin V walks hand-over-hand: single fluorophore imaging with 1.5-nm localization. Science 300(5628):2061–2065
17. Shaner NC, Patterson GH, Davidson MW (2007) Advances in fluorescent protein technology. J Cell Sci 120(24):4247–4260
18. McKinney SA, Murphy CS, Hazelwood KL, Davidson MW, Looger LL (2009) A bright and photostable photoconvertible fluorescent protein. Nat Methods 6(2):131–133
19. van de Linde S, Wolter S, Heilemann M, Sauer M (2010) The effect of photoswitching kinetics and labeling densities on super-resolution fluorescence imaging. J Biotechnol 149(4):260–266
20. Shroff H, Galbraith CG, Galbraith JA, Betzig E (2008) Live-cell photoactivated localization microscopy of nanoscale adhesion dynamics. Nat Methods 5(5):417–423
21. Bates M, Huang B, Zhuang X (2008) Super-resolution microscopy by nanoscale localization of photo-switchable fluorescent probes. Curr Opin Chem Biol 12(5):505–514
22. Shroff H et al (2007) Dual-color superresolution imaging of genetically expressed probes within individual adhesion complexes. Proc Natl Acad Sci U S A 104(51):20308–20313
23. Subach FV et al (2009) Photoactivatable mCherry for high-resolution two-color fluorescence microscopy. Nat Methods 6(2):153–159
24. Subach FV, Patterson GH, Renz M, Lippincott-Schwartz J, Verkhusha VV (2010) Bright monomeric photoactivatable red fluorescent protein for two-color super-resolution sptPALM of live cells. J Am Chem Soc 132(18):6481–6491
25. Fernandez-Suarez M, Ting AY (2008) Fluorescent probes for super-resolution imaging in living cells. Nat Rev Mol Cell Biol 9(12):929–943
26. Henriques R, Mhlanga MM (2009) PALM and STORM: what hides beyond the Rayleigh limit? Biotechnol J 4(6):846–857
27. Patterson G, Davidson M, Manley S, Lippincott-Schwartz J (2010) Superresolution imaging using single-molecule localization. Annu Rev Phys Chem 61:345–367
28. Biteen JS et al (2008) Super-resolution imaging in live Caulobacter crescentus cells using photo-switchable EYFP. Nat Methods 5(11):947–949
29. Moerner WE, Fromm DP (2003) Methods of single-molecule fluorescence spectroscopy and microscopy. Rev Sci Instrum 74(8):3597–3619
30. Greenfield D et al (2009) Self-organization of the Escherichia coli chemotaxis network imaged with super-resolution light microscopy. PLoS Biol 7(6):e1000137
31. Axelrod D (2001) Total internal reflection fluorescence microscopy in cell biology. Traffic 2(11):764–774

32. Tokunaga M, Imamoto N, Sakata-Sogawa K (2008) Highly inclined thin illumination enables clear single-molecule imaging in cells. Nat Methods 5(2):159–161
33. Wiedenmann J et al (2004) EosFP, a fluorescent marker protein with UV-inducible green-to-red fluorescence conversion. Proc Natl Acad Sci U S A 101(45):15905–15910
34. Ando R, Mizuno H, Miyawaki A (2004) Regulated fast nucleocytoplasmic shuttling observed by reversible protein highlighting. Science 306(5700):1370–1373
35. Patterson GH, Lippincott-Schwartz J (2002) A photoactivatable GFP for selective photolabeling of proteins and cells. Science 297:1873–1877
36. Chudakov DM et al (2004) Photoswitchable cyan fluorescent protein for protein tracking. Nat Biotechnol 22(11):1435–1439
37. Andresen M et al (2008) Photoswitchable fluorescent proteins enable monochromatic multi-label imaging and dual color fluorescence nanoscopy. Nat Biotechnol 26(9):1035–1040
38. Garini Y, Young IT, McNamara G (2006) Spectral imaging: principles and applications. Cytometry A 69(8):735–747
39. Keshava N, Mustard JF (2002) Spectral unmixing. IEEE Signal Process Mag 19(1):44–57
40. Zimmermann T (2005) Spectral imaging and linear unmixing in light microscopy. Adv Biochem Eng 95:245–265
41. Gunewardene MS et al (2011) Superresolution imaging of multiple fluorescent proteins with highly overlapping emission spectra in living cells. Biophys J 101(6):1522–1528
42. Piatkevich KD et al (2010) Monomeric red fluorescent proteins with a large Stokes shift. Proc Natl Acad Sci U S A 107(12):5369–5374
43. Gurskaya NG et al (2006) Engineering of a monomeric green-to-red photoactivatable fluorescent protein induced by blue light. Nat Biotechnol 24(4):461–465
44. Huang B, Wang W, Bates M, Zhuang X (2008) Three-dimensional super-resolution imaging by stochastic optical reconstruction microscopy. Science 319(5864):810–813
45. Juette MF et al (2008) Three-dimensional sub-100 nm resolution fluorescence microscopy of thick samples. Nat Methods 5(6):527–529
46. Pavani SRP et al (2009) Three-dimensional, single-molecule fluorescence imaging beyond the diffraction limit by using a double-helix point spread function. Proc Natl Acad Sci U S A 106(9):2995–2999
47. Tang J, Akerboom J, Vaziri A, Looger LL, Shank CV (2010) Near-isotropic 3D optical nanoscopy with photon-limited chromophores. Proc Natl Acad Sci U S A 107(22):10068–10073
48. Foelling J et al (2008) Fluorescence nanoscopy by ground-state depletion and single-molecule return. Nat Methods 5(11):943–945
49. Vaziri A, Tang J, Shroff H, Shank CV (2008) Multilayer three-dimensional super resolution imaging of thick biological samples. Proc Natl Acad Sci U S A 105(51):20221–20226
50. York AG, Ghitani A, Vaziri A, Davidson MW, Shroff H (2011) Confined activation and sub-diffractive localization enables whole-cell PALM with genetically expressed probes. Nat Methods 8(4):327–333
51. Manley S, Gillette JM, Lippincott-Schwartz J (2010) Single-particle tracking photoactivated localization microscopy for mapping single-molecule dynamics. Methods Enzymol 475:109–120
52. Hedde PN, Fuchs J, Oswald F, Wiedenmann J, Nienhaus GU (2009) Online image analysis software for photoactivation localization microscopy. Nat Methods 6(10):689–690
53. Wolter S et al (2010) Real-time computation of subdiffraction-resolution fluorescence images. J Microsc 237(1):12–22
54. Krizek P, Raska I, Hagen GM (2011) Minimizing detection errors in single molecule localization microscopy. Opt Express 19(4):3226–3235
55. Rogers SS, Waigh TA, Zhao X, Lu JR (2007) Precise particle tracking against a complicated background: polynomial fitting with Gaussian weight. Phys Biol 4(3):220–227
56. Smith CS, Joseph N, Rieger B, Lidke KA (2010) Fast, single-molecule localization that achieves theoretically minimum uncertainty. Nat Methods 7(5):373–375
57. Henriques R et al (2010) QuickPALM: 3D real-time photoactivation nanoscopy image processing in ImageJ. Nat Methods 7(5):339–340
58. Matsuda A et al (2010) Condensed mitotic chromosome structure at nanometer resolution using PALM and EGFP-histones. PLoS One 5(9):1–12
59. Niu L, Yu J (2008) Investigating intracellular dynamics of FtsZ cytoskeleton with photoactivation single-molecule tracking. Biophys J 95(4):2009–2016
60. Högbom JA (1974) Aperture synthesis with a non-regular distribution of interferometer baselines. Astron Astrophys Suppl 15:417–426
61. Holden SJ, Uphoff S, Kapanidis AN (2011) DAOSTORM: an algorithm for high-density super-resolution microscopy. Nat Methods 8(4):279–280

62. Mlodzianoski MJ et al (2011) Sample drift correction in 3D fluorescence photoactivation localization microscopy. Opt Express 19(16): 15009–15019
63. Rasband, W.S., ImageJ, U. S. National Institutes of Health, Bethesda, Maryland, USA, http://imagej.nih.gov/ij/, 1997-2012
64. Blanpied TA, Kerr JM, Ehlers MD (2008) Structural plasticity with preserved topology in the postsynaptic protein network. Proc Natl Acad Sci U S A 105(34):12587–12592
65. Newpher TM, Ehlers MD (2009) Spine microdomains for postsynaptic signaling and plasticity. Trends Cell Biol 19(5):218–227
66. Penzes P, Cahill ME, Jones KA, VanLeeuwen JE, Woolfrey KM (2011) Dendritic spine pathology in neuropsychiatric disorders. Nat Neurosci 14(3):285–293
67. Sheng M, Hoogenraad CC (2007) The postsynaptic architecture of excitatory synapses: a more quantitative view. Annu Rev Biochem 76:823–847
68. Dani A, Huang B, Bergan J, Dulac C, Zhuang X (2010) Superresolution imaging of chemical synapses in the brain. Neuron 68(5): 843–856
69. Wilt BA et al (2009) Advances in light microscopy for neuroscience. Annu Rev Neurosci 32:435–506
70. Gaietta G et al (2002) Multicolor and electron microscopic imaging of connexin trafficking. Science 296(5567):503–507
71. Shu X et al (2011) A genetically encoded tag for correlated light and electron microscopy of intact cells, tissues, and organisms. PLoS Biol 9(4):e1001041
72. Watanabe S et al (2011) Protein localization in electron micrographs using fluorescence nanoscopy. Nat Methods 8(1):80–84
73. Hoze N et al (2012) Heterogeneity of AMPA receptor trafficking and molecular interactions revealed by superresolution analysis of live cell imaging. Proc Natl Acad Sci U S A 109(42): 17052–17057
74. Derkach VA, Oh MC, Guire ES, Soderling TR (2007) Regulatory mechanisms of AMPA receptors in synaptic plasticity. Nat Rev Neurosci 8(2):101–113
75. Kandel ER, Schwarts JH, Jessell TM (2000) Principles of neural science. McGraw-Hill Medical, New York
76. Matsuzaki M, Honkura N, Ellis-Davies GCR, Kasai H (2004) Structural basis of long-term potentiation in single dendritic spines. Nature 429(6993):761–766
77. Malenka RC, Bear MF (2004) LTP and LTD: an embarrassment of riches. Neuron 44(1): 5–21
78. Okamoto KI, Nagai T, Miyawaki A, Hayashi Y (2004) Rapid and persistent modulation of actin dynamics regulates postsynaptic reorganization underlying bidirectional plasticity. Nat Neurosci 7(10):1104–1112
79. Vanharreveld A, Fifkova E (1975) Swelling of dendritic spines in fascia dentata after stimulation of perforant fibers as a mechanism of post-tetanic potentiation. Exp Neurol 49(3):736–749
80. Holtmaat A, Svoboda K (2009) Experience-dependent structural synaptic plasticity in the mammalian brain. Nat Rev Neurosci 10(9): 647–658
81. Kopec CD, Li B, Wei W, Boehm J, Malinow R (2006) Glutamate receptor exocytosis and spine enlargement during chemically induced long-term potentiation. J Neurosci 26(7):2000–2009
82. Naegerl UV, Willig KI, Hein B, Hell SW, Bonhoeffer T (2008) Live-cell imaging of dendritic spines by STED microscopy. Proc Natl Acad Sci U S A 105(48):18982–18987
83. Urban NT, Willig KI, Hell SW, Naegerl UV (2011) STED nanoscopy of actin dynamics in synapses deep inside living brain slices. Biophys J 101(5):1277–1284
84. Izeddin I et al (2011) Super-resolution dynamic imaging of dendritic spines using a low-affinity photoconvertible actin probe. PLoS One 6(1):e15611
85. Wilson CJ, Groves PM, Kitai ST, Linder JC (1983) 3-Dimensional structure of dendritic spines in the rat neostriatum. J Neurosci 3(2):383–398
86. Harris KM, Jensen FE, Tsao B (1992) 3-Dimensional structure of dendritic spines and synapses in rat hippocampus (CA1) at postnatal day-15 and adult ages—implications for the maturation of synaptic physiology and long-term potentiation. J Neurosci 12(7): 2685–2705
87. Naegerl UV, Bonhoeffer T (2010) Imaging living synapses at the nanoscale by STED microscopy. J Neurosci 30(28):9341–9346
88. Halpain S, Hipolito A, Saffer L (1998) Regulation of F-actin stability in dendritic spines by glutamate receptors and calcineurin. J Neurosci 18(23):9835–9844
89. Honkura N, Matsuzaki M, Noguchi J, Ellis-Davies GCR, Kasai H (2008) The subspine organization of actin fibers regulates the structure and plasticity of dendritic spines. Neuron 57(5):719–729
90. Hotulainen P et al (2009) Defining mechanisms of actin polymerization and depolymerization during dendritic spine morphogenesis. J Cell Biol 185(2):323–339
91. Manley S et al (2008) High-density mapping of single-molecule trajectories with photoacti-

vated localization microscopy. Nat Methods 5(2):155–157

92. Frost NA, Shroff H, Kong H, Betzig E, Blanpied TA (2010) Single-molecule discrimination of discrete perisynaptic and distributed sites of actin filament assembly within dendritic spines. Neuron 67(1):86–99

93. Tatavarty V, Das S, Yu J (2012) Polarization of actin cytoskeleton is reduced in dendritic protrusions during early spine development in hippocampal neuron. Mol Biol Cell 23(16): 3167–3177

94. Tatavarty V, Kim E-J, Rodionov V, Yu J (2009) Investigating sub-spine actin dynamics in rat hippocampal neurons with super-resolution optical imaging. PLoS One 4(11):e7724

95. Hirokawa N (1998) Kinesin and dynein superfamily proteins and the mechanism of organelle transport. Science 279(5350): 519–526

96. Morfini GA et al (2009) Axonal transport defects in neurodegenerative diseases. J Neurosci 29(41):12776–12786

97. Nakata T, Niwa S, Okada Y, Perez F, Hirokawa N (2011) Preferential binding of a kinesin-1 motor to GTP-tubulin-rich microtubules underlies polarized vesicle transport. J Cell Biol 194(2):245–255

Chapter 6

Data Analysis for Single-Molecule Localization Microscopy

Steve Wolter, Thorge Holm, Sebastian van de Linde, and Markus Sauer

Abstract

We review single-molecule localization microscopy techniques with a focus on computational techniques and algorithms necessary for their use. The most common approach to single-molecule localization, Gaussian fitting at positions pre-estimated from local maxima, is illustrated in depth and techniques for two- and three-dimensional data analysis are highlighted. After an introduction explaining the principle requirements of single-molecule localization microscopy, we discuss and contrast novel approaches such as maximum likelihood estimation and model-less fitting. Finally, we give an overview over the existing, scientifically available software and show how these techniques can be combined to quickly and easily obtain super-resolution images.

Key words Super-resolution imaging, Localization microscopy, *d*STORM, PALM, Data analysis, rapi*d*STORM

1 Introduction

The key to super-resolution imaging is the temporal separation of fluorescence emission using either stochastic or deterministic approaches. Stochastic methods, termed single-molecule localization microscopy (SMLM) here, are defined by three factors: stochastic fluorescence emission, precise position determination of individual fluorescent probes (localization), and far-field fluorescence microscopy. That is, stochastic single-molecule localization methods such as photoactivated localization microscopy (PALM) [1], fluorescence photoactivation localization microscopy (FPALM) [2], stochastic optical reconstruction microscopy (STORM) [3], direct stochastic optical reconstruction microscopy (*d*STORM) [4, 5], and related methods [6–9] enable the user access to a reconstructed high resolved image of an object through the stochastically isolated observation of the fluorescence emission of single fluorescent probes in time. Stochastic single-molecule localization methods can be distinguished from the closely related super-resolution imaging methods such as stimulated emission depletion (STED) [10] and structured illumination

(SIM) [11] that separate fluorescence emission of fluorophores in time deterministically using, e.g., a phase mask.

The general principle of all localization microscopy methods is identical: The target of interest is labeled with fluorescent probes and imaged using a wide-field fluorescence microscope. Stochastic transitions between fluorescent *on*- and non-fluorescent *off*-states of the fluorescent probes are induced by irradiation with light of appropriate wavelength and a temporal sequence of images (stack) typically consisting of tens of thousands of images is recorded. Each of these images contains the same part of the sample, but only a very sparse subset of the total fluorophore population is expected to be active, i.e., to reside in their *on*-states, in each image. It is therefore impossible to see any structure in any single image of a stack, but by localizing the precise position of each fluorophore in each image, the positions of all fluorophores can be determined and a single, high-resolution image reconstructed.

Experimentally one is confronted with the problem to reversibly photoswitch, photoactivate, or photoconvert synthetic fluorophores or fluorescent proteins (FPs). Ideally the rate of activation is sufficiently low to ensure that the majority of active, i.e., fluorophores residing in the *on*-state, are spaced further apart than the resolution limit. The fluorescence of individual fluorophores is read out and the fluorophores are photobleached or photoswitched to the *off*-state before the next subset of fluorophores is activated. All localization microscopy methods share this modus operandi, differing mainly in the employed type of fluorophore. *d*STORM and related methods use standard, organic fluorescent probes, and PALM and FPALM employ photoactivatable fluorescent proteins (PA-FPs). The latter enable stoichiometric protein labeling with efficiencies of nearly 100 %, but FPs emit only a few hundred detectable photons before bleaching. The former are mostly used with chemical staining, which has a lower labeling efficiency, but exhibit a higher brightness of more than 1,000 photons per cycle [5, 12, 13] and a higher photostability [14], thereby permitting a higher localization precision [15–17]. Furthermore, their photoswitching rates can be controlled by external means [4, 5, 12, 13, 18, 19]. In addition, carbocyanine dyes such as Cy5, Cy5.5, Cy7, Alexa Fluor 647, Alexa Fluor 680, and Alexa Fluor 750 as well as oxazine and rhodamine dyes (ATTO and Alexa dyes) spanning the entire visible spectral range can be operated as reversible photoswitches in aqueous solvents simply by adding millimolar concentrations of reducing thiols [5, 18, 20].

In this respect *d*STORM with standard fluorescent probes volunteers as reliable and easy SMLM method. Among other things, *d*STORM has been used to study the number, distribution, and density of cellular or membrane proteins in fixed cells and the in vitro dynamics of molecular motor proteins [21–28]. Since living cells contain reducing agents such as glutathione at millimolar concentration levels, live-cell SMLM is possible without addition of external reagents. This fact has been exploited to study the distribution and dynamics of different proteins in living cells applying standard fluorophores with trimethoprim (TMP) and SNAP tags, respectively [22, 26, 29, 30]. Thus *d*STORM and related methods can be combined with (F)PALM to perform multicolor super-resolution imaging experiments in living cells, profiting from orthogonal labeling strategies. The tunability of photoswitching of synthetic fluorophores under physiological conditions allows frame rates of 10–1,000 Hz at excitation intensities between 0.5 and 50 kW cm^{-2}, such that data acquisition takes only seconds to a few minutes depending on structural demands.

Even though (F)PALM and *d*STORM experiments using PA-FP and standard fluorescent probes, respectively, can be performed under similar experimental conditions, the initial situation of a super-resolution imaging experiment is completely different. PA-FPs are essentially non-fluorescent at the beginning of the experiment, allowing the density of fluorescent molecules to be tightly controlled by the irradiation intensity of a 405 nm laser. In contrast, one of the key steps of single-molecule-based localization microscopy with synthetic fluorophores is the efficient transfer of the fluorophores to a reversible, relatively stable *off*-state at the beginning of the experiment whereas photodamage of the sample and photobleaching of the fluorophores have to be avoided. Thus, a highly reliable and, in the case of synthetic fluorophores, reversible photoswitching mechanism and appropriate photoswitching rates (i.e., stable *off*-states) are of utmost importance for SMLM. Furthermore, highly efficient and dense labeling with photoswitchable fluorescent probes is important for all super-resolution imaging methods. Whereas optical resolution is defined as the minimal distance resolvable between two emitters, a high experimental structural resolution affords high labeling densities according to the Nyquist–Shannon sampling theorem [31].

The ease and speed of data acquisition afforded by the various SMLM methods increase the computational requirement concerning fast and reliable data processing and image reconstruction. The general problem of processing stochastic SMLM data is characterized by a high quantity of data, typical in the range of gigabytes, uncertainty in the exact size of the point spread function (PSF), high background noise as common in single-molecule experiments, and stochastically occurring insufficient spatial separation between simultaneously emitting fluorophores. There are two general approaches to data processing: Most often, fluorophore emission is initially assumed to be generated by single fluorophores, which are first localized on a pixel scale by finding maxima in each input image followed by localization on a nanometer scale by fitting a model of the PSF to the experimental data. In a different approach, the super-resolution optical fluctuation imaging (SOFI) algorithm does not rely on single-molecule emission but computes the temporal autocorrelation of the fluorescence signal for each pixel [32].

In this chapter we will focus our attention on the computational tools and processing options for SMLM by following the logical processing chain from finding the spots of fluorescent emission in unannotated images, fitting the emitter location in these spots with nanometer resolution, and finally generating a super-resolved image from all localizations. We will continue with a discussion of the advanced techniques of gaining z position information in spot fitting (3D imaging). Finally, we will focus our attention on the photoswitching mechanism of *d*STORM and the importance of the photoswitching ratio, i.e., the importance of the formation of stable *off*-states, and its influence on the quality of localization microscopy data.

2 Experimental Setup for dSTORM Measurements

Precise position determination of single fluorophores requires high fluorescence signals, optimally combined with a very low background. Typically, an inverted fluorescence microscope is equipped with an oil-immersion objective, with a high numerical aperture (NA > 1.45) operated in objective-type total internal reflection fluorescence (TIRF) microscopy mode [5]. TIRF microscopy limits fluorophore excitation to a thin evanescent field (100–200 nm) and minimizes background light. The scheme can be extended to three-dimensional or whole-cell imaging by using a wide-field configuration, albeit at the cost of a lower signal-to-noise ratio (SNR) and consequently a lower localization precision and final image resolution.

The fluorophores are excited with multiple laser lines, e.g., at 488, 514, 532, 568, or 647 nm to induce fluorescence and at 405 nm for photoswitching from the *off*- to the *on*-state. Excitation intensities vary in the range of 0.5–50 kW cm^{-2} for excitation and are below 0.1 kW cm^{-2} for *off*-to-*on* photoswitching. The fluorescence light in the detection path is filtered using suitable bandpass filters and imaged with a sensitive camera, e.g., an electron-multiplying charge-coupled device (EMCCD) camera with quantum yields of 80–90 % in the visible range. To preserve most of the position information in the fluorescence signal data while reducing shot noise to a minimum, a pixel size of ~0.4× the Abbe resolution limit of the optical system is generally used [16]. A typical image pixel is therefore ~100 nm wide, which is achieved by the objective and additional lenses in the detection path.

Labeling techniques for in vivo and in vitro SMLM include photoactivatable fusion proteins and organic fluorophores coupled to antibodies, DNA oligonucleotides, or small specifically binding peptides such as phalloidin. A wealth of probes labeled with Alexa Fluor and ATTO fluorophores is commercially available. Typically, a tandem of a primary (against the antigen of interest) and a secondary antibody (against the first antibody) is used for labeling. For fixed cell experiments, we recommend the use of Alexa Fluor 647 or Cy5 for single-color *d*STORM experiments because they can be used at very high excitation intensities (allowing short acquisition times [21, 34]) and because concurrent readout at 641 or 647 nm and reactivation at 488, 514, or 532 nm do not induce strong photodamage. For live-cell *d*STORM, the TMP-tag method [22] and SNAP tags [26] have been successfully used. Figure 1 shows a *d*STORM image of the mitochondrial protein cytochrome *c* oxidase in a mammalian COS-7 cell. This image was acquired by using an sCMOS camera (Andor), which is the currently most promising detection device to image whole cells at high frame rates with so far unmatched resolution.

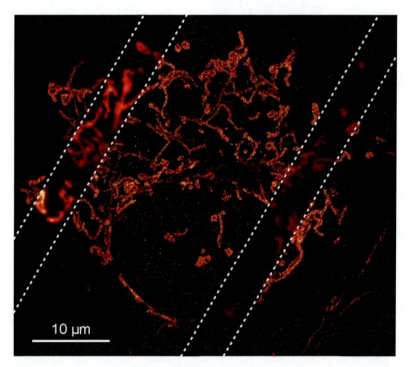

Fig. 1 Qualitative comparison of *d*STORM and wide-field fluorescence imaging of mitochondrial structures in COS-7 cells. Cytochrome *c* oxidase was immunostained using Alexa Fluor 647-labeled antibodies, immersed in a switching buffer containing 100 mM β-mercaptoethylamine (MEA) and an oxygen scavenging system at pH 7.4. Fluorophores were excited with an intensity of 3 kW cm^{-2} at 644 nm. The fluorescence signal was detected with an Andor Neo sCMOS camera at a frame rate of 100 Hz. The image was reconstructed using rapi*d*STORM [34]. A part of the conventional wide-field fluorescence image is shown in the two stripes between the *white dashed lines*

3 Spot Finding

The first task of computational post-processing after image acquisition is spot finding (Fig. 2). Spot finding is the computational task of finding the spots, i.e., probable locations of fluorophore activity, in an unannotated wide-field microscopy image. Spot finding is generally [33, 34] performed by finding local maxima, which can be done quite efficiently using the Neubeck algorithm [35]. Since local maxima are also found in random background noise,

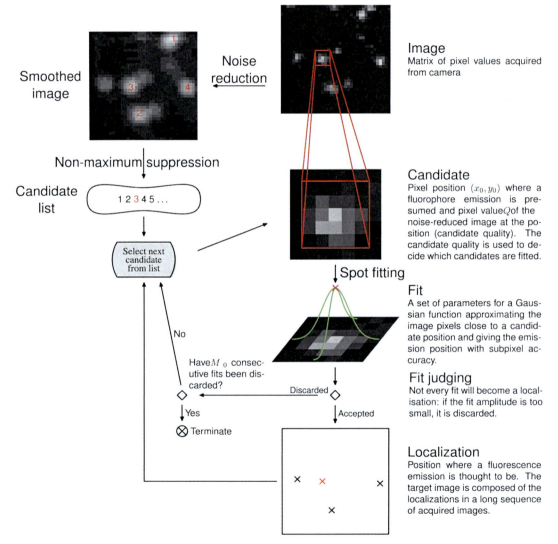

Fig. 2 Visualization of a typical data processing scheme for localization microscopy. From *top* to *bottom*, the data are processed from raw image form to a set of likely candidates for emission positions (*spots*), fitted by a Gaussian point spread function model and, given that this fit yields acceptable values, considered localizations. These localizations can be visualized in a result image or analyzed quantitatively (with permission from ref. [34])

a discrimination is usually made either through simple thresholding [36] of the maximum or through dynamic thresholding [34] by sorting the maxima by their intensity and fitting consecutively until a given number of false localizations is reached.

When noise conditions are unusually high, which can happen in a number of circumstances such as high autofluorescence, very high single pixels are quite common and will foil a straightforward local maxima approach. In this case, low-pass filters have been used successfully, and several works have been published that compare different low-pass filters [34, 37]. Most of the successful and widely used filters are simple kernel operators such as an average mask [33, 34], or a Gaussian low-pass [34, 36, 37]. Morphological operators such as erosion or fill-hole transforms have also been considered [34].

Smoothing is prone to both under- and over-smoothing. To understand this predicament, it is useful to consider the process in frequency space, i.e., the effect of smoothing operators in Fourier space. Any smoothing operator acts as a low-pass filter in terms of frequency and thus removes high frequencies from the image. On the one hand, this is useful to remove noise since noise is usually spatially uncorrelated and thus only represented in high frequencies. On the other hand, the separation of close-by spots is encoded in the high frequencies. Thus, it is important to smooth the image to remove noise, but not to over-smooth in order to keep as much position information as possible. An average mask with a width of one PSF full width at half maximum (FWHM) has been reported to be successful [34].

4 Spot Fitting

Spot fitting is the computational task of finding the number and most likely positions of fluorophores in a small sub-image given an initial rough estimate of the parameters. Since this task is central to localization microscopy, there are a number of approaches and variations to this problem, and we will describe them by first giving an overview over the most common and basic method of fixed-form Gaussian least-squares (XGLS) fitting and then showing where other methods differ.

When using XGLS, the PSF is modelled with a two-dimensional Gaussian function as given in Eq. 1. The parameters modelling emitter position $(\vec{x}_0)$, signal strength (A), and background (B) are initialized using rough estimates such as the position of the brightest pixel in the sub-image or overall intensity. The covariance matrix Σ is provided from prior knowledge of the optical system. The initial estimates are then improved using nonlinear least-squares parameter optimization, that is, the sum of squared deviations of the data and the model (χ^2) is minimized iteratively.

The fitted emitter position $(\vec{x}_0)$ gives the subpixel-precise emitter position, and the fitted amplitude [34] or the sum of the remaining squared deviations is a good observable for deciding whether the fitted sub-image actually contained an emitting fluorophore.

$$PSF(\vec{x}_0, A, B) = \frac{A}{2\pi\sqrt{|\Sigma|}} \exp\left[-0.5(\vec{x}-\vec{x}_0)^T \Sigma^{-1}(\vec{x}-\vec{x}_0)\right] + B \quad (1)$$

Fitting with XGLS relies on three assumptions:

1. The PSF is sufficiently close to a Gaussian function.
2. The noise is sufficiently close to white Gaussian noise.
3. There is at most a single fluorophore in the diffraction-limited region (DLR).

Each of these assumptions has been questioned in literature and improvements suggested, though no solution exists to our knowledge where all three assumptions can be dropped. In practice, the *d*STORM method has mostly been used with straight XGLS fitting and very good results have been obtained with it. The most common failure of these assumptions happens when imaging densely emitting samples, i.e., samples where the density of fluorophores is so high that the assumption of having only a single active fluorophore per DLR is invalid sufficiently often to degrade the overall image quality. Such samples have been super-resolved [28, 33, 36, 38] by replacing the single Gaussian in Eq. 1 with a sum of Gaussian kernels and comparing the improvements in χ^2 values for increasing number of kernels. The highest number of kernels for which χ^2 still decreases significantly compared to the next-lowest number is considered the most likely number of active fluorophores.

Another common variant seen in *d*STORM imaging is the use of free-form fitting, that is, fitting the entries of the covariance matrix Σ individually to each spot. While it has been shown [34] that this approach decreases resolution, probably through the added degrees of freedom in fitting, it has proven useful in calibrating and using techniques for three-dimensional imaging. Advanced thresholding techniques seem to be necessary for deciding the presence of a fluorophore with such data.

Several researchers [33, 39] have also pointed out that the assumption that underlies least-squares fitting, i.e., the noise being additive white Gaussian noise, is wrong. The dominating noise in SMLM is Poisson shot noise, and least-squares estimation yields suboptimal results for such noise. Using true maximum likelihood estimators for shot noise has been reported to improve precision on simulated data and fluorescent beads, but to our knowledge no improvement has been published for switchable fluorescent probes.

Lastly, some researchers have challenged the fundamental assumption of modelling the PSF with a Gaussian function. While the theoretical PSF of a wide-field microscope is described by a Besselian function of first kind and first order, experimental deviations from the theoretical PSFs are unavoidable. Mlodzianoski and Bewersdorf [40] have addressed this problem for the especially problematic case of 3D microscopy by using an approach based on Fourier analysis of the experimentally measured PSF, while Aguet et al. [41] have approached the problem by performing maximum likelihood estimation on the raw integral (which is extremely time-consuming, but gives very precise results in distances from focal planes that are unattainable by other methods). Chromophore dipole orientation [42] can also have an effect on dSTORM imaging.

Despite the criticisms, XGLS fitting is still widely used in software implementations of dSTORM and other methods. The good performance and easy implementation of least-squares fitting and the computationally cheap use of the Gaussian function continue to be large advantages in this field.

5 Image Generation

The image generation part of localization microscopy is the computational task of converting the set of localizations into a super-resolved image that closely resembles the imaged structure. While this task seems trivial, it is complicated by several factors. The number of localizations available is limited by the measurement time and by fluorophore stability and thus the image formation is subject to Poisson noise. This noise is amplified by differences in the local chemical nanoenvironment: Some fluorophores are much more active than others, generating a huge dynamic range in the numbers of localizations per pixel.

Different localization microscopy methods exhibit these problems with different severity. dSTORM measurements show excellent reversibility and photostability [4] and thus have little problems with photobleaching. However, the local chemical environment is extremely important because a quickly recovering fluorophore can generate hundreds of localizations in the course of a measurement [34]. Other methods such as PALM can usually rely on a more predictable localization count per fluorophore, but at the price of far fewer localizations. The simple and usually sufficient approach to dSTORM imaging is to construct a high-resolution density map of the sample with a small pixel size (~10 nm) and count the number of localizations in each pixel, using interpolation to conserve some subpixel position information. This density map is then converted into an image by applying a color map [34]. The contrast issues arising in dSTORM have been addressed [34] with weighted histogram normalization. This technique offers a tradeoff between linearity, i.e., a straightforward linear mapping from the number of localizations represented by a pixel to the grey value of that pixel, and contrast. Classical histogram normalization would ensure maximum contrast by reassigning brightness values in the image in such a way that each brightness value would be populated equally, completely disregarding linearity. Weighted histogram normalization allows control of the degree of histogram normalization, making automated generation of images with high contrast and good linearity possible.

The high Poisson noise in the image has been approached by displaying resolutions with a Gaussian function with a width equal to the localization precision. While this obviously degrades resolution [43], it results in an overall smoother image. This method is commonly used with PALM images, but generally unnecessary with dSTORM. If the number of localizations is small, Baddeley et al. [43] have suggested several alternative methods.

6 Three-Dimensional Fitting

The PSF of a high numerical aperture (NA) microscope conveys considerable information about the axial position of an emitter. Upon defocusing, the detected PSF quickly deviates from the ideal Besselian function of first kind and first order, both growing wider and gaining additional interference minima and maxima [44]. Close to the focal plane, i.e., up to 1 µm away, the PSF can still be approximated by a Gaussian function, but the width of this Gaussian is a parameter of the axial deviation of the emitter from the focal plane.

This deviation can be functionalized using a quadratic function [45] or a Euclidean norm approach [46] and then used to modify the PSF to include the Z position as a parameter. Equations 2 and 3 show the 3D PSF using the quadratic approach, with x_0 and y_0 as the components of $\vec{x}_0$, Σ_0 giving the best-focused covariance matrix, a denoting a widening constant dependent on the NA, and $\vec{\delta}_z$ the distance from the best-focused plane for each coordinate.

$$PSF(\vec{x}_0, z, A, B) = \frac{A}{2\pi\sigma_x(z)\sigma_y(z)} \exp\left[-0.5\left(\frac{(x-x_0)^2}{\sigma_x^2(z)} + \frac{(y-y_0)^2}{\sigma_y^2(z)}\right)\right] + B \quad (2)$$

$$\sigma_i(z) = a\delta_{z,i}^2 + \Sigma_{0,ii} \quad (3)$$

Close inspection of this PSF reveals two problems: Firstly, if the $\delta_{z,I}$ are identical for both dimensions, the PSF is symmetric in z, thus allowing no conclusion whether a given emitter is below or above the focal plane. Secondly, the derivative of the PSF with respect to z depends linearly on $\delta_{z,i}$ and is thus very small close to the focal plane, allowing no precise Z position measurement.

Both problems can be alleviated by measuring the PSF width at two different defocusing levels. The first solution, astigmatic imaging [46], separates the $\delta_{z,I}$ by using a cylindrical lens, thus focusing x and y at different planes. The second solution, biplane imaging [47], uses two detectors with two different focal planes for the same effect. By separating the focal planes, two advantages are gained: Firstly, an asymmetry in z is observable, and secondly, there is no point with vanishing z derivative since the minima for the z derivatives of the x and y dimensions (or for the two detectors) do not coincide. Naturally, the distance between the focal planes has to be chosen carefully such that the Gaussian approximation still holds for both planes concurrently, but still having sufficient focal plane distance to break PSF symmetry and derivative minima.

In practical terms the functional form of the depth information in the PSF can be exploited by fitting the z position alongside the x and y parameters. If this is infeasible, alternative methods include fitting the PSF with a diagonal covariance matrix Σ and using a calibration curve for the ratio or the difference between the diagonal elements σ_x^2 and σ_y^2. It must be emphasized that the parameter a in the PSF model confers a degree of freedom into the result image: The scale of the z axis cannot be inferred from the scale of the x and y axes. Therefore, it must be independently calibrated, e.g., by the use of a piezo-stage.

A completely alternate approach is the engineering of a microscope with a different, more depth-sensitive PSF. Examples of this approach are the double-helix PSF suggested by Pavani et al. [48] or interferometric localization microscopy suggested by Shtengel et al. [49]. However, PSF engineering is a methodically difficult approach requiring extensive calibration.

7 Photoswitching Ratio and High Spot Densities

SMLM can achieve resolution improvements of drastically higher magnitude than classical microscopy. This improvement is possible because of the central assumption of fluorophore isolation: Fluorophores in SMLM are fluorescent so rarely that each spot can be safely assumed to be the emission pattern of a single fluorescent molecule. Thereby, the deconvolution problem, which is known in computer science to be computationally hard and cannot be solved generally, can be solved reliably.

The isolation assumption can only hold if the fluorophore population is small or if the ratio of non-fluorescent inactive time to fluorescent active time is very large for each fluorophore, i.e., fluorophores are in a non-fluorescent *off*-state most of the time.

A small fluorophore population is generally not an option: The sampling theorem [31] dictates that the average distance between sampling points, i.e., fluorophores, must be half the size of the lowest reliably resolvable level of detail. Consequently, when imaging a two-dimensional structure at a structural resolution of 50 nm, one fluorophore must be placed every 25 nm or at least 1,600 fluorophores per square micrometer. Assuming that the PSF is small enough that a statistical average of one active molecule per square micrometer suffices to make the assumption of isolated fluorophores reasonable, each fluorophore must statistically be non-fluorescent 1,600 times longer than fluorescent.

Reversibly photoswitchable fluorophores as used in *d*STORM can be switched multiple times from a fluorescent *on*-state to a non-fluorescent *off*-state with rate k_{off} and vice versa with rate k_{on}. The number of molecules residing in the *on*-state N_{on} is given by the differential rate equation (4), which is idealized in terms of photostability. The ratio of photoswitching rates $k_{\text{off}}/k_{\text{on}}$ can be used to give the ratio between *on*-state and *off*-state population sizes at equilibrium, i.e., for $dN_{\text{on}}/dt = 0$. The switching rates control the active fluorophore density and together with the processing algorithm they determine the ability to resolve a structure [50]. Both rates are usually given as the average number of transitions per molecule and time. Giving a rate is in this context equivalent to giving a lifetime of a state, e.g., $\tau_{\text{on}} = k_{\text{off}}^{-1}$ is the lifetime of the *on*-state and gives the average time a fluorescent molecule stays fluorescent. The photoswitching ratio r is defined in terms of these rates or, equivalently, lifetimes (5) and can be influenced by using oxidizing agents [51, 52] or applying direct excitation of the non-fluorescent *off*-state [5, 19, 20, 53].

$$\frac{dN_{on}}{dt} = N_{off}k_{on} - N_{on}k_{off} \qquad (4)$$

$$r = \frac{k_{off}}{k_{on}} = \frac{\tau_{off}}{\tau_{on}} \qquad (5)$$

High label densities therefore require that the majority of fluorophores must be kept in the *off*-state whereas only a small subset stays fluorescent. Hence, k_{off} must exceed k_{on} by several orders of magnitude, i.e., a photoswitching ratio of $r \geq 1{,}000$ is often required. The lifetime τ_{off} thereby gives an upper limit to the reasonably achievable photoswitching ratio; when τ_{off} is in the order of 10–100 ms, τ_{on} has to be in the order of 10–100 µs. Such short lifetimes necessitate on the one hand very high irradiation intensities and prevent live-cell imaging, and on the other hand require faster cameras than are generally available today. Thus, longer fluorescent *off*-times in the order of seconds are necessary to achieve high photoswitching ratios and

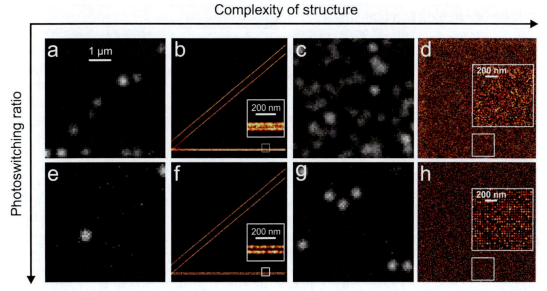

Fig. 3 Simulated data demonstrating the influence of the ratio of photoswitching rates, r, on different complexities. With increasing complexity of the imaged structure, the number of fluorophores present per area increases. The photoswitching ratio is mainly determined by the stability of the *off*-state. Images shown in (**a**, **c**, **e**, and **g**) are snapshots from the localization microscopy stacks from which the reconstructed super-resolution images (**b**, **d**, **f**, and **h**) were generated. Panels (**b**) and (**f**) show pairs of *lines* separated by 300 and 50 nm with a fluorophore attached every 8.5 nm, whereas (**d**) and (**h**) show a lattice with 40 nm spacing. (**a–d**) Images were simulated with $r=250$. (**e–h**) Images were simulated with $r=3,000$. It becomes clear that the necessary photoswitching ratio depends strongly on the structure imaged; when the photoswitching ratio is too low (**c**, **d**) more than one fluorophore may fluoresce per diffraction-limited area and subsequent fitting of the PSF can result in false localizations and image artifacts. On the contrary, high photoswitching ratios result in a small number of localizations per image and thus prolong the acquisition time unnecessarily. Note that the density of fluorescence emissions is roughly the same for the images shown in (**a**) and (**g**), even though they are simulated for different values of r. Scale bars, 1 μm for (**a**) (valid for all images); 200 nm for *insets* in (**b**), (**d**), (**f**), and (**h**) (with permission from ref. [5])

have been reported by several groups [7, 12, 13, 53]. Figure 3 summarizes the effect of different switching kinetics and label densities on the achievable resolution.

The assumption of isolated fluorophores has been reported to hold up to roughly 0.1–1 fluorophores per square micrometer [28, 50]. For up to ten fluorophores per square micrometer, multispot fitting routines [28, 33] have been reported to be successful. These fitting approaches have the option to explain a spot as a sum of multiple emitting fluorophores. While the idea is older than super-resolution microscopy [38], it came to prominence in super-resolution only very recently with the focus on live-cell imaging and the imaging of cellular structures different from filaments.

8 Photoswitching Mechanism

SMLM Methods that use irreversibly photoactivatable or -convertible fluorescent proteins [1, 2] need to activate a sparse subset of fluorophores. The structure of interest is dark at the beginning and upon photoactivation (e.g., at 405 nm) single FPs are activated stochastically and imaged until photobleaching terminates the switching process. In contrast, by using organic dyes as reversibly photoswitchable fluorophores [3, 4, 6–8, 51, 54, 55], all fluorophores reside in the fluorescent *on*-state at the beginning of the experiment; the structure therefore appears bright. In the first step of the experiment, the majority of fluorophores must be switched reversibly to a non-fluorescent *off*-state and only small subpopulations of fluorophores, separated in time, are allowed to be fluorescent.

The formation of metastable dark states has its origin in prolonging the triplet state lifetime of fluorophores, e.g., in polymer films [56]. For resolving only small fluorophore densities, even prolonged triplet state lifetimes can serve as dark states. However, for resolving complex structures, i.e., high fluorophore densities, triplet state lifetimes are generally too short. In aqueous solution, triplet states only exhibit lifetimes in the microsecond to millisecond range, but they can serve as a platform for further redox reactions, e.g., by a reducing and oxidizing system (ROXS) which has been successfully used to minimize photobleaching of standard organic dyes [57]. The refined understanding of depopulating excited states by reducing agents has been used to generate stable *off*-states [7, 18, 20, 53] with lifetimes ranging from hundreds of milliseconds to several seconds.

Using thiols as weak electron donors, it has been demonstrated that upon irradiation rhodamine and oxazine dyes are reversibly switched to a radical or a fully reduced state, respectively [5, 20, 58]. If dye radicals are oxidized or directly irradiated, the ground state is recovered and fluorescence emission occurs again. Since cells naturally contain reducing and oxidizing agents, e.g., glutathione and oxygen [59, 60], the photoinduced formation of dye radicals and further reduced species can be exploited in live-cell super-resolution microscopy experiments [22].

9 Existing Software

QuickPALM is an ImageJ-based plug-in for super-resolution by Henriques et al. [24]. While fairly fast, it is based on center of mass computations, which are known to be quite suboptimal for particle localization [15]. However, practical results with QuickPALM are usually acceptable, and the method has the advantage of having few free parameters. Features include 2D and 3D localization and embedded laser control software. As it is based on ImageJ, it allows on the one hand a wide range of image formats to be used and a good integration into a huge tool database, but on the other hand is hard to automate and shares the harsh memory requirements of ImageJ, which can only process files that fully fit into the computer's main memory. QuickPALM is available as free open-source software (FOSS) from http://code.google.com/p/quickpalm/.

Palm3d is a Python program by York et al. [61]. It is based on cross-correlating prerecorded instances of the PSF with the located image data, which has the advantage of being able to work with a very aberrated PSF, but the distinct disadvantage of necessary lengthy calibration and of lower precision than full Gaussian fitting [15]. Palm3d can perform 2D and 3D localization. The computation time is acceptable for slow acquisition methods, and the program is very concise and sufficiently documented to allow easy extension. Palm3d is available as FOSS from http://code.google.com/p/palm3d/.

LivePalm is a Matlab suite by Hedde et al. [62]. It is based on the fluoroBancroft algorithm, which is according to the tests published with the program only slightly less precise than Gaussian fitting. LivePalm can only perform 2D localization. The run time seems to be acceptable, but the program depends on the proprietary and expensive Matlab software suite and is itself closed source, thus making further analysis complicated. LivePalm is available from the supporting information of the article by Hedde et al. [62].

rapidSTORM is a standalone program by Wolter et al. [34]. It employs Gaussian least-squares fitting and makes this computation (which has often been considered to be too expensive) feasible and even real-time capable. From the mentioned programs, it seems to be the largest and by far the most complete and configurable analysis suite, can perform 2D, 3D, and multicolor localization, and has a large number of additional features (*see* Fig. 4). However, its size, complexity of infrastructure, and build process require in-depth knowledge for further extension or adaptation, and the learning curve is fairly steep. rapidSTORM is available as FOSS from http://www.super-resolution.biozentrum.uni-wuerzburg.de/home/rapidstorm/.

daoSTORM is a Python program by Holden et al. [28] based on the image reduction and analysis facility (IRAF). It adapts the Daophot module of IRAF, which specializes in astronomical crowded-field photometry, and applies it to microscopy data. daoSTORM is specialized on such crowded-field acquisitions, i.e., those with high fluorophore densities, and a prototype, consequently not the best choice for more sparse images. However, the code is concise and documented and should be easily extensible. daoSTORM can only perform 2D localization and has a quite long runtime. daoSTORM is available as FOSS from the supporting information of the article by Holden et al. [28].

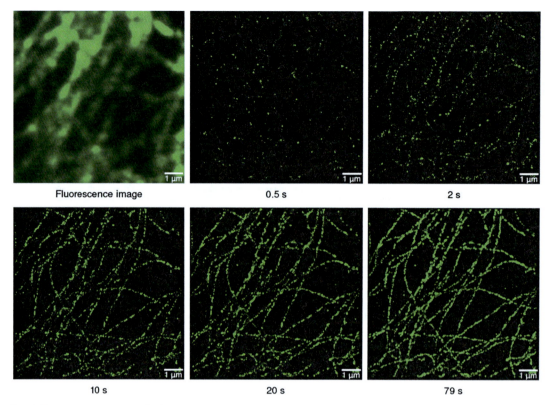

Fig. 4 Time-resolved view of *d*STORM image reconstruction on a real sample. The microtubulin network of a mammalian cell was labeled via immunocytochemistry [4] using a secondary antibody carrying the photo-switchable fluorophore Alexa Fluor 647. Images were acquired at frame rate of ~1 kHz and were processed in real-time with rapi*d*STORM. Over 730,000 fluorophore localizations were found in the 70,000 images acquired (with permission from ref. [34])

Overall, new researchers interested in learning and adapting the technique should, in our opinion, choose rapi*d*STORM for its richness of features, speed, and relative stability.

10 Concluding Remarks

SMLM is a powerful and easily applied technique for resolution enhancement in fluorescence microscopy. The algorithms for SMLM have received considerable attention in the last years and are continuing to do so. This chapter has described the basic algorithmic approach for computational fluorescence super-resolution imaging, consisting of a coarse spot finding stage which selects likely positions of fluorescence emission, and a spot fitting stage, where these estimates are checked and refined to nanometer precision. The spot finding stage is usually implemented via a smoothing algorithm that reduces the measurement noise to a level at which coarse localization becomes confident. These coarse position estimates are then refined by one of a multitude of localization algorithms, which mostly differ by the assumptions they make about the shape of the PSF and the nature of the background noise. For many current applications, the general choice of a least-squares optimal Gaussian fit to the data is still optimal and widely used. The precise position estimates found by the fitting stage can then be displayed by combining them into a single density map, which serves as a super-resolution image. We have seen that a number of software implementations exist for this process.

Straightforward SMLM processing can be performed reliably and with little user interactions. However, there are several special cases in which user interaction and advanced processing is necessary. Firstly, three-dimensional processing needs careful selection of PSF models and calibration of z defocus behavior. Secondly, an appropriate photoswitching rate must be achieved for a sufficiently low average spot density, or the number of correctly localized spots and localization precision will decrease considerably. Thirdly, sufficient knowledge about the PSF that is put into the computation process, e.g., by fixing the covariance matrix of the fitted Gaussian function, considerably helps computation and improves localization precision.

Overall, it should be stressed that algorithmic processing is a critical part of SMLM and should be given appropriate consideration when designing an experiment. While maybe intimidating to a new researcher in the field, the power and wealth of options in this post-processing is one of the great strengths of SMLM and should be treated as such.

Acknowledgment

We would like to thank the Biophotonics Initiative of the BMBF for financial support (Grants #13N11019 and #13N12507).

References

1. Betzig E, Patterson GH, Sougrat R, Lindwasser OW, Olenych S, Bonifacino JS, Davidson MW, Lippincott-Schwartz J, Hess HF (2006) Imaging intracellular fluorescent proteins at nanometer resolution. Science 313(5793):1642–1645
2. Hess ST, Girirajan TP, Mason MD (2006) Ultra-high resolution imaging by fluorescence photoactivation localization microscopy. Biophys J 91(11):4258–4272
3. Rust MJ, Bates M, Zhuang X (2006) Sub-diffraction-limit imaging by stochastic optical reconstruction microscopy (storm). Nat Methods 3(10):793–795
4. Heilemann M, van de Linde S, Schüttpelz M, Kasper R, Seefeldt B, Mukherjee A, Tinnefeld P, Sauer M (2008) Subdiffraction-resolution fluorescence imaging with conventional fluorescent probes. Angew Chem Int Ed 47(33): 6172–6176
5. van de Linde S, Löschberger A, Klein T, Heidbreder M, Wolter S, Heilemann M, Sauer M (2011) Direct stochastic optical reconstruction

microscopy with standard fluorescent probes. Nat Protoc 6(7):991–1009, ISSN1754-2189
6. Fölling J, Bossi M, Bock H, Medda R, Wurm CA, Hein B, Jakobs S, Eggeling C, Hell SW (2008) Fluorescence nanoscopy by ground-state depletion and single molecule return. Nat Methods 5(11):943–945, ISSN 1548-7091
7. Vogelsang J, Cordes T, Forthmann C, Steinhauer C, Tinnefeld P (2009) Controlling the fluorescence of ordinary oxazine dyes for single-molecule switching and superresolution microscopy. Proc Natl Acad Sci U S A 106(20):8107–8112
8. Lemmer P, Gunkel M, Weiland Y, Müller P, Baddeley D, Kaufmann R, Urich A, Eipel H, Amberger R, Hausmann M, Cremer C (2009) Using conventional fluorescent markers for far-field fluorescence localization nanoscopy allows resolution in the 10-nm range. J Microsc 235(2):163–171
9. Dedecker P, Hotta J-I, Flors C, Sliwa M, Uji-i H, Roeffaers MBJ, Ando R, Mizuno H, Miyawaki A, Hofkens J (2007) Subdiffraction imaging through the selective donut mode depletion of thermally stable photoswitchable fluorophores: numerical analysis and application to the fluorescent protein Dronpa. J Am Chem Soc 129(51):16132–16141
10. Hell SW, Wichmann J (1994) Breaking the diffraction resolution limit by stimulated emission: stimulated-emission-depletion fluorescence microscopy. Opt Lett 19(11):780
11. Gustafsson MGL (2005) Nonlinear structured-illumination microscopy: wide-field fluorescence imaging with theoretically unlimited resolution. Proc Natl Acad Sci U S A 102(37):13081–13086
12. Heilemann M, Margeat E, Kasper R, Sauer M, Tinnefeld P (2005) Carbocyanine dyes as efficient reversible single-molecule optical switch. J Am Chem Soc 127(11):3801–3806
13. Bates M, Blosser TR, Zhuang X (2005) Short-range spectroscopic ruler based on a single-molecule optical switch. Phys Rev Lett 94(10):108101
14. Shaner NC, Lin MZ, McKeown MR, Steinbach PA, Hazelwood KL, Davidson MW, Tsien RY (2008) Improving the photostability of bright monomeric orange and red fluorescent proteins. Nat Methods 5(6):545–551, ISSN 1548-7091
15. Cheezum MK, Walker WF, Guilford WH (2001) Quantitative comparison of algorithms for tracking single fluorescent particles. Biophys J 81(4):2378–2388
16. Thompson RE, Larson DR, Webb WW (2002) Precise nanometer localization analysis for individual fluorescent probes. Biophys J 82(5):2775–2783
17. Yildiz A, Forkey JN, McKinney SA, Ha T, Goldman YE, Selvin PR (2003) Myosin v walks hand-over-hand: single fluorophore imaging with 1.5-nm localization. Science 300(5628):2061–2065
18. Heilemann M, van de Linde S, Mukherjee A, Sauer M (2009) Super-resolution imaging with small organic fluorophores. Angew Chem Int Ed 48(37):6903–6908
19. van de Linde S, Wolter S, Heilemann M, Sauer M (2010) The effect of photoswitching kinetics and labeling densities on superresolution fluorescence imaging. J Biotechnol 149(4):260–266, ISSN 0168-1656
20. van de Linde S, Krstić I, Prisner T, Doose S, Heilemann M, Sauer M (2011) Photoinduced formation of reversible dye radicals and their impact on superresolution imaging. Photochem Photobiol Sci 10:499–506
21. Endesfelder U, van de Linde S, Wolter S, Sauer M, Heilemann M (2010) Subdiffraction resolution fluorescence microscopy of myosin-actin motility. Chemphyschem 11(4):836–840
22. Wombacher R, Heidbreder M, van de Linde S, Sheetz MP, Heilemann M, Cornish VW, Sauer M (2010) Livecell super-resolution imaging with trimethoprim conjugates. Nat Methods 7(9):717–719, ISSN 1548-7091
23. Owen DM, Rentero C, Rossy J, Magenau A, Williamson D, Rodriguez M, Gaus K (2010) Palm imaging and cluster analysis of protein heterogeneity at the cell surface. J Biophotonics 3(7):446–454, ISSN 1864-0648
24. Henriques R, Lelek M, Fornasiero EF, Valtorta F, Zimmer C, Mhlanga MM (2010) QuickPALM: 3D real-time photoactivation nanoscopy image processing in ImageJ. Nat Methods 7(5):339–340, ISSN 1548-7091
25. Williamson DJ, Owen DM, Rossy J, Magenau A, Wehrmann M, Gooding JJ, Gaus K (2011) Pre-existing clusters of the adaptor lat do not participate in early T cell signaling events. Nat Immunol 12(7):655–662, ISSN 1529-2908
26. Klein T, Löschberger A, Proppert S, Wolter S, van de Linde S, Sauer M (2011) Live-cell dSTORM with SNAP-tag fusion proteins. Nat Methods 8(1):7–9, ISSN 1548-7091
27. Izeddin I, Specht CG, Lelek M, Darzacq X, Triller A, Zimmer C, Dahan M (2011) Super-resolution dynamic imaging of dendritic spines using a low-affinity photoconvertible actin probe. PLoS One 6(1):e15611
28. Holden SJ, Uphoff S, Kapanidis AN (2011) Daostorm: an algorithm for high-density superresolution microscopy. Nat Methods 8(4):279–280, ISSN 1548-7091
29. Testa I, Wurm CA, Medda R, Rothermel E, von Middendorf C, Fölling J, Jakobs S,

Schönle A, Hell SW, Eggeling C (2010) Multicolor fluorescence nanoscopy in fixed and living cells by exciting conventional fluorophores with a single wavelength. Biophys J 99(8):2686–2694, ISSN0006-3495

30. Jones SA, Shim S-H, He J, Zhuang X (2011) Fast, three-dimensional super-resolution imaging of live cells. Nat Methods 8(6):499–505
31. Shannon CE (1984) Communication in the presence of noise (reprinted). Proc IEEE 72(9):1192–1201, ISSN 0018-9219
32. Dertinger T, Colyer R, Iyer G, Weiss S, Enderlein J (2009) Fast, background-free, 3D super-resolution optical fluctuation imaging (SOFI). Proc Natl Acad Sci U S A 106(52): 22287–22292
33. Huang F, Schwartz SL, Byars JM, Lidke KA (2011) Simultaneous multiple-emitter fitting for single molecule super-resolution imaging. Biomed Opt Express 2(5):1377–1393
34. Wolter S, Schüttpelz M, Tscherepanow M, van de Linde S, Heilemann M, Sauer M (2010) Real-time computation of subdiffraction-resolution fluorescence images. J Microsc 237(1):12–22
35. Neubeck A, Van Gool L (2006) Efficient non-maximum suppression. In: ICPR'06: proceedings of the 18th international conference on pattern recognition. IEEE Computer Society, Washington, DC, pp 850–855. ISBN 0-7695-2521-0
36. Thomann DM (2003) Algorithms for detection and tracking of objects with super-resolution in 3D fluorescence microscopy. PhD thesis, ETH Zürich.
37. Křížek P, Raška I, Hagen GM (2011) Minimizing detection errors in single molecule localization microscopy. Opt Express 19(4):3226–3235
38. Thomann D, Dorn J, Sorger PK, Danuser G (2003) Automatic fluorescent tag localization II: improvement in superresolution by relative tracking. J Microsc 211(Pt 3):230–248, ISSN 0022-2720
39. Bobroff N (1986) Position measurement with a resolution and noise-limited instrument. Rev Sci Instrum 57(6):1152–1157
40. Mlodzianoski MJ, Bewersdorf J (2009) 3D-resolution in FPALM/PALM/STORM. Biophys J 96(3 suppl 1):636–637, ISSN 0006-3495
41. Aguet F, van de Ville D, Unser M (2005) A maximum-likelihood formalism for sub-resolution axial localization of fluorescent nanoparticles. Opt Express 13:10503–10522
42. Stallinga S, Rieger B (2010) Accuracy of the Gaussian point-spread-function model in 2D localization microscopy. Opt Express 18(24): 24461–24476
43. Baddeley D, Cannell MB, Soeller C (2010) Visualization of localization microscopy data. Microsc Microanal 16(1):64–72
44. Zhang B, Zerubia J, Olivo-Marin JC (2007) Gaussian approximations of fluorescence microscope point-spread function models. Appl Optics 46(10):1819–1829
45. Huang B, Wang W, Bates M, Zhuang X (2008) Three-dimensional super-resolution imaging by stochastic optical reconstruction microscopy. Science 319(5864):810–813, ISSN 1095-9203
46. Holtzer L, Meckel T, Schmidt T (2007) Nanometric three-dimensional tracking of individual quantum dots in cells. Appl Phys Lett 90(5):053902
47. Juette MF, Gould TJ, Lessard MD, Mlodzianoski MJ, Nagpure BS, Bennett BT, Hess ST, Bewersdorf J (2008) Three-dimensional sub-100 nm resolution fluorescence microscopy of thick samples. Nat Methods 5(6):527–529, ISSN 1548-7091
48. Pavani SRP, Thompson MA, Biteen JS, Lord SJ, Liu N, Twieg RJ, Piestun R, Moerner WE (2009) Three-dimensional, single-molecule fluorescence imaging beyond the diffraction limit by using a double-helix point spread function. Proc Natl Acad Sci U S A 106(9):2995–2999
49. Shtengel G, Galbraith JA, Galbraith CG, Lippincott-Schwartz J, Gillette JM, Manley S, Sougrat R, Waterman CM, Kanchanawong P, Davidson MW, Fetter RD, Hess HF (2009) Interferometric fluorescent super-resolution microscopy resolves 3D cellular ultrastructure. Proc Natl Acad Sci U S A 106(9):3125–3130
50. Wolter S, Endesfelder U, van de Linde S, Heilemann M, Sauer M (2011) Measuring localization performance of superresolution algorithms on very active samples. Opt Express 19(8):7020–7033
51. Steinhauer C, Forthmann C, Vogelsang J, Tinnefeld P (2008) Superresolution microscopy on the basis of engineered dark states. J Am Chem Soc 130(50):16840–16841
52. Cordes T, Strackharn M, Stahl SW, Summerer W, Steinhauer C, Forthmann C, Puchner EM, Vogelsang J, Gaub HE, Tinnefeld P (2010) Resolving single-molecule assembled patterns with superresolution blink-microscopy. Nano Lett 10(2):645–651, PMID:20017533
53. van de Linde S, Kasper R, Heilemann M, Sauer M (2008) Photoswitching microscopy with standard fluorophores. Appl Phys B 93(4): 725–731
54. Bates M, Huang B, Dempsey GT, Zhuang X (2007) Multicolor super-resolution imaging with photo-switchable fluorescent probes. Science 317(5845):1749–1753

55. Flors C, Ravarani CNJ, Dryden DTF (2009) Super-resolution imaging of DNA labelled with intercalating dyes. Chemphyschem 10(13): 2201–2204
56. Weston KD, Carson PJ, DeAro JA, Buratto SK (1999) Single-molecule detection fluorescence of surface-bound species in vacuum. Chem Phys Lett 308:58–64
57. Vogelsang J, Kasper R, Steinhauer C, Person B, Heilemann M, Sauer M, Tinnefeld P (2008) A reducing and oxidizing system minimizes photobleaching and blinking of fluorescent dyes. Angew Chem Int Ed 47(29):5465–5469
58. Kottke T, van de Linde S, Sauer M, Kakorin S, Heilemann M (2010) Identification of the product of photoswitching of an oxazine fluorophore using Fourier transform infrared difference spectroscopy. J Phys Chem Lett 1(21):3156–3159
59. Schafer FQ, Buettner GR (2001) Redox environment of the cell as viewed through the redox state of the glutathione disulfide/glutathione couple. Free Radic Biol Med 30(11): 1191–1212, ISSN 0891-5849
60. Sies H (1999) Glutathione and its role in cellular functions. Free Radic Biol Med 27(9–10): 916–921
61. York AG, Ghitani A, Vaziri A, Davidson MW, Shroff H (2011) Confined activation and subdiffractive localization enables whole-cell palm with genetically expressed probes. Nat Methods 8(4):327–333, ISSN 1548-7091
62. Hedde PN, Fuchs J, Oswald F, Wiedenmann J, Nienhaus GU (2009) Online image analysis software for photoactivation localization microscopy. Nat Methods 6(10):689–690, ISSN 1548-7091

Chapter 7

Super-Resolution Fluorescence Microscopy Using Structured Illumination

Kai Wicker

Abstract

The resolution of far-field fluorescence microscopy is limited by the Abbe diffraction limit. Making use of the moiré effect, structured illumination microscopy circumvents this limit by projecting fine patterns of light into the sample. From several diffraction limited raw images taken for different pattern positions and orientations, a high resolution image can be calculated. This way, linear structured illumination can enhance the resolution by a factor of about two. Employing nonlinearities such as fluorescence saturation, the resolution can be enhanced even further. In this article, a conceptual as well as a mathematical introduction to the technique is provided, as well as several examples of applications.

Key words Super-resolution, Photo-switchable fluorophores, Wide-field microscopy

1 Introduction

With the discovery of the Green Fluorescent Protein (GFP, 2008 Nobel Prize in Chemistry) [1–3] and the subsequent development of a whole rainbow-like assortment of fluorescent markers, fluorescence microscopy has developed into an indispensable tool of modern cell biology. The ability to label specific cellular components, e.g., through genetic expression of fluorescent proteins or antibody-labeling, combined with the relatively low invasiveness of light microscopy, allows the imaging of both structure and function, as well as the observation of dynamic processes in living cells, with unprecedented clarity. While for conventional wide-field microscopes these images may suffer from out-of-focus light, which contributes a blurry background and thus reduces image contrast in the focal plane, the confocal microscope [4] suppresses this unwanted light through a combination of laser-scanning point illumination and a detection pinhole, which predominantly (but unfortunately not only) blocks light stemming from out-of-focus regions. The thus acquired confocal images exhibit optical sectioning and contain more information about the three-dimensional structure of the sample.

The last decade has witnessed the emergence of numerous novel microscopy techniques. Nevertheless, the confocal microscope remains the workhorse of the biological community. But as a scanning technique confocal microscopy is inherently limited in its acquisition speed: the larger the field of view, the longer a scan will take. This may lead to problems when trying to image fast cellular processes. Although scan speeds can be increased, the reduced pixel dwell time (the time the illumination is incident on each sample point) also leads to a reduction of the already weak signal, leading to noisy images. This can partly be countered by increasing the illumination intensity. However, as the intensities needed for confocal scanning are usually very high, a further increase will eventually lead to saturation of fluorescence, so that the emitted signal cannot be increased arbitrarily. Furthermore, the necessary laser source and scanning optics put a rather higher price tag on confocal microscopes, compared to the much simpler wide-field microscope.

Structured illumination microscopy (SIM) can achieve optical sectioning in a wide-field microscope. To this end the sample is incoherently illuminated with a harmonic line pattern, i.e., a grating, rather than with a homogeneous light distribution (or other patterns, e.g., point patterns can also be used). From three (or more) images acquired for different positions of the light pattern, an optically sectioned image much like a confocal one can be calculated [5]. Besides the potential of much faster image acquisition, these microscopes are much simpler and cheaper than confocal ones.

But apart from optical sectioning, the biggest potential of structured illumination lies in high resolution imaging [6, 7], and this is the aspect of SIM this chapter will mainly focus on. In a wide-field fluorescence microscope the resolution, i.e., the smallest sample detail about which information can be captured by the microscope, can be no smaller than half the wavelength (in a particular medium) of the light emitted by the fluorescent molecules. This is the famous Abbe resolution limit. Smaller details simply cannot be captured by the microscope objective. Various techniques have been devised to circumvent this limit. As a scanning technique, stimulated emission depletion (STED) microscopy [8] reduces the size of the effective excited volume (the volume from which captured fluorescence light may have emanated) by de-exciting fluorophores around the center of illumination through stimulated emission. Localization techniques such as photoactivated localization microscopy (PALM) [9], stochastic optical reconstruction microscopy (STORM) [10], and fluorescence photoactivation localization microscopy (fPALM) [11] manage to sequentially image individual fluorescent molecules. These can then be localized with a precision far surpassing the Abbe resolution limit.

SIM uses yet another approach to circumvent the Abbe limit: the moiré effect. For coherent imaging similar ideas were proposed as early as 1966 [12]. This technique does not break the Abbe limit. It rather combines the individual limits of illumination and detection and in this way can improve the resolution by a factor of about two—very much like the confocal microscope, which also combines the resolution limits of illumination and detection, but does so very inefficiently. Furthermore, due to its finite pinhole size, the confocal starts off with a relatively bad detection resolution, so that the overall resolution improvement is almost negligible.

Any fluorescence microscopy technique attempting to push this limit further, and truly circumvent the Abbe limit, must induce a nonlinear relation between fluorescence excitation and emission [13, 14]. By using, e.g., photo-switchable fluorophores [15, 16], or by exploiting the nonlinear fluorescence response near saturation [17, 18], SIM can in principle be used to image samples with theoretically unlimited resolution. In reality the resolution will always be limited for signal-to-noise reasons.

2 How Does SIM Work?

The resolution of fluorescence microscopes is limited by diffraction. According to the Abbe diffraction limit, only spatial frequencies below a maximum frequency of $\mathbf{k}_{max} = 4\pi NA/\lambda$ (in the lateral direction) can be captured, where NA is the numerical aperture of the microscope objective and λ is the wavelength of the fluorescence light. All higher frequencies will be lost. This maximum frequency corresponds to a minimum resolvable distance of $d_{min} = \lambda/2NA$. Finer detail than this simply cannot be captured.

Using various different approaches, all super-resolution techniques attempt to circumvent the Abbe limit and extend the range of accessible spatial frequencies.

2.1 The Moiré Effect

Structured illumination microscopy makes use of the moiré effect to improve the resolution of fluorescence microscopy. When two fine (transparent) patterns are overlaid coarser patterns become visible (Fig. 1).

This effect can be readily observed in everyday life, e.g., in the brightness modulation visible in a semi-transparent curtain folding back on itself. This principle is exploited in high resolution SIM. By overlaying the sample, which contains very fine details not detectable by the microscope (left part of Fig. 1), with a very fine illumination pattern (right part of Fig. 1), coarse patterns in the light distribution emitted by the sample are created (middle of Fig. 1), which can be captured by the microscope. From the coarse information recorded under several different positions of the illumination pattern and with sufficient knowledge of the illumination

Fig. 1 The moiré effect. When a high-frequency pattern (*left*) is multiplied with another high frequency, regular pattern (*right*), the resulting pattern will exhibit both high and low frequencies (*middle*)

pattern itself, one can deduce what the corresponding finer sample details must have been, and thus improve the resolution of the microscope.

2.2 SIM in a Nutshell

When trying to understand the working principle of SIM, it helps to look at it in Fourier space. A sample's fluorophore distribution containing fine detail (Fig. 2a) will contain high spatial frequencies (Fig. 2b, the zero frequency is in the center). As mentioned above, the frequencies which can be captured by the microscope are limited by the Abbe diffraction limit. Only frequencies within the support of the optical transfer function (OTF), marked by a cyan circle in Fig. 2d, will be available, resulting in a blurry image (Fig. 2c), in which much of the finer detail is lost.

In order to improve the resolution of the microscope, a sinusoidal light pattern (Fig. 2c, close-up shown in cyan circle) is projected into the sample. In Fourier space this light pattern has three peaks, marked by the cyan arrows in Fig. 2f: two peaks for the sinusoidal modulation, and one central peak for a general offset of the pattern, as its intensity cannot be negative. The light emitted by a fluorescent sample illuminated with such a pattern corresponds to a product of the sample fluorophore distribution with the illumination pattern (Fig. 3a). Note the coarse moiré patterns visible in this light distribution (one example is marked by a cyan arrow). In Fourier space this product of sample and illumination is described by a convolution of the two corresponding Fourier transforms. A convolution of two functions can be depicted as using one of the functions as a brush and using this brush to repaint the other function. In our example this means taking the Fourier transformed sample and placing it at each of the Fourier

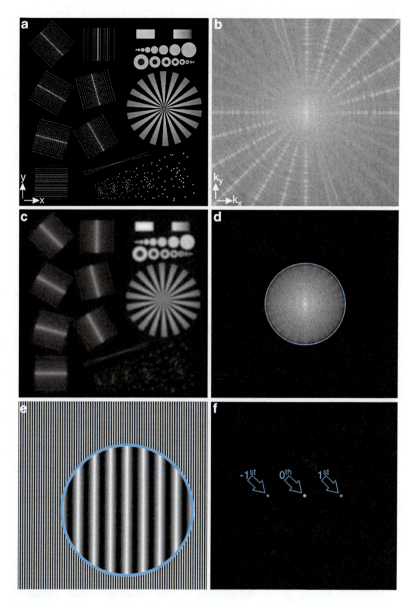

Fig. 2 A sample containing fine detail (**a**) will have very high spatial frequencies in its Fourier representation (**b**). (**d**) In the imaging process only frequencies within the support region of the OTF will be transmitted. (**c**) This leads to a blurring of features and a loss of detail in the image. (**e**) To recover some high spatial frequency information, SIM illuminates the sample with a fine sinusoidal line pattern of light, a close-up of which is shown in the *cyan circle*. (**f**) This can be represented in Fourier space by peaks at three spatial frequencies

transformed illumination's peaks, which yields the Fourier transformed emitted fluorescence shown in Fig. 3b. Here we can clearly see how the sample's central Fourier peak has been placed at each of the illumination peaks, again marked by cyan arrows. Note that

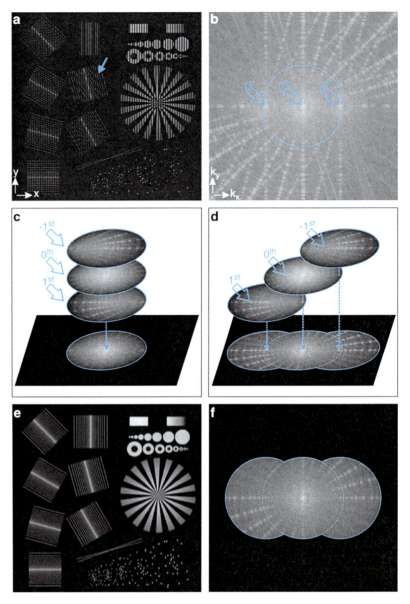

Fig. 3 (**a**) The patterned illumination will leave some parts of the sample dark, whereas others will be excited and emit fluorescence. The emission from the sample corresponds to a product of sample fluorophore density and the illumination pattern. (**b**) In Fourier space, this product translates into a convolution of Fourier transformed sample and light pattern; the sample information is placed once at each of the illumination peaks. (**c**) Again, imaging discards all information outside the support region. However, this region now contains three superimposed components stemming from different regions in frequency space. (**d**) After acquiring three images for different pattern positions, these components can be computationally separated and shifted back to their true origin in Fourier space. After recombination the final image has an improved spatial resolution (**e**) stemming from the effective increase of the support region in frequency space (**f**)

the fluorescence emission (Fig. 3a) and its Fourier transform (Fig. 3b) still contain fine detail and high frequencies. Imaging this distribution once again discards these high frequencies and only transmits those inside the OTF support. However, because the Fourier transformed sample was placed at each of the pattern's Fourier peaks, the low frequency information accessible will actually contain three different components of information, two of which carry information corresponding to higher sample frequencies (Fig. 3c). If at least three (the number of images required depends on the number of components to separate and therefore on the number of peaks in the Fourier transformed illumination pattern) images are acquired for different pattern positions, the information contained in these images allows us to separate the three information components superimposed in the individual Fourier images (Fig. 3c). Once separated, these components can be shifted back to their true origins and then recombined (Fig. 3d). This way frequencies from outside the OTF support are recovered, the effective support is increased (Fig. 3f) and the resolution improved (Fig. 3e).

As this process will only increase the support in the direction of the illumination pattern, it will have to be repeated for several orientations of the pattern. A total of three directions is usually enough for near-isotropic resolution.

For sake of simplicity, the above explanation was limited to two-dimensional samples and images. The general idea remains valid for three-dimensional imaging.

2.3 Image Formation and Reconstruction in SIM

After this conceptual explanation, we will now take a more detailed look at the maths behind SIM. For this we will first analyze the image formation and then see how these images can be used to reconstruct high resolution SIM images.

2.3.1 Image Formation

In fluorescence microscopy the acquired data $D(\mathbf{r})$ can be written as a convolution ($\otimes$) of the emitted fluorescence intensity $E(\mathbf{r}')$ with the microscope's point spread function (PSF) $h(\mathbf{r}')$: $D(\mathbf{r}) = [E \otimes h](\mathbf{r})$. Here $\mathbf{r}$ denotes the image coordinate, whereas the dashed coordinate $\mathbf{r}'$ refers to sample space. For simplicity of the equations we assume a magnification of one. If the fluorophores' response to the illumination light is linear, the emitted intensity can be written as a product of the sample's fluorophore density $S(\mathbf{r}')$ and the illumination intensity $I(\mathbf{r}')$ (omitting constant scaling factors): $E(\mathbf{r}') = S(\mathbf{r}')I(\mathbf{r}')$. This yields the acquired data

$$D(\mathbf{r}) = [(SI) \otimes h](\mathbf{r}). \quad (1)$$

The Illumination Pattern

There are a great variety of light patterns which can be used for structured illumination: one-dimensional line patterns, two-dimensional dot-patterns, both of which can be either sparse or dense, and many more. In this chapter we will limit ourselves to one-dimensional patterns which fulfil the three following conditions, described by Gustafsson et al. [19]:

1. They should comprise a finite number M of components which are separable into an axial and a lateral part, i.e.,
$$I(\mathbf{r}') = \sum_m I_m(z') J_m(\mathbf{r}'_{xy}) \;.$$
2. Each lateral part—we will henceforth refer to these as illumination *orders*—should be a harmonic wave, containing only a single spatial frequency $\mathbf{p}_m$, i.e., $J_m(\mathbf{r}'_{xy}) = \exp\{i(2\pi \mathbf{p}_m \cdot \mathbf{r}'_{xy} + \phi_m)\}$, where ϕ_m is the phase of the mth harmonic, and $i = \sqrt{(-1)}$. If we assume a periodic pattern with reflection symmetry, both the spatial frequencies and the phases of the individual illumination orders will be multiples of a fundamental frequency and phase: $\mathbf{p}_m = m\mathbf{p}$ and $\phi_m = m\phi$. As the illumination intensity has to be real valued, each illumination order has to have a complex conjugate partner $J_m^*(\mathbf{r}')$. This can be achieved by counting orders from $m = -(M-1)/2$ to $(M-1)/2$ whereby the definition of the lateral orders automatically yields $J_{-m}(\mathbf{r}') = J_m^*(\mathbf{r}')$.
3. The axial parts of the illumination pattern should remain fixed with respect to the focal plane of the microscope objective, and *not* move with the sample when refocussing. As we will see, this allows us to treat the axial illumination components as part of the microscope's OTF.

Data Acquisition

Using such an illumination pattern in the image formation Eq. (1) yields
$$\begin{aligned}D(\mathbf{r}) &= [(SI) \otimes h](\mathbf{r}) \\ &= \sum_m \int S(\mathbf{r}') I_m(z') J_m(\mathbf{r}'_{xy}) h(\mathbf{r} - \mathbf{r}') d^3\mathbf{r}'\end{aligned}$$

However, in this equation the illumination pattern is fixed with respect to the sample rather than the microscope's objective, when refocussing the sample. If we want to keep the illumination fixed with respect to the objective, we have to change this equation to
$$\begin{aligned}D(\mathbf{r}) &= \sum_m \int S(\mathbf{r}') I_m(z'-z) J_m(\mathbf{r}'_{xy}) h(\mathbf{r}-\mathbf{r}') d^3\mathbf{r}' \\ &= \sum_m [(SJ_m) \otimes (I'_m h)](\mathbf{r}),\end{aligned} \quad (2)$$

where in the last step we defined the mirrored axial illumination components $I'_m(z) := I_m(-z)$. We can now incorporate the axial illumination components into the PSF by defining new PSFs $h_m(\mathbf{r}') = I'_m(z') h(\mathbf{r}') = I_m(-z') h(\mathbf{r}')$.

Shifting the Pattern

The above describes the image acquisition for one particular pattern position. For the reconstruction of a high resolution SIM image, we need at least M images—the number of lateral illumination orders—with different pattern positions (note that there are less straightforward approaches, which allow a reconstruction from fewer raw images [20, 21], which utilize the fact that the raw images have redundancies in the acquired information). As the pattern is to remain rigid under translation (i.e., it does not change its inherent shape), we can describe this translation simply by varying the phase ϕ. We do this by introducing the index n for image number. We also introduce the upper index (d) to account for the D different orientations of the pattern required for isotropic resolution enhancement. The lateral illumination orders then become

$$J_{m,n}^{(d)}(\mathbf{r}') = \exp\{i(2\pi m \mathbf{p}^{(d)} \cdot \mathbf{r}' + m\phi_n^{(d)})\}$$
$$= \exp\{im\phi_n^{(d)}\}\exp\{i(2\pi m \mathbf{p}^{(d)} \cdot \mathbf{r}')\}$$

where $\phi_n^{(d)}$ now denotes the pattern phase (i.e., position) of the nth image for the $(d)^{\text{th}}$ pattern orientation.

Matrix Notation

With these illumination phases Eq. (2) becomes

$$D_n^{(d)}(\mathbf{r}) = \sum_m [(e^{im\phi_n^{(d)}} e^{i(2\pi m \mathbf{p}^{(d)} \cdot r')} S) \otimes h_m](\mathbf{r}),$$

or in Fourier space

$$\tilde{D}_n^{(d)}(\mathbf{k}) = \sum_m e^{im\phi_n^{(d)}} \tilde{S}(\mathbf{k} - m\mathbf{p})\tilde{h}_m(\mathbf{k}). \quad (3)$$

This equation shows how each image consists of a superposition of different components $\tilde{\Omega}_m^{(d)}(\mathbf{k}) = \tilde{S}(\mathbf{k} - m\mathbf{p}^{(d)})\tilde{h}_m(\mathbf{k})$. If we see the individual Fourier images $\tilde{D}_n^{(d)}(\mathbf{k})$ as elements of a vector $\tilde{\boldsymbol{D}}^{(d)}(\mathbf{k})$ of Fourier images and likewise the components $\tilde{\Omega}_m^{(d)}(\mathbf{k})$ as elements of a vector $\tilde{\boldsymbol{\Omega}}^{(d)}(\mathbf{k})$ of components, we can write Eq. (3) in matrix notation

$$\tilde{\boldsymbol{D}}^{(d)}(\mathbf{k}) = \mathbf{M}^{(d)}\tilde{\boldsymbol{\Omega}}^{(d)}(\mathbf{k}), \quad (4)$$

with the mixing matrix

$$\mathbf{M}_{m,n}^{(d)} = e^{im\phi_n^{(d)}}.$$

This describes the image formation for the $(d)^{\text{th}}$ orientation of the pattern. This process has to be repeated for at least $D = 3$ pattern orientations in order to acquire the necessary information for near-isotropic resolution enhancement.

2.3.2 Image Reconstruction

Component Separation

In order to reconstruct a high resolution SIM image from the acquired Fourier data $\tilde{\boldsymbol{D}}^{(d)}(\mathbf{k})$ (Eq. 4), we need to computationally separate the different Fourier components $\tilde{\Omega}_m^{(d)}(\mathbf{k})$. This can be achieved through a simple matrix inversion in Eq. (4):

$$\tilde{\boldsymbol{\Omega}}^{(d)}(\mathbf{k}) = \mathbf{M}^{(d)^{-1}}\tilde{\boldsymbol{D}}^{(d)}(\mathbf{k}), \quad (5)$$

where $\mathbf{M}^{(d)-1}$ is the inverse mixing matrix. When $\mathbf{M}^{(d)}$ is not square, i.e., when there are more raw images than components to separate, $\mathbf{M}^{(d)-1}$ can be calculated as the Moore–Penrose pseudo inverse [22] $\mathbf{M}^{(d)-1} := (\mathbf{M}^{\dagger}\mathbf{M})^{-1}\mathbf{M}^{\dagger}$, where the dagger symbol † denotes the Hermitian transpose.

The separated Fourier components then have to be shifted by $-m\mathbf{p}^{(d)}$ to their true origins in Fourier space:

$$\tilde{\Omega}_m^{(d)}(\mathbf{k} + m\mathbf{p}^{(d)}) = \tilde{S}(\mathbf{k})\tilde{h}_m(\mathbf{k} + m\mathbf{p}^{(d)}).$$

These shifted components then have to be merged in order to reconstruct a higher-frequency representation of the sample information $\tilde{S}(\mathbf{k})$.

Combining the Shifted Components

The simplest way to recombine the separated and shifted components is a summation. The reconstructed image thus becomes

$$\begin{aligned}\tilde{F}_{\text{sum}}(\mathbf{k}) &= \sum_{m,d}\tilde{\Omega}_m^{(d)}(\mathbf{k} + m\mathbf{p}^{(d)}) \\ &= \tilde{S}(\mathbf{k})\sum_{m,d}\tilde{h}_m(\mathbf{k} + m\mathbf{p}^{(d)}) \\ &= \tilde{S}(\mathbf{k})\tilde{h}_{\text{sum}}(\mathbf{k}),\end{aligned}$$

which corresponds to imaging the sample with an effective OTF

$$\tilde{h}_{\text{sum}}(\mathbf{k}) = \sum_{m,d}\tilde{h}_m(\mathbf{k} + m\mathbf{p}^{(d)}).$$

Weighted Averaging in Fourier Space

While a summation of the individual components does yield a reconstructed (Fourier) image $\tilde{F}_{\text{sum}}$ with improved frequency support of the effective OTF $\tilde{h}_{\text{sum}}$, this recombination is by no means ideal when considering noisy images. Assuming a constant mean intensity in all acquired raw images, the corresponding Fourier images will exhibit white (albeit correlated) noise with a standard deviation σ. After unmixing components using Eq. (5) the individual separated components will also exhibit white noise. Their noise will have a standard deviation of $\sqrt{N}\sigma$, as the components are a summation of N Fourier images, whose noise will add in quadrature.

So while all the Fourier components have the same noise of standard deviation $\sqrt{N}\sigma$, their information transfer strength (the level of the corresponding shifted OTF $\tilde{h}_m(\mathbf{k} + m\mathbf{p}^{(d)})$) will differ from component to component. When adding a component of very weak (or even no) transfer strength to a component of high transfer strength, one adds only little (or no) information, while the noise of this additional component will be added and thus increases the overall noise. In this case it would be beneficial to use mainly (or only) the information of the stronger component.

This is the general idea of weighted averaging: letting each component contribute to the final reconstructed image in such a way, that the ratio of total transfer strength to total noise is maximal.

The ideal recombination assuming uncorrelated white noise is a weighted-averaging approach,

$$\tilde{F}_{\text{wa}}(\mathbf{k}) = \frac{\sum_{m,d} \tilde{w}_m^{(d)}(\mathbf{k}) \tilde{\Omega}_m^{(d)}(\mathbf{k} + m\mathbf{p}^{(d)})}{\sum_{m,d} \tilde{w}_m^{(d)}(\mathbf{k})},$$

with component and frequency-dependent weights

$$\tilde{w}_m^{(d)}(\mathbf{k}) = \tilde{h}_m(\mathbf{k} + m\mathbf{p}^{(d)}).$$

The reconstructed image can then be written as

$$\tilde{F}_{\text{wa}}(\mathbf{k}) = \tilde{S}(\mathbf{k}) \tilde{h}_{\text{wa}}(\mathbf{k}), \qquad (6)$$

with an effective weighted-averaging OTF

$$\tilde{h}_{\text{wa}}(\mathbf{k}) = \frac{\sum_{m,d} \tilde{h}_m^2(\mathbf{k} + m\mathbf{p}^{(d)})}{\sum_{m,d} \tilde{h}_m(\mathbf{k} + m\mathbf{p}^{(d)})}. \qquad (7)$$

This reconstruction no longer has a white noise distribution, but rather noise with a frequency-dependent standard deviation of

$$\sigma_{\text{wa}}(\mathbf{k}) = \frac{\sqrt{N}\sigma \sqrt{\sum_{m,d} \tilde{h}^2(\mathbf{k} + m\mathbf{p}^{(d)})}}{\sum_{m,d} \tilde{h}_m(\mathbf{k} + m\mathbf{p}^{(d)})}. \qquad (8)$$

Wiener Filter Deconvolution Assuming uncorrelated white noise, the above weighted averaging is the optimum way of recombining noisy components. It is, however, not the optimum way of displaying the final reconstructed image.

We are finally interested in a representation of our data, which most closely resembles the original sample information. Naïvely one might think that this could be achieved by dividing $\tilde{F}_{\text{wa}}$ by its effective OTF $\tilde{h}_{\text{wa}}$ where the latter is not zero, and setting it to zero otherwise. This way the information present within the support of the thus deconvolved Fourier image would just be the sample information $\tilde{S}$, free from any OTF apodization. While this approach would indeed yield the best results in the absence of noise, it utterly fails in the presence of noise: where the weighted-averaging OTF is low, dividing by $\tilde{h}_{\text{wa}}$ unduly magnifies noise present in these frequencies.

Wiener filter deconvolution [23] attempts to reverse the apodization of information, which is a result of multiplication with an OTF, while avoiding an over-amplification of noise.

Equation (6) really only tells us the expectation value of the reconstructed weighted-averaging image. If we want to account for noise in that equation, we have to write

$$\tilde{F}_{wa}(\mathbf{k}) = \tilde{S}(\mathbf{k})\tilde{h}_{wa}(\mathbf{k}) + \tilde{n}(\mathbf{k}),$$

where $\tilde{n}$ is the Fourier noise with a standard deviation of $\sigma_{wa}(\mathbf{k})$ (Eq. 8). The Wiener filter is a linear multiplicative filter $\tilde{W}(\mathbf{k})$ which has to be chosen such that the filtered image

$$\tilde{F}_{wien}(\mathbf{k}) = \tilde{F}_{wa}(\mathbf{k})\tilde{W}(\mathbf{k})$$

has a minimal mean square error $\epsilon(\mathbf{k}) = \langle |\tilde{S}(\mathbf{k}) - \tilde{F}_{wien}(\mathbf{k})|^2 \rangle$ as compared to the true sample information, which is equivalent to minimizing the mean square error of the real space image because of Parseval's theorem. Here $\langle \rangle$ denotes the expectation value. This is fulfilled for a Wiener filter of

$$\tilde{W}(\mathbf{k}) = \frac{\tilde{h}_{wa}^*(\mathbf{k})|\tilde{S}(\mathbf{k})|^2}{|\tilde{h}_{wa}|^2(\mathbf{k})|\tilde{S}(\mathbf{k})|^2 + \langle |\tilde{n}(\mathbf{k})|^2 \rangle},$$

where $|S|^2$ is the sample's power spectrum.

Incorporating the Wiener filter in the component recombination yields a Wiener filtered reconstructed image

$$\tilde{F}_{wien}(\mathbf{k}) = \frac{\sum_{m,d} \tilde{h}_m^*(\mathbf{k} + m\mathbf{p}^{(d)}) \tilde{\Omega}_m^{(d)}(\mathbf{k} + m\mathbf{p}^{(d)})}{\sum_{m,d} |\tilde{h}_m|^2(\mathbf{k} + m\mathbf{p}^{(d)}) + \sqrt{N}\sigma/|\tilde{S}|^2(\mathbf{k})}.$$

As the sample power spectrum will in most cases be unknown, we need to make further assumptions at this point. For lack of a better estimate, the power spectrum is usually chosen to be constant, and the thus constant Wiener parameter $\sqrt{N}\sigma/|\tilde{S}|^2$ is adjusted somewhat empirically.

Apodization of the Wiener Filtered Image

Although the Wiener filter minimizes the mean square error, the resulting effective PSF may exhibit ringing and even negative values. In the absence of noise for example, the Wiener filter is the inverse weighted-averaging OTF. As a result, the Wiener filtered image's final OTF will be one anywhere inside the OTF's support and zero outside. This will cause ringing artifacts in the final image, which should be avoided.

Furthermore, natural samples tend to have power spectra which decrease for higher frequencies. Assuming instead the power spectrum and thus the Wiener parameter to be constant therefore puts an undue emphasis on higher frequencies, also amplifying noise in these.

Both the above issues are countered by apodizing the Wiener filtered Fourier image with an apodization function. Several different approaches can be used for apodization. Gustafsson et al. use a three-dimensional triangular function, decreasing linearly from unity in the center to zero at the edges of support of the effective OTF [19].

Our approach is based on the Euclidean distance transform $d(\tilde{h}_{wa}(\mathbf{k}))$ of the OTF, which is defined as the closest distance from the point $\mathbf{k}$ to the support edge of the OTF, normalized to a maximum value of one. As an apodization function we then use the apodization function $\tilde{A}(\mathbf{k}) = d^{\kappa}(\tilde{h}_{wa}(\mathbf{k}))$, whereby through the choice of κ we can selectively emphasize either lower frequency ($\kappa > 1$) or higher frequency ($\kappa < 1$) information ($\kappa = 0.5$ usually being a good choice).

This leads us to a resulting final Fourier image

$$\tilde{F}_{final}(\mathbf{k}) = \frac{\sum_{m,d} \tilde{h}^{*}(\mathbf{k} + m\mathbf{p}^{(d)}) \tilde{\Omega}_{m}^{(d)}(\mathbf{k} + m\mathbf{p}^{(d)})}{\sum_{m,d} |\tilde{h}_{m}|^{2}(\mathbf{k} + m\mathbf{p}^{(d)}) + const.} \tilde{A}(\mathbf{k}), \quad (9)$$

which corresponds to an image $\tilde{S}(\mathbf{k})\tilde{h}_{final}(\mathbf{k})$ with the effective final OTF

$$\tilde{h}_{final}(\mathbf{k}) = \frac{\sum_{m,d} |\tilde{h}_{m}|^{2}(\mathbf{k} + m\mathbf{p}^{(d)})}{\sum_{m,d} |\tilde{h}_{m}|^{2}(\mathbf{k} + m\mathbf{p}^{(d)}) + const.} \tilde{A}(\mathbf{k}). \quad (10)$$

The boxes on pages 59 and 60 summarize the maths behind the image formation and reconstruction in SIM. While this way of SIM reconstruction is the one most commonly applied, there recently have been promising approaches to SIM reconstruction based on maximum likelihood deconvolution [21, 24].

3 Experimental Realizations

The previous section gave an introduction into the maths behind structured illumination microscopy. In this section we will look at various different experimental realizations of SIM. For the most part, we will be looking at two variants of epi-fluorescence SIM. These are two-beam and three-beam illumination. Both employ a very similar setup, which is depicted schematically in Fig. 4.

Laser light is incident on a spatial light modulator (SLM), displaying a phase grating. Alternatively, a physical phase grating can be placed in the light path. This grating diffracts the light into

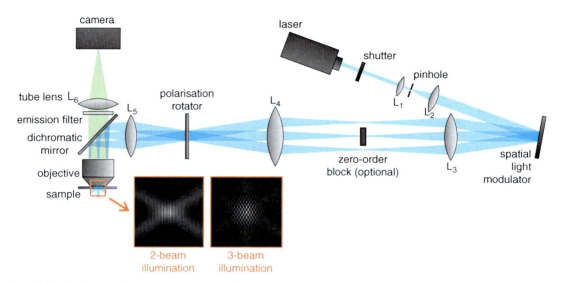

Fig. 4 A typical setup for two- or three-beam SIM in epi-fluorescence. A spatial light modulator diffracts laser light into several diffraction orders. The ± 1st—and for three-beam illumination also the 0th—diffraction orders are polarized perpendicular to the pattern vector and via a dichromatic reflector are relayed into the back focal plane of the microscope objective. They interfere in the sample to generate a structured illumination pattern. Light emitted from the sample is captured by the objective and after transmission through dichromatic reflector and emission filters is focussed by the tube lens to form an image on the CCD camera

several diffraction orders. A lens (L_3) images (i.e., Fourier transforms) these diffraction orders into a plane conjugate to the back focal plane (BFP) of the objective. In this plane each diffraction order forms a focused spot. Any second or higher orders—and, if the setup is used in the two-beam SIM configuration, the zero-order— are blocked in this plane. A polarization rotating device ensures that the light polarization is always orthogonal to the pattern vector **p** of the grating. In high NA systems this ensures that, upon interfering inside the sample, the diffraction orders will all have polarizations which are parallel to each other, thereby guaranteeing a high contrast in the modulated illumination. Via a dichromatic reflector the diffraction orders are focused into the BFP of the objective, and will interfere in the sample to generate the desired illumination pattern.

The emitted fluorescence is captured by the objective, transmitted through the dichromatic reflector, and imaged onto a camera.

The illumination pattern is translated by displaying shifted patterns on the SLM, or by physically translating the phase grating. After the required number of images have been acquired, the grating (either displayed or real) as well as the polarization rotator are turned and the acquisition process is repeated for this new pattern orientation. When imaging three dimensional samples, the sample is now translated axially and the acquisition process repeated for a new focal slice. This order of raw image acquisition is but one example; it can be varied from system to system and is usually adjusted to minimize total acquisition time. The thus acquired information can then be reconstructed into a two- or three-dimensional high resolution SIM image.

3.1 Two-Beam Illumination

The conceptually simplest realization of SIM is two-beam illumination. As we will see, however, two-beam illumination has to compromise between lateral resolution and optical sectioning. When trying to exploit two beam SIM's full potential with regard to lateral resolution enhancement, the missing cone in Fourier space (i.e., the region around the k_z-axis where no information is transmitted, shown in orange in Fig. 7b) will not be filled, and the image will exhibit no optical sectioning. For this reason, two-beam illumination is mainly employed when imaging very thin samples, or when the sectioning can be achieved through other means, e.g., by using SIM in combination with total internal reflection fluorescence (TIRF) microscopy. In these conditions two-beam illumination may be beneficial, as it requires a lower number of raw images than three-beam SIM and can thus achieve faster acquisition times. Two-beam SIM employing TIRF illumination also has the greatest resolution enhancement, as the illumination pattern can be finer than in non-TIRF SIM.

3.1.1 Illumination

As mentioned above, in two-beam illumination two diffraction orders from the SLM or grating are focussed into the back focal plane of the microscope's objective. The BFP is a projection of McCutchen's generalized aperture [25] onto the lateral (k_{xy}) frequency plane in Fourier space (Fig. 5). While the Ewald sphere [26] contains all possible spatial frequencies for a single wavelength of light, the McCutchen aperture comprises all those frequencies which can be captured by the microscope objective (or for that matter also the frequencies which can be used for illumination). Reversing this idea, projecting the two focused diffraction orders from the BFP back up onto the McCutchen aperture (i.e., the Ewald sphere) tells us which spatial frequencies will contribute to the illumination pattern in the sample.

For two-beam illumination, these frequencies are shown as cyan spots in Figs. 5 and 6a. An inverse Fourier transform of these spots, which describe an amplitude frequency distribution $\tilde{E}(\mathbf{k})$, yields the amplitude field distribution $E(\mathbf{r})$ in the sample. As the fluorophores' response depends on the illumination intensity rather than on the amplitude, we are interested in just that: the intensity $I(\mathbf{r}) = |E(\mathbf{r})|^2 = E(\mathbf{r})E^*(\mathbf{r})$, which is also shown in Fig. 6a. It is the result of two interfering plane waves (indicated by blue arrows). Going back to Fourier space and applying the convolution theorem, we get the Fourier transformed intensity pattern, $\tilde{I}(\mathbf{k}) = \tilde{E}(\mathbf{k}) \otimes \tilde{E}^*(-\mathbf{k})$ (which is the same as an auto-correlation of

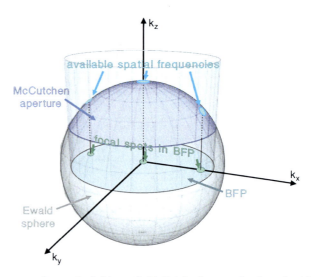

Fig. 5 For monochromatic light any field distribution can be described by superpositions of plane waves of the same wavelength propagating in all possible directions in space. In Fourier space these plane waves' k-vectors all lie on a sphere of radius $2\pi\,n/\lambda$, the Ewald sphere. Out of these, only a limited number can be transmitted through the optical system (i.e., microscope objective). These are given by the McCutchen generalized aperture, which describes the amplitude and phase of the transmission coefficients for all frequencies. The support of this generalized aperture is generally a cap of the Ewald sphere, which can be determined by projecting the back focal plane (BFP) along the k_z-direction onto the Ewald sphere. Focussing three diffraction orders into the BFP leads to a light distribution whose frequencies can also be determined by projecting these foci onto the Ewald sphere

$\tilde{E}(\mathbf{k})$) telling us which spatial frequencies are present in the pattern. These frequencies are shown as red dots in Fig. 6b, with the amplitude frequencies again indicated in cyan.

3.1.2 Data Acquisition

The light emitted by the fluorescent sample is captured by the microscope objective. The opening angle of the objective defines the McCutchen aperture—or the coherent OTF (denoted as $\tilde{a}$ in this chapter)—telling us which fraction of frequencies on the Ewald sphere the microscope can capture, as shown in Figs. 5 and 7a. These are amplitude frequencies however, whereas we are interested in intensity. An auto-correlation of the coherent OTF (Fig. 7a) yields the familiar wide-field OTF (Fig. 7b), which defines which frequencies will be present in the acquired images and which exhibits the typical missing cone, i.e., it transmits no information for frequencies around the k_z-axis and therefore cannot achieve optical sectioning.

The illumination intensity pattern has three orders (-1, 0, and $+1$) in Fourier space. The acquired images therefore contain

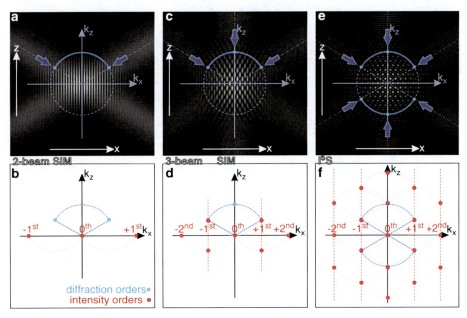

Fig. 6 (**a**) In two-beam SIM two diffraction orders interfere to generate a fine grid of light. The Fourier components of this amplitude distribution are shown as *cyan* spots. Because they both have the same k_z-component, their interference pattern will not vary along the z-direction. (**b**) To determine the Fourier components of the intensity distribution, an autocorrelation of the amplitude frequency distribution is calculated, resulting in three spots, shown in *red*. (**c**) For three-beam SIM three diffraction orders interfere to generate a light pattern which does vary along z. (**d**) An autocorrelation of the amplitude spots yields seven intensity orders. Out of these, only five modulate independently with translation of the pattern. (**e**) In I^{5}S two opposing objectives are used to project three illumination orders each into the sample, leading to a total of six amplitude orders. Because of the greater range of frequencies along k_z the resulting interference pattern becomes finer in the z-direction. (**f**) An auto-correlation of the six amplitude orders yields 19 intensity orders, out of which again only five are independent

three components, which in Fourier space correspond to the sample information being shifted (laterally) to the respective orders and multiplied with their corresponding OTF (see Eq. 11). As none of the three intensity orders have a k_z-component, their corresponding OTFs will all be identical and are simply the microscope's wide-field OTF.

Three images have to be taken for three different pattern positions, so that the three superimposed components can be computationally separated.

3.1.3 Image Reconstruction

After separation, the separated Fourier components are shifted laterally, so that the information contained in these components will be moved to the correct frequencies in Fourier space. This is illustrated in Fig. 8a. The separated 0th-order band is shown in blue, the separated and shifted + 1st-order band in yellow. The minus first-order is symmetrical to the + 1st-order band and is not

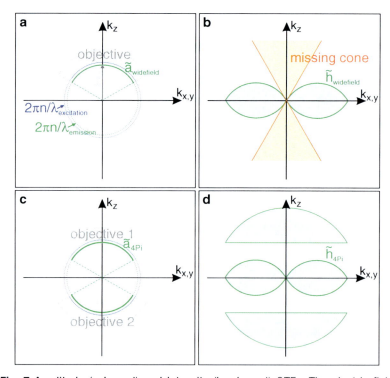

Fig. 7 Amplitude (coherent) and intensity (incoherent) OTFs. The electric field distribution that can be generated by the microscope can be described by the microscope's amplitude OTF, which corresponds to the McCutchen aperture. (**a**) For a conventional epi-fluorescence microscope this corresponds to the cap of a sphere, which is defined by the opening angle of the microscope objective. Its radius depends on the wavelength of light (in medium) and therefore is slightly different for illumination and detection. (**b**) The intensity OTF is an auto-correlation of this amplitude OTF, which exhibits the typical missing cone, i.e., it contains no information around the k_z-axis. (**c**) In a 4Pi or I^{5}M configuration the microscope has two opposing objectives, leading to a coherent OTF comprising two opposing spherical caps. (**d**) An auto-correlation of this coherent OTF yields the 4Pi intensity OTF, which has significantly improved axial support and partly fills the missing cone. However, it still leaves unfilled gaps around the k_z-axis

shown. The final SIM OTF resulting from combining all separated and shifted components is shown in green. It has a lateral frequency support which is just over two times higher than that of the widefield OTF (blue).

This combined OTF still does not fill the missing cone and thus cannot achieve optical sectioning. This could be remedied by choosing a coarser illumination pattern: in Fourier space this would lead to intensity orders of lower spatial frequency. Shifting the separated bands to these lower frequencies would lead to the ± 1st-order OTFs filling the missing cone. This of course would result in a less powerful resolution enhancement, as the lateral support is not extended by the same amount as for fine patterns, i.e., high-frequency intensity orders. Two-beam SIM is therefore mostly

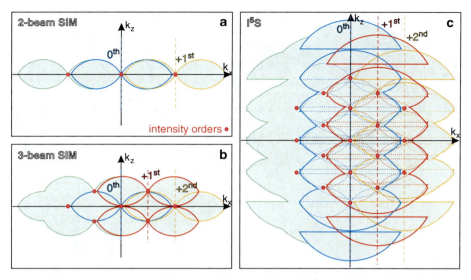

Fig. 8 During image reconstruction the individual Fourier components are computationally separated and shifted to their true origins in Fourier space. As each component has its own associated OTF, the final reconstructed OTF's frequency support will correspond to a combination of all shifted OTFs' regions of support. (**a**) Two-beam SIM images have three Fourier components per grating orientation, all of which have the same support. Shifting the + 1st (indicated in *yellow*) and − 1st component increases the lateral support, but does not necessarily fill the missing cone. (**b**) Three-beam SIM images have an additional two components. Their OTFs (the new + 1st component's OTF is indicated in *red*) exhibit an extended axial support and when shifted to their proper place in Fourier space (as in this figure) fill the missing cone. (**c**) I^{5}S combines SIM with the benefits of illumination and detection through two opposing objectives. This leads to dramatically increased support of all components' OTFs

3.1.4 Example Images

used when optical sectioning can be achieved by other means, e.g., through TIRF, or when imaging very thin samples.

As an example of two-beam SIM in TIRF configuration Fig. 9 shows microtubules expressing EGFP-alpha tubulin in a live HeLa [27] cell. Experimental parameters were: excitation wavelength $\lambda_{ex} = 488$ nm; emission wavelength $\lambda_{em} > 500$ nm; numerical aperture $NA = 1.45$, oil immersion objective; refractive index in sample $RI = 1.33$. The bottom left corners of the image and the close-up show a TIRF wide-field image. The top right corners show the TIRF SIM image. The grating constant in the sample was $d_{grat} = 2\pi / |\mathbf{p}| = 181$ nm.

3.2 Three-Beam SIM

As we saw in Sect. 3.1 SIM employing two-beam illumination has to compromise between filling the missing cone and maximum improvement of lateral resolution. Employing three (or more)-beam illumination, both benefits can be achieved at the same time. Three-beam SIM furthermore improves the axial support of the microscope's OTF (Fig. 8b), leading to finer axial resolution in addition to the sectioning effect.

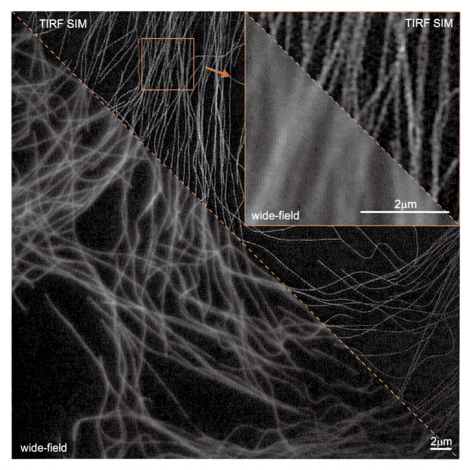

Fig. 9 Microtubules expressing EGFP-alpha tubulin in a live HeLa [27] cell. *Bottom left half* of main image and close-up: TIRF wide-field. *Top right half* of main image and close-up: TIRF SIM. Images were kindly provided by Reto Fiolka, Janelia Farm Research Campus, Howard Hughes Medical Institute, Ashburn, Virginia, USA

3.2.1 Illumination

Besides the two diffraction orders used in two-beam illumination, three-beam SIM uses an additional third diffraction order in the center of the objective's BFP. The light distribution in the sample (shown in Fig. 6c) is therefore the interference pattern of three plane waves (again indicated by blue arrows). Projecting these Fourier-orders focused into the BFP onto the Ewald sphere (Fig. 5) yields the amplitude frequencies shown as cyan dots in Fig. 6c. Auto-correlating these (i.e., convolving them with their mirrored, complex conjugated copy) yields the intensity frequencies shown as red dots in Fig. 6d (again with the amplitude frequencies indicated in cyan). The corresponding intensity distribution is shown in Fig. 6c. Unlike the two-beam intensity distribution, it exhibits a modulation in the z-direction stemming from the Talbot effect [28], which is responsible for the improved axial resolution.

3.2.2 Data Acquisition

As two-beam SIM, three-beam SIM also uses one objective in epi-fluorescence and the microscope's OTF (without SIM) is therefore the conventional wide-field OTF (Fig. 7b).

Unlike in two-beam illumination, the ± 1st orders in three-beam illumination do have a k_z (or axial) component. As shown in Eq. (2), these illumination orders should be described by their lateral component only, and their axial components be incorporated into a modified PSF $h_{\pm 1}(\mathbf{r}) = I_{\pm 1}(z)h(\mathbf{r})$. The corresponding OTF $\tilde{h}_{\pm 1}(\mathbf{k}) = [\tilde{I}_{\pm 1} \otimes \tilde{h}](\mathbf{k})$ can be found by placing the wide-field OTF at each of the ± 1st order peaks, as outlined in red for $\tilde{h}_{+1}(\mathbf{k})$ in Fig. 8b. As the 0th and ± 2nd orders have no axial frequency components their OTFs correspond to the conventional wide-field OTF (modified by the respective order strength), as outlined in blue (0th) and yellow (+ 2nd). Note that the modified OTFs in this figure have already been laterally shifted and that for data acquisition they would be centered around $\mathbf{k} = 0$.

As there are now five independent Fourier components (- 2 to 2), at least five images have to be acquired to allow the components' separation.

3.2.3 Image Reconstruction

As was the case for two-beam illumination, the individual components are computationally separated from the acquired data. They are then shifted (laterally) to their true origin in Fourier space and recombined using weighted averaging and the filters described in Sect. 2.3.2. The main difference to two-beam illumination lies in the modified 3D OTFs of the ± 1st orders, which have to be accounted for in the weights for recombination. The final SIM OTF resulting from combining all separated and shifted components is shown in green in Fig. 8b. While the lateral support is as high as for two-beam illumination, the three-beam SIM OTF approximately doubles the axial support as well as filling the missing cone. The resulting high axial resolution and optical sectioning make three-beam SIM the method of choice for three-dimensional SIM imaging.

3.2.4 Example Images

Three-beam illumination is arguably the most useful variant of SIM—it is most certainly the one that is used the most and it is commercially available from a number of manufacturers. With its sectioning capability but superior resolution it is a strong competitor to confocal microscopy, with further advantages regarding speed, light efficiency and therefore photo bleaching.

To illustrate the power of three-beam SIM Fig. 10 shows images of a live mixed cell culture of hippocampal neurons and glia. The sample expresses cytosolic GFP (green) and is actin-labeled with tdTomato-LifeAct (red). The images show a maximum intensity projection of a three-dimensional stack of 24 focal slices, with a distance of 125 nm between slices. Figure 10a shows an SIM overview over a neuron, with the top right showing the

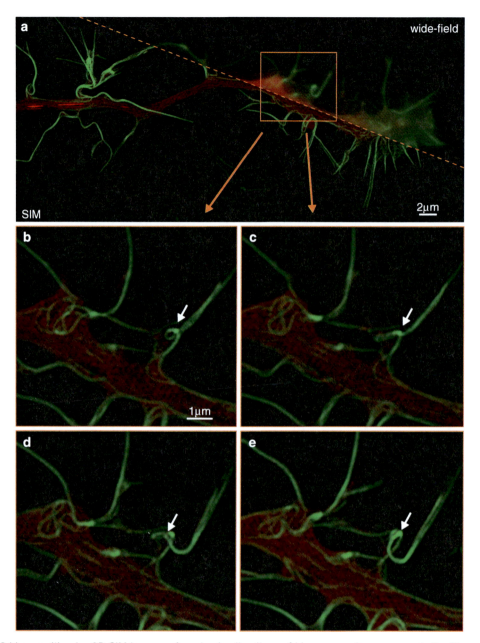

Fig. 10 Live, multi-color 3D SIM images of a mixed cell culture of hippocampal neurons and glia. The sample expresses cytosolic GFP (*green*) and is actin-labeled with tdTomato-LifeAct (*red*). The images show a maximum intensity projection of a three-dimensional stack of 24 focal slices, with a distance of 125 nm between slices. (**a**) A SIM overview of a neuron, with the *top right* showing the conventional wide-field for comparison. (**b**)–(**e**) SIM close-ups of the *orange box* in (**a**), taken 17 s apart. A lot of movement can be observed: as a prominent example the *white arrows* indicate the coiling of a filopodium. Images were provided by Reto Fiolka, Janelia Farm Research Campus, Howard Hughes Medical Institute, Ashburn, Virginia, USA. Modified with permission from [29]

conventional wide-field for comparison. Figures 10b–e show SIM close-ups of the orange box in Fig. 10a, taken 17 s apart. A lot of movement can be observed: as a prominent example the white arrows indicates the coiling of a filopodium. Experimental parameters were: excitation wavelength GFP $\lambda_{ex} = 488$ nm, tdTomato-Life-Act $\lambda_{ex} = 561$ nm; numerical aperture $NA = 1.2$, water immersion objective. The light patterns were generated, rotated and translated using a ferroelectric SLM, which allowed the extremely fast acquisition of raw images. Images were provided by Reto Fiolka, Janelia Farm Research Campus, Howard Hughes Medical Institute, Ashburn, Virginia, USA, and have previously been published in [29].

3.3 I^5S

Although the previous examples impressively demonstrate the power of structured illumination microscopy, the technique can be pushed even further, by combining it with illumination and detection from two sides, i.e., through two opposing objectives [30]. This approach is known from microscopy techniques such as 4Pi microscopy [31, 32] for scanning and I^5M [33, 34] (a combination of image interference microscopy, I^2M, and incoherent interference illumination microscopy, I^3M) for wide-field imaging. I^5S combines structured illumination with I^5M. If carefully aligned to extreme precision, the two objectives do not merely each detect the illumination light independently, but rather act as one objective with a much extended aperture. Their combined coherent OTF is depicted in Fig. 7c and leads to an intensity OTF with dramatically increased axial frequency support and a partly filled missing cone, shown in Fig. 7d. Even without the added complication of structured illumination, the coherent combination of two microscope objectives in an I^5M configuration is technically extremely demanding. The feat of combining both approaches was achieved by the group around Mats Gustafsson [30], and—although too complex for everyday lab work—demonstrates the astonishing possibilities of SIM.

3.3.1 Illumination

Besides improving the detection, the two objectives can be used to improve the illumination patterns of SIM. Rather than illuminating with three diffraction orders in the BFP of one objective, this illumination is mirrored for the second objective, leading to an effective six-beam illumination, and thus a much finer modulation along the axial direction, as seen in Fig. 6e. The frequency distribution of this intensity pattern is again shown as red dots in Fig. 6f, with the amplitude frequencies indicated in cyan.

3.3.2 Data Acquisition

I^5S uses two opposing objectives, each contributing with opposing McCutchen apertures to the microscope's total coherent OTF (Fig. 7c). An auto-correlation of these two spherical caps yields the microscope's intensity OTF, which is the well-known 4Pi detection OTF shown in Fig. 7d. Although this OTF leaves gaps in the missing cone region, it does have very large axial frequency support yielding very high axial resolution.

As before in three-beam illumination, illumination orders with identical lateral frequency components are combined into one order with a corresponding modified PSF/OTF. For I^{5}S this is the case for all intensity orders. The mth modified OTF $\tilde{h}_m(\mathbf{k})$ can be found by placing this time the 4Pi OTF at each of the mth order peaks. The different bands' modified OTFs are outlined in blue for $\tilde{h}_0(\mathbf{k})$, red for $\tilde{h}_{+1}(\mathbf{k})$, and yellow for $\tilde{h}_{+2}(\mathbf{k})$ in Fig. 8c. Note that, as before, the OTFs shown in the figure have already been laterally shifted, while the OTFs relevant for data acquisition are centered around $\mathbf{k} = 0$.

3.3.3 Image Reconstruction

As before, the individual components are computationally separated from the acquired data, shifted (laterally) to their true origin in Fourier space and recombined using weighted averaging and the filters described in Sect. 2.3.2. The final I^{5}S OTF resulting from combining all separated and shifted components is shown in green in Fig. 8c. While the lateral support is as large as for two- and three-beam illumination, the I^{5}S OTF has a dramatically enhanced axial support and axial resolution.

3.3.4 Example Images

I^{5}S has been demonstrated imaging fluorescent beads as well as biological specimen (microtubules in HeLa cells and meiotic cells from C. elegans) [30]. As an example of the resolving power of I^{5}S, Fig. 11 shows a lateral (Fig. 11a) and an axial section (Fig. 11b) of a fluorescent bead sample. The left part of the figure shows the wide-field image, the right the I^{5}S image. With a full width at half the maximum (FWHM) of 119 nm the lateral resolution improvement of I^{5}S corresponds to that of two- and three-beam illumination. However, the axial FWHM resolution of 116 nm exhibits a tremendous improvement over that of microscopy techniques using only one objective. Experimental parameters were: excitation wavelength $\lambda_{\text{ex}} = 532$ nm; emission wavelength $\lambda_{\text{em}} = 605 \pm 25$ nm; numerical aperture $NA = 2 \times 1.4$, oil immersion objectives. The grating constant in the sample was $d_{\text{grat}} = 2\pi / |\mathbf{p}| = 176$ nm.

3.3.5 Comparison of Two-Beam SIM, Three-Beam SIM, and I^{5}S

Figure 12 shows a comparison of the frequency support for the various SIM techniques' OTFs as well as that of wide-field microscopy. Figure 13 shows a side view of these OTFs.

The wide-field OTF (blue) has a far smaller lateral support than any of the SIM OTFs and does not fill the missing cone (as can be seen in the 3D visualization in Fig. 12). The two-beam TIRF SIM OTF (red) has a much greater lateral support than the wide-field OTF; in fact, its lateral support is slightly greater even than that of the other SIM techniques, as the TIRF

Super-Resolution Fluorescence Microscopy Using Structured Illumination 157

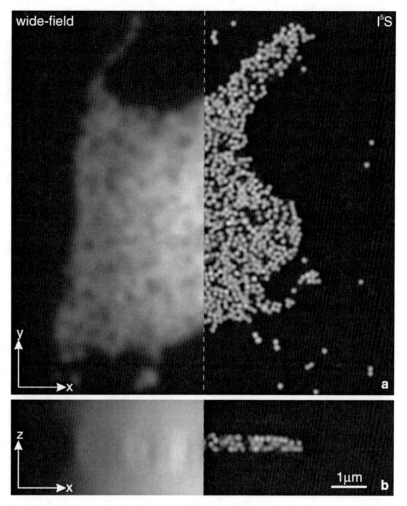

Fig. 11 (a) A lateral and **(b)** an axial image section of a bead sample acquired with I^{5}S. The *left half* of the images show the wide-field image, the *right* the I^{5}S image. The FWHM of the I^{5}S PSF was measured to be 119 nm in the lateral and 116 nm in the axial direction. Data was provided by Lin Shao, Janelia Farm Research Campus, Howard Hughes Medical Institute, Ashburn, Virginia, USA. Modified with permission from [30]

illumination allows illumination patterns beyond the limit possible for non-TIRF illumination. As the wide-field OTF, the two-beam SIM OTF does not fill the missing cone, but optical sectioning is achieved through the low penetration depth of TIRF illumination (not depicted). Three-beam SIM achieves nearly the same lateral support as two-beam TIRF SIM but has twice the axial support and fills the missing cone. While having the same lateral support as three-beam SIM, the axial support of I^{5}S is dramatically improved over that of any of the other OTFs.

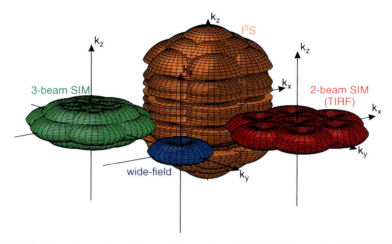

Fig. 12 Three-dimensional view of the various SIM OTFs. The wide-field OTF is shown in *blue*, the two-beam TIRF SIM OTF in *red*, three-beam SIM in *green* and I^{5}S in *orange*

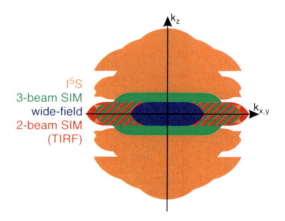

Fig. 13 Two-dimensional projections of the various SIM OTFs. The wide-field OTF is shown in *blue*, the two-beam TIRF SIM OTF in *red*, three-beam SIM in *green* and I^{5}S in *orange*

4 Nonlinear SIM: An Outlook

The three variants of SIM described in the previous section do not surpass the Abbe resolution limit, but rather combine the respective limits of illumination and detection to yield an extended overall support. In spite of this, both illumination and detection remain diffraction limited, restricting the possible resolution enhancement to a factor of about two.

Introducing and exploiting nonlinearities in the fluorophores' response to the illumination light, it is possible to circumvent the Abbe limit for the (effective) illumination and therefore for SIM in general [17]. This way, the possible resolution enhancement can in

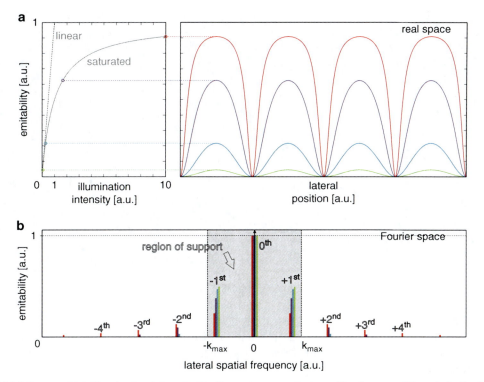

Fig. 14 (**a**) For very high illumination intensity the fluorescence emission will not respond linearly to the illumination intensity. This is mathematically equivalent to the fluorophores responding linearly to an *effective* illumination, which incorporates this saturation of emission. In SIM the effective pattern corresponding to a harmonic illumination pattern need no longer be harmonic, but will exhibit flattened peaks. (**b**) This distortion of the effective illumination pattern leads to higher harmonics and thus more orders in Fourier space. These can lie outside the support of the illumination OTF and can therefore help extend the resolution beyond the Abbe limit

principle be made arbitrarily high. In practice, however, it will still be limited by signal-to-noise issues and experimental imperfections.

Nevertheless, lateral resolution below 50 nm has been demonstrated this way [16, 18].

4.1 Fluorescence Saturation

The conceptually simplest way to achieve a nonlinear fluorescence response is fluorescence saturation [17]. This effect is illustrated in the left part of Fig. 14a. At low illumination intensities the number of emitted photons will be linearly proportional to the illumination intensity. But as illumination intensity increases, this relation will not remain linear as a fluorophore cannot emit photons at a rate higher than its inherent fluorescence decay rate; fluorescence is thus saturated.

If a homogeneous fluorescent plane is illuminated with a harmonic (sinusoidal) light pattern, its emission need not be harmonic, as shown in the right part of Fig. 14a. For intensities high enough to cause fluorescence saturation, the emitted light pattern will be distorted. While a Fourier transform of the harmonic

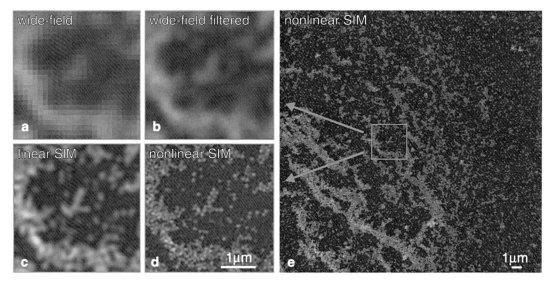

Fig. 15 50 nm fluorescent beads imaged with nonlinear SIM employing fluorescence saturation to generate a nonlinear sample response. (**a**) The wide-field image. (**b**) The wide-field image, filtered and resampled to a higher pixel number. (**c**) Linear SIM shows clear improvement of the lateral resolution. (**d**) Nonlinear SIM using three additional harmonics shows further resolution improvement over linear SIM. (**e**) As a wide-field technique (nonlinear) SIM can be used to image large fields of view. Images modified from [18]. Copyright (2005) National Academy of Sciences, USA

pattern will have only three orders, the distorted patterns will also have higher orders, which can even lie outside the frequency support of the illumination OTF, as can be seen in Fig. 14b.

Although the sample is really responding nonlinearly to a harmonic illumination distribution, this is mathematically equivalent to the sample responding linearly to an *effective* illumination which is proportional to this distorted, inharmonic emission pattern. Saturated SIM can therefore be treated the same way as linear SIM, but with an illumination pattern that will have more orders and no longer needs to be diffraction limited.

This way it would in principle be possible to extend the frequency support of the SIM OTF indefinitely. However, while stronger saturation leads to higher orders appearing in the illumination pattern, these orders themselves become weaker and weaker. In practice the resolution improvement is therefore limited by signal-to-noise levels of the raw images, as the Fourier orders will eventually disappear in a constant, white noise floor stemming from Poisson noise in the acquired images.

The concept of nonlinear SIM as well as using fluorescence saturation as one possible way of achieving the necessary nonlinearity was first proposed by Heintzmann et al. in 2002 [17]. It was demonstrated in 2005 by Mats Gustafsson [18], who imaged a field of 50 nm fluorescent beads using two-beam illumination. His results are shown in Fig. 15. Figure 15a shows a wide-field image of the sample. Figure 15b shows the same wide-field image, digitally filtered and resampled to a higher pixel number. Linear SIM (Fig. 15c) already shows a clear improvement in resolution; however, individual beads still cannot be resolved. In the nonlinear SIM image (Fig. 15d, e), which reconstructs three additional harmonics in the illumination pattern, individual beads are clearly resolved. In this image the averaged bead profiles have an FWHM of 58.6 ± 0.5 nm. After accounting for the size of the beads, this yields an FWHM resolution of 48.8 ± 0.5 nm, a factor 5.5 better than the unfiltered wide-field resolution of 265 nm. Experimental parameters were: excitation wavelength $\lambda_{ex} = 532$ nm; 3.6 J pulses with 640 ps duration and a repetition rate of 6 kHz, leading to a pulse energy density of 5.3 mJ/cm^2; emission filter band $\lambda_{em} = \{580-630 \text{ nm}\}$; numerical aperture $NA = 1.4$, oil immersion objective. The grating constant in the sample was $d_{grat} = 2\pi / |\mathbf{p}| = 200$ nm.

While Gustafsson's experiment demonstrates an impressive resolution enhancement, the high local intensities necessary to achieve fluorescence saturation may lead to increased photo bleaching and to date the technique has not been demonstrated for imaging biological samples.

4.2 Photo-Switchable Fluorophores

Fortunately nonlinearities in the fluorescence response can be introduced by other means than saturation.

Photo-switchable fluorophores are markers that can be switched between a dark non-fluorescent off-state and a bright fluorescent on-state through irradiation with particular wavelengths of light. Only in the on-state do they behave as a fluorescent molecule, can be excited and will emit light.

There are many different photo-switchable fluorophores and their switching mechanisms may vary. Their common feature is the possibility of activation/deactivation. These markers are famous for their use in super-resolution microscopy based on stochastic activation and localization of individual molecules (i.e., PALM) and have long been a candidate for other super-resolution approaches [13]. In nonlinear SIM they can be used to introduce nonlinearities between illumination and emission. If a homogenous plane of switchable fluorophores in the off-state are activated using a sinusoidal illumination pattern of low intensity, the number of fluorophores in the on-state will (at first) be linearly proportional to the sinusoidal illumination. However, activation is a stochastic process. As the irradiation with the activation light continues, fluorophores which have thus far not been activated will have another chance to be activated. Fluorophores which already have been activated however, cannot be activated again. While initially rising at a rate proportional to the activation intensity, the fluorophores' probability of being activated can never exceed 100 % and will saturate somewhat similarly to fluorescence saturation shown in the right part of Fig. 14a (while saturation follows a hyperbolic curve, the curve for activation/deactivation is exponential).

For signal-to-noise reasons it is beneficial to first activate all fluorophores and then use sinusoidal illumination to deactivate the fluorophores. Rather than driving activation into saturation this drives deactivation into saturation. The resulting activation patterns then correspond to the difference between one and the curves shown in the right part of Fig. 14a and for strong saturation the illumination pattern becomes a series of fine lines.

Unlike fluorescence saturation this saturation of activation/deactivation does not depend on momentary illumination intensity but rather on the total light dosage received. It can thus also be achieved with low intensity and is therefore more suitable for the imaging of biological specimen.

Recently Rego et al. demonstrated nonlinear SIM using the photo-switchable protein Dronpa [16]. Using TIRF illumination they imaged various biological samples. As an example Fig. 16 shows a SIM image of nuclear pores in a human embryonic kidney (HEK293, [35]) cell nucleus transiently transfected with POM121-Dronpa. Figure 16a shows the entire nucleus; the bottom left part shows the wide-field TIRF image, whereas the top right shows the image acquired with nonlinear SIM. Figure 16b–e show a close-up of a single pore. While linear SIM (Fig. 16c) already significantly enhances the resolution over the wide-field image (Fig. 16b), it is outperformed by nonlinear SIM reconstructing one (Fig. 16d) or two (Fig. 16e) additional harmonics. With this technique ring-like structures with an inner diameter of around 40 nm could be resolved. Experimental parameters were: activation wavelength $\lambda_{act} = 405$ nm; excitation as well as deactivation wavelength $\lambda_{ex} = 488$ nm; emission filter band $\lambda_{em} = \{505-545 \text{ nm}\}$; numerical aperture $NA = 1.46$, oil immersion objective; TIRF illumination. The grating constant in the sample was $d_{grat} = 2\pi / |\mathbf{p}| = 166$ nm.

5 Discussion

While the theory of image formation and reconstruction in SIM is relatively straightforward, there are many pitfalls on the way to high quality SIM images.

Insufficient knowledge of many experimental parameters may lead to artifacts, e.g., through bad component unmixing (Eq. 5). The quality of reconstruction can be significantly degraded by fluctuations in illumination intensity, sample drift or bleaching, imprecise knowledge of the pattern period or direction, unknown or even varying contrast of the illumination pattern, imprecise knowledge of the pattern position in the individual raw images, etc.

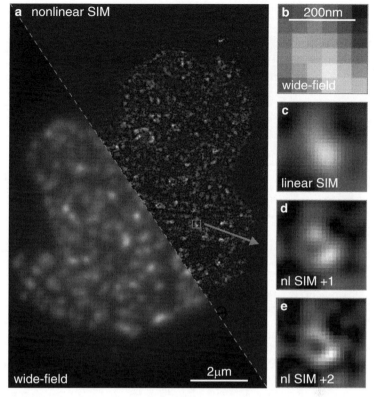

Fig. 16 Nuclear pores of a human embryonic kidney imaged with nonlinear SIM employing photo-switchable fluorophores to generate a nonlinear sample response. (**a**) An overview of the sample, the *bottom left* showing the wide-field image, the *top right* the nonlinear SIM image. (**b**)–(**e**) Close-ups of one nuclear pore in (**b**) wide-field, (**c**) linear SIM, and nonlinear SIM with (**d**) one and (**e**) two additional harmonics. Modified with permission from [16]

SIM therefore requires very precise control over all these experimental parameters. This can be facilitated by using sophisticated reconstruction algorithms which manage to extract the required knowledge of these parameters a posteriori from the acquired data [36–40].

Recent years have seen many advances in both SIM hardware and reconstruction and the technique is becoming more and more widespread, with several microscope manufacturers now offering commercial systems. While the resolution of linear SIM does not quite reach that of other super-resolution techniques such as STED, PALM or STORM, its simplicity and large field of view nevertheless makes it attractive to users in the biomedical sciences. Unlike these other techniques, linear SIM does not require specialized sample preparation and works with any dye used in conventional or confocal fluorescence microscopy. While the confocal microscope remains the workhorse of the biomedical field for the time being, SIM may well become a strong competitor with its improved resolution and light efficiency.

Significant progress has also been made concerning acquisition speed. Using ferroelectric liquid crystal SLMs, which can switch (i.e., translate and rotate) the displayed patterns with rates of more than 1 kHz, 2D SIM has been carried out at video rate [41]. 3D SIM image stacks of whole cells have been acquired in as little as 5 s. As a wide-field technique the field of view barely influences the acquisition time (the readout speed of the cameras may be affected), unlike for scanning techniques such as STED or confocal microscopy. SIM therefore allows fast live cell imaging of large samples.

While nonlinear SIM yet has to be improved before it will become a useful tool in the biomedical sciences, it shows huge potential as a technique which will allow fast live cell imaging with a resolution below 50 nm and thus make SIM a serious player among other super-resolution techniques.

References

1. Shimomura O, Johnson F, Saiga Y (1962) Extraction, purification and properties of aequorin, a bioluminescent protein from the luminous hydromedusan, aequorea. J Cell Comp Physiol 59:223–239
2. Chalfie M, Tu Y, Euskirchen G, Ward W, Prasher D (1994) Green fluorescent protein as a marker for gene expression. J Cell Comp Physiol 263:802
3. Tsien R (1988) The green fluorescent protein. Annu Rev Biochem 67:509–544
4. Amos WB, White JG, Fordham M (1987) Use of confocal imaging in the study of biological structures. Appl Opt 26:3239–3243
5. Neil MAA, Juskaitis R, Wilson T (1997) Method of obtaining optical sectioning by using structured light in a conventional microscope. Opt Lett 22(24):1905–1907
6. Heintzmann R, Cremer C (1999) Laterally modulated excitation microscopy: Improvement of resolution by using a diffraction grating. In: Proceedings of the Society of Photographic Instrumentation Engineers, vol 3568, pp 185–196, 1999
7. Gustafsson MGL (2000) Surpassing the lateral resolution limit by a factor of two using structured illumination microscopy. J Microsc 198:82–87
8. Hell SW, Wichmann J (1994) Breaking the diffraction resolution limit by stimulated emission: Stimulated-emission-depletion fluorescence microscopy. Opt Lett 19:780–782
9. Betzig E, Patterson GH, Sougrat R, Lindwasser W, Olenych S, Bonifacino JS, Davidson MW, Lippincott-Schwartz J, Hess HF (2006) Imaging intracellular fluorescent proteins at nanometer resolution. Science 313:1642–1645
10. Rust MJ, Bates M, Zhuang X (2006) Sub-diffraction-limit imaging by stochastic optical reconstruction microscopy (STORM). Nat Method 3:793–796
11. Hess ST, Girirajan TPK, Mason MD (2006) Ultra-high resolution imaging by fluorescence photoactivation localization microscopy. Biophys J 91:4258–4272
12. Lukosz W (1966) Optical systems with resolving powers exceeding the classical limit. J Opt Soc Am 56:1463–1472
13. Hofmann M, Eggeling C, Jakobs S, Hell SW (2005) Breaking the diffraction barrier in fluorescence microscopy at low light intensities by using reversibly photoswitchable proteins. Proc Natl Acad Sci USA 102:17565–17569
14. Heintzmann R, Gustafsson MGL (2009) Subdiffraction resolution in continuous samples. Nat Photon 3:363–364
15. Hirvonen L (2008) Structured illumination microscopy using photoswitchable fluorescent proteins. PhD thesis, King's College London, UK
16. Rego EH, Shao L, Macklin J, Winoto L, Johansson GA, Kamps-Hughes N, Davidson MW, Gustafsson MGL (2012) Nonlinear structured-illumination microscopy with a photoswitchable protein reveals cellular structures at 50-nm resolution. Proc Natl Acad Sci USA 109:E135–E143
17. Heintzmann R, Jovin TM, Cremer C (2002) Saturated patterned excitation microscopy – a concept for optical resolution improvement. J Opt Soc Am A 19:1599–1609

18. Gustafsson MGL (2005) Nonlinear structured-illumination microscopy: Wide-field fluorescence imaging with theoretically unlimited resolution. Proc Natl Acad Sci USA 102: 13081–13086
19. Gustafsson MGL, Shao L, Carlton PM, Wang CJR, Golubovskaya IN, Cande WZ, Agard DA, Sedat JW (2008) Three-dimensional resolution doubling in wide-field fluorescence microscopy by structured illumination. Biophys J 94:4957–4970
20. Heintzmann R (2003) Saturated patterned excitation microscopy with two-dimensional excitation patterns. Micron 34:283–291
21. Orieux F, Sepulveda E, Loriette V, Dubertret B, Olivo-Marin J-C (2012) Bayesian estimation for optimized structured illumination microscopy. IEEE Trans Image process 21: 601–614
22. Penrose R (1955) A generalized inverse for matrices. In: Proceedings of the Cambridge Philosophical Society, vol 51, pp 406–413, 1955
23. Wiener N (1949) Extrapolation, interpolation and smoothing of stationary time series with engineering applications. MIT Press, Cambridge; Wiley, New York; Chapman & Hall, London
24. Arigovindan M, Elnatan D, Fung J, Branlund E, Sedat J, Agard D (2012) High resolution restoration of 3D structures from extreme low dose widefield images. In: Focus on Microscopy 2012 abstract book, 2012
25. McCutchen CW (1964) Generalized aperture and the three-dimensional diffraction image. J Opt Soc Am 54(2):240–244
26. Ewald PP (1969) Introduction to the dynamical theory of X-ray diffraction. Acta Crystallogr A 25(1):103–108
27. Scherer WF, Syverton JT, Gey GO (1953) Studies on the propagation in vitro of poliomyelitis viruses IV. Viral multiplication in a stable strain of human malignant epithelial cells (strain HeLa) derived from an epidermoid carcinoma of the cervix. J Exp Med 97: 695–710
28. Talbot HF (1836) Facts relating to optical science. Lond Edinb Philos Mag J Sci 9: 401–407
29. Fiolka R, Shao L, Rego EH, Davidson MW, Gustafsson MGL (2012) Time-lapse two-color 3D imaging of live cells with doubled resolution using structured illumination. Proc Natl Acad Sci USA 109:5311–5315
30. Shao L, Isaac B, Uzawa S, Agard DA, Sedat JW, Gustafsson MGL (2008) I^{5}S: Wide-field light microscopy with 100-nm-scale resolution in three dimensions. Biophys J 94: 4971–4983
31. Hell SW, Stelzer EHK (1992) Fundamental improvement of resolution with a 4Pi-confocal fluorescence microscope using two-photon excitation. Opt Commun 93:277–282
32. Hell SW, Stelzer EHK (1992) Properties of a 4Pi-confocal fluorescence microscope. J Opt Soc Am A 9:2159–2166
33. Gustafsson MGL, Agard DA, Sedat JW (1995) Sevenfold improvement of axial resolution in 3D widefield microscopy using two objective lenses. Proc SPIE 2412:147–156
34. Gustafsson MGL, Agard DA, Sedat JW (1999) I5M: 3D widefield light microscopy with better than 100 nm axial resolution. J Microsc 195:10–16
35. Graham FL, Smiley J (1977) Characteristics of a human cell line transformed by dna from human adenovirus type 5. J Gen Virol 36: 59–72
36. Schaefer LH, Schuster D, Schaffer J (2004) Structured illumination microscopy: Artefact analysis and reduction utilizing a parameter optimization approach. J Microsc 216(2): 165–174
37. Shroff SA, Fienup JR, Williams DR (2009) Phase-shift estimation in sinusoidally illuminated images for lateral superresolution. J Opt Soc Am 26:413–424
38. Wicker K (2010) Increasing resolution and light efficiency in fluorescence microscopy. PhD thesis, King's College London, UK
39. Wicker K, Mandula O, Best G, Fiolka R, Heintzmann R (2013) Phase optimisation for structured illumination microscopy. Opt Express 21:2032–2049
40. Wicker K (2013) Non-iterative determination of pattern phase in structured illumination microscopy using auto-correlations in Fourier space. Opt Express 21:24692–24701
41. Kner P, Chhun BB, Griffis ER, Winoto L, Gustafsson MGL (2009) Super-resolution video microscopy of live cells by structured illumination. Nat Method 6:339–342

Chapter 8

Application of Three-Dimensional Structured Illumination Microscopy in Cell Biology: Pitfalls and Practical Considerations

Daniel Smeets, Jürgen Neumann, and Lothar Schermelleh

Abstract

Super-resolution fluorescence microscopy techniques have paved the way to address cell biological questions with unprecedented spatial resolution. Of these, three-dimensional structured illumination microscopy (3D-SIM) reaches a nearly eightfold increased volumetric resolution compared to conventional diffraction-limited methods and allows multicolor optical sectioning of standard fluorescently labeled fixed or live samples. Owing to its broad application spectrum, 3D-SIM is likely to become a key method in cell biological far-field imaging, complementing more specialized higher-resolving techniques, such as single molecule localization and cryo-electron microscopy. To fully explore the potential of 3D-SIM, however, considerably greater care needs to be taken with regard to the preparation of the sample, calibration of the instrument, post-processing of the data, and extraction of valid quantitative measurements. In this chapter we discuss technical problems typically encountered and provide guidelines for troubleshooting.

Key words Super-resolution imaging, 3D structured illumination microscopy, Immunofluorescence, Sample preparation, Colocalization

1 Introduction

For decades fluorescence light microscopy has provided an invaluable tool for cell biologists and contributed substantially to our current knowledge of how cells and organisms function [1, 2]. With today's toolset of modern fluorescent dyes, advanced immunocytochemistry approaches, fluorescent protein tags, and in situ hybridization methods it is possible to label and visualize literally any protein or DNA sequence of interest [3–6]. This enables scientists to study and understand the spatial organization and time-dependent dynamics of fundamental cellular processes on the single cell level, and to complement population-wide biochemical and genetic approaches. However, the majority of biological structures of interest range in size from tens to a few hundreds of nanometers and thus escape the resolution of classical light microscopy imposed by the diffraction limit of visible light. This limit is wavelength dependent and reaches ~200–300 nm in lateral (x, y) and ~500–800 nm in axial (z) directions (see ref. [7]).

In the past decade, this conceptual limitation has been overcome with the emergence of groundbreaking optical microscopy techniques that combine the advantages of fluorescence light microscopy with the promise of theoretically unlimited resolution. Three major concepts of these collectively called super-resolution techniques are now maturing into commercially available turnkey instruments: single molecule localization microscopy approaches (e.g., PALM, STORM) [7–9], stimulated emission depletion (STED) microscopy [10–12], and super-resolution structured illumination microscopy (SIM) [13–15]. While all approaches rely on different principles that have distinctive advantages and impose specific constraints, they all share the revolutionary ability to circumvent the long thought fundamental physical barrier of the diffraction limit (for reviews see [16–20]).

In this chapter we will focus on the application of three-dimensional SIM (3D-SIM) that provides optical sectioning with improved resolution of a factor of two in all three spatial dimensions (i.e., a resolution of 100–130 nm in lateral dimension and 250–350 nm along the optical axis), equivalent to an eightfold improvement in volumetric resolution compared to conventional light microscopy (Fig. 1). While the gain in lateral resolution may seem modest compared to other super-resolution techniques, 3D-SIM features some important benefits, such as the simultaneous imaging of multiple colors using a broad range of standard fluorescent dyes, and the ability to acquire three-dimensional image stacks of up to 20 μm in depth and with hugely improved contrast. Moreover, live-cell applicability of 2D- and 3D-SIM has been recently demonstrated [21–24] and has become a standard feature in the latest generation of turnkey instruments. To date, a growing number of publications demonstrate the broad applicability and versatility of super-resolution SIM to address a wide range of cell biological questions [25–36].

Here, we explain in detail practical considerations associated with the application of 3D-SIM, discuss potential pitfalls that need to be considered from a user's perspective, and provide guidelines on how to counteract them. While our experiences rest on the usage of a DeltaVision OMX V3 prototype (Applied Precision, Inc.), the elaborated principles also apply to other commercial systems (e.g., from Zeiss or Nikon).

Application of Three-Dimensional Structured Illumination Microscopy in Cell Biology...

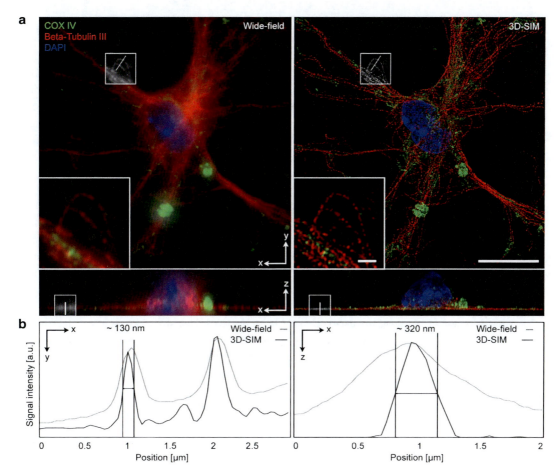

Fig. 1 Enhanced lateral and axial resolution and improved optical sectioning quality by 3D-SIM. (**a**) Comparison of wide-field versus 3D-SIM imaging of a murine neuronal cell differentiated from an embryonic stem cell (ESC) immunostained with antibodies against the mitochondrial inner membrane protein COX IV (*green*) and the neuronal marker protein beta-tubulin III (*red*). Secondary antibodies were conjugated to Alexa 488 and Alexa 594, respectively, and DNA was counterstained with 4′,6-diamidino-2-phenylindole (DAPI, *blue*). The pseudo wide-field image was generated by average projection of the five phase-shifted SI-raw images recorded for each single plane of the first angle, hence allowing a direct comparison of structural features and signal intensities. Maximum intensity projections of optical serial sections in lateral (*x–y, upper panel*) and orthogonal (*x–z, lower panel*) directions are shown. *Scale bar* represents 10 μm (*inset* 0.5 μm). (**b**) Intensity line profile plots through Alexa 594-labeled tubulin fibers from the assigned region in (**a**), and measurements of peak width (FWHM) demonstrate the twofold improvement in lateral (*left panel*) as well as in axial direction (*right panel*) for the according wavelength (excitation at 592 nm, emission at >620 nm). Importantly, the physical size of the assessed structure (including covering antibodies) is about 50 nm, and thus well below the resolution limit of the approach

2 3D-SIM: Setup and Basic Principles

The 3D-SIM technology was first implemented on the OMX (Optical Microscope eXperimental) platform, originally designed by John Sedat and coworkers [25, 37]. It is based on a wide-field illumination light microscope, equipped with nano-precision stage controls for minimal drift, up to four laser lines and dedicated high-sensitivity electron multiplying charge-coupled device (EMCCD), or scientific complementary metal–oxide–semiconductor (sCMOS) cameras for multicolor acquisition (Fig. 2; for detailed description of the OMX setup see ref. [38]). Instead of illuminating the sample uniformly, a three-dimensional illumination pattern with a spatial frequency just above the diffraction limit is used. This is typically achieved by passing a collimated laser beam through a diffraction grating. The central three beams (diffraction orders −1, 0, 1) are collected by a lens and targeted to the back aperture of the objective, where they combine to generate a modulated pattern of excitation, visible as stripes of around 200 nm width in the focal plane. This modulation interferes with higher spatial frequency information (i.e., finer structures) emitted from the sample generating a lower frequency interference pattern (Moiré effect), coarse enough to be transmitted back through the objective. The detected raw images therefore encode higher frequency information of otherwise unresolved sample structures, but shifted to a lower frequency. This information can be decoded by computational post-processing using dedicated algorithms to reconstruct a super-resolution image, analogous to deconvolution (Fig. 3). Notably, this linear structured illumination approach doubles the frequencies transmitted by the employed optical system and is thus by definition still diffraction-limited. This is in contrast to other super-resolution techniques that either exploit nonlinear effects when driving fluorochromes into saturation (STED, saturated SIM), or generate a pointillist projection of the positions of hundred thousands of temporally separated single emitting fluorophores (PALM/STORM). The resolution or localization accuracy of these methods is only dependent on the energies applied or the number of photons detected, and not limited by diffraction [39].

In 3D-SIM, in order to capture the interference information in all regions and spatial directions the illumination pattern needs to be shifted between camera exposures in five lateral phase steps and rotated in three angles for every focal plane. A stack size of at least eight planes (equivalent to 1 μm z-height) is required to reconstruct an optical section with super-resolution in all three axes. The imaging depth is limited depending on the nature of the specimen to 10–20 μm due to refractive index mismatch-induced optical aberrations (see below). To meet the Nyquist–Shannon sampling theorem in the axial direction, the focal planes are typically acquired with a z-distance of 0.125 μm. This means that to cover the volume of a typical mammalian cultured cell (e.g., an 8 μm stack height), a raw dataset of 1,000 16-bit images (>500 MB) per color channel is required. In multicolor experiments, the specific channels are collected sequentially by switching the exciting lasers for every phase step and angular position of the SI-pattern and at every

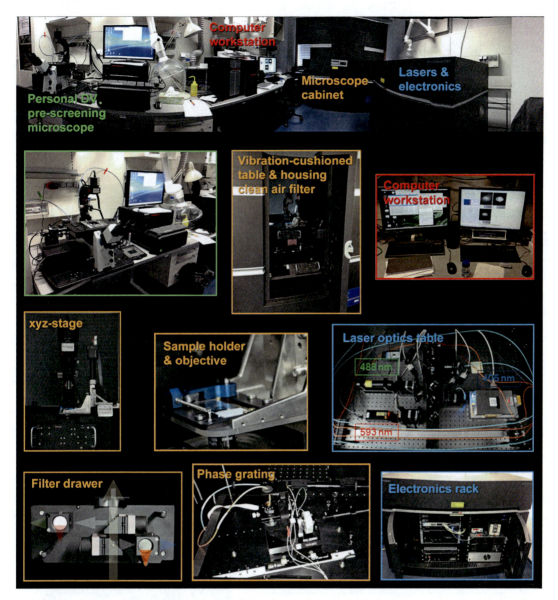

Fig. 2 OMX V3 prototype setup. The OMX V3 prototype system (API/GE Healthcare) is composed of (an optional) conventional wide-field deconvolution microscope (*green*), which allows sample pre-screening through the eye-piece to visually judge sample quality. Its stage position is calibrated to the OMX stage allowing users to store positions of interest for subsequent revisiting once the same sample is mounted on the OMX stage. The super-resolution imaging system is configured with three high-power diode lasers controlled by high-speed shutters and neutral density filters mounted on an accessible breadboard within a separate laser enclosure (*purple*). This also harbors all electronic equipment necessary for hardware control. The laser light is directed via a switchable mirror to two optical fibers, either for the SI or for the wide-field light path, which then lead to the double-walled insulated cabinet harboring the OMX microscope stage with three EMCCD cameras assembled on a vibration-damped table (*yellow*). The laser light is directed through a polarizer and an optical phase grating mounted on a rotary stage to generate the structured illumination pattern. Sample holder, objective holder, filter drawer, and optical light path are designed to provide maximum mechanical and thermal stability and detection sensitivity. The image data is stored and processed on a high-performance computer workstation (*red*) using the SoftWorx 6.0 software package

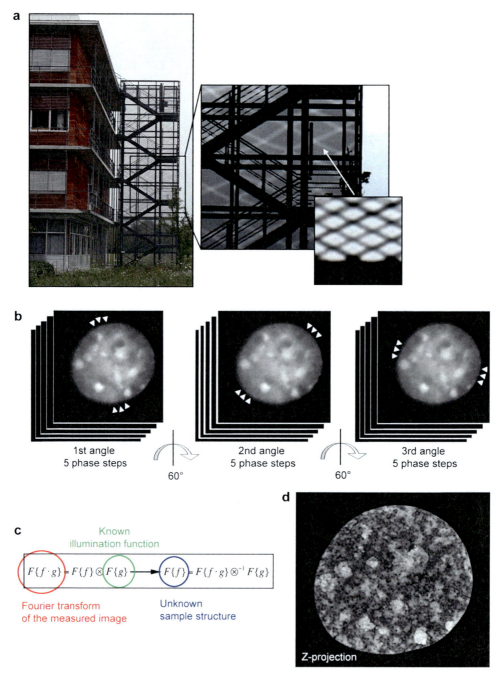

Fig. 3 Principle of 3D-SIM reconstruction exploits the Moiré effect in the interference between known stripe pattern and unknown sample structure. (a) Macroscopic example for a typical Moiré effect (seen at an emergency staircase outside the LMU Biocenter research building). The structure of the interference pattern reflects the checkered structure of the superimposing metal trellises which are non-resolvable in the larger image (small *inset* shows magnification of the trellis structure). (b) Illustration of the 3D-SIM acquisition procedure with lateral phase-shifting and rotating the stripe pattern to acquire a total of 15 raw images per *z*-section. (c) Simplified formula that describes the very basic principle of the reconstruction procedure. The unknown sample structure can be resolved from the measured image and the known illumination frequency. (d) *Z*-projection of the DAPI-stained cell nucleus shown in (b) after the reconstruction process

z-position in order to avoid detection of bleed through signal of the shorter wavelength emitting fluorophore onto the longer wavelength camera when simultaneously exciting with all laser lines. With exposure times of typically 10–200 ms and delay times of 50–100 ms in between exposures, the total acquisition time (full frame, 512 by 512 pixels) may vary from a few tens of seconds to a few minutes, depending on the stack size.

While the first SIM implementations relied on mechanical movement of a transmission phase grating, new approaches have been devised, e.g., using liquid crystal spatial light modulators or galvanometer-driven mirror clusters (OMX Blaze) that allow a much faster pattern change. In combination with sCMOS cameras that enable much faster read-out, the acquisition speed can be accelerated by at least an order of magnitude, a vital prerequisite for live-cell imaging [23, 24].

3 Sample Preparation

Painstaking sample preparation is the first crucial step and a key to optimal, artifact-free microscopic images. While this is true for conventional high-resolution imaging methods, it is even more so when it comes to super-resolution microscopy [40, 41]. The following aspects are of importance and need to be considered: (1) the structural preservation of the specimen by fixation, permeabilization, and labeling procedure; (2) the detection quality, given by the labeling specificity, labeling density, and signal-to-background ratio; (3) the choice of fluorophores in terms of brightness and photostability; and (4) optical aberrations and effects of refractive index mismatch.

3.1 Structural Preservation

The reconstructed image acquired from a fixed cell specimen should ideally reflect the closest representation possible of the unperturbed living cell. However, super-resolution live-cell imaging is still in its infancy and pioneering studies have been restricted to rather specific applications [41–45]. In fact, the resolution gain comes at the price of depositing higher energies (i.e., more photons) onto the sample, by increasing either the laser power (STED, SSIM) or the number of exposures (STORM, SIM). This increases the sampling time and poses special challenges to the live-cell fluorophore, in most cases fluorescent proteins (e.g., EGFP, mCherry) fused to a protein of interest. Moreover, depending on the time scale of observation and the deposited energy, the concomitant phototoxicity increasingly compromises the general physiology and viability of the observed specimen. Fast progressing technical advances, such as the use of pulsed lasers and faster and more sensitive detectors, will make it possible to reduce the energy loads and shorten sampling times in order to be able to capture faster cellular processes, and open the way to a broader application of live-cell super-resolution imaging.

Still, some trade-offs will persist. For many reasons indirect labeling methods of fixed and permeabilized specimens using photostable organic dyes will continue to be a prime application for super-resolution imaging, e.g., when immunodetecting endogenous proteins that may carry for instance a posttranslational modification or when labeling specific DNA/RNA sequences by in situ hybridization. In these cases it is important to maintain the structural integrity of the specimen to the best possible extent, at least at the level of the observable resolution. Of note, many routine methods have a substantial effect on the maintenance of cellular morphology. For instance, dehydration by methanol fixation or embedding with hardening mounting media can lead to a significant flattening of the specimen. Chemical fixation with the strong bivalent cross-linker glutaraldehyde, routinely used for electron microscopy, on the other hand, can sometimes impair epitopes and affect the integrity of certain cellular structures [46]. In our experience, the best morphological and structural preservation for fluorescence detection at the light microscopy level is given by cross-linking with formaldehyde [47]. Because of the potential influence on epitope accessibility, the optimal concentration for IF applications may vary between 1 and 4 % depending on the antibodies used. From experience, fixation with 2 % formaldehyde (freshly prepared from paraformaldehyde, from sealed ampules of 16 % solution, or from 37 % solution stabilized with 10 % methanol) in PBS for 10 min at room temperature typically followed by a permeabilization step with 0.2–0.5 % Triton X-100 in PBS for 10 min works well for most applications.

3.2 Labeling and Detection Quality, Signal-to-Background

The ratio of the fluorescence signal specific to the target structure to the surrounding (nonspecific) background, i.e., the signal contrast, is a key criterion for fluorescence imaging applications. This is particularly true for 3D-SIM as the signal contrast directly dictates the modulation (stripe) contrast and thus the quality of interference, which translates into the level of extractable high-frequency information. The signal contrast depends often largely on the staining procedure and is therefore in the users' hands. It must not be confused with the signal-to-noise ratio (SNR) of the microscope setup, although the inherent noise characteristics of the camera (shot noise, dark noise, fixed pattern noise, read-out noise) contribute to the overall background (see below). Additional contribution of stray light contamination should be excluded by careful microscope design, e.g., by enclosing the microscope, optimizing apertures, and using light traps. Other contributions to the overall background include the autofluorescence of the specimen and the embedding media, nonspecifically bound or unbound fluorescent molecules, and out-of-focus blur as discussed below.

One can distinguish several types of labeling-induced background, which can be counteracted by various means:

1. Distinct, immobile signal spots caused by nonspecific binding to specimen and glass surfaces, or by specific cross-reactivity of the antibody to other epitopes. Nonspecific binding may be counteracted by a quenching of non-reacted aldehyde (e.g., 100 mM glycine/PBS) and incubation with blocking reagents, such as serum albumin, fish skin gelatin, casein, or mixtures thereof, which saturate nonspecific binding sites before and while incubating with antibodies. The efficacy of blocking agents varies depending on the antibodies used and should be tested. Specific cross-reactivity, e.g., to a similar epitope of a different protein, is an inherent property of the antibodies of choice. Therefore these need to be thoroughly tested and selected for high affinity and specificity [40, 48]. The optimal antibody dilution should be determined for every batch and cell type to find the best trade-off between epitope coverage, signal strength, and low background. Too high antibody concentration disproportionally increases nonspecific background signals and thus could be as detrimental as a very low concentration. Often disregarded is the benefit of extensive washing after antibody incubations in large volumes of PBS containing small amount of detergent (e.g., 0.02 % Tween-20). In addition, for some antibodies, stringent washing with, e.g., a high salt solution (300 mM NaCl in PBS) can further help to reduce cross-reactivity to low affinity off-target epitopes. Of note, while these artifacts are often blurred out in conventional imaging, with 3D-SIM they become more obvious as spurious spots that superimpose the signals originating from the actual structure of interest and may hamper the statistical evaluation and correct interpretation of the results.

2. Moving fluorescent particles, e.g., antibody aggregates or cell debris, can lead to local reconstruction artifacts within a confined region of movement during image acquisition. Usage of well-tested high affinity antibodies, thorough washing, and post-fixation with formaldehyde after the last antibody incubation usually prevent this source of background. Antibodies sometimes aggregate upon prolonged storage. In these cases it may be useful to spin down the stock solution ($11,000 \times g$ for 20 min) and transfer the supernatant to a new vial. Filtering a working dilution in blocking solution through a syringe filter with 0.5 μm pore size may also help to reduce precipitates and other debris.

3. For the overall reconstruction quality, the most critical factor is the level of diffuse signal background permeating the structures of interest. As discussed above, the SI principle relies on the detection of the Moiré interference of three-dimensional

illumination and the light of variable frequencies emitted from structures of interest. The amount of mathematically recoverable higher frequency information depends on the local contrast, i.e., the ratio of specific signal in-focus to the nearby surrounding background. Hence, sample-induced background, typically associated with indirect (immuno-)fluorescence labeling approaches, should be minimized by all aforementioned means (blocking, washing, reagent concentration, low autofluorescing media). Direct or covalent labeling approaches, e.g., by click-chemistry [49], are typically less prone to background-associated artifacts and are a superior alternative [50, 51]. In this respect, densely packed structures of extended volumes (e.g., nuclear chromatin or large filament bundles) that inherently generate high levels of out-of-focus blur pose a greater challenge compared to small solitary structures and require adjustments in the acquisition settings (discussed in Sect. 4).

3.3 Choice of Fluorophores

Sufficiently high signal intensities and bleaching stability are important criteria, which depend on the choice of fluorophores as well as on the system settings. The utilized dyes not only have to match their spectral properties to the laser and the filter setup of the imaging system but also must withstand the extended number of exposures required for SI imaging (see Sect. 4). Most modern organic dyes, e.g., of the Alexa or ATTO family are sufficiently bright (as defined by the product of quantum yield and quantum efficiency) and photostable [52], and are well suited for 3D-SIM imaging, while imaging of GFP, mRFP, and other fluorescent proteins, due to their propensity for bleaching, typically requires post-labeling with specific antibodies [28].

3.4 Optical Quality of the Sample

All matter encountered by light after exiting the front lens of the objective is an integral part of the optical system and can have substantial influence on the output. Spherical aberration caused by the refractive index mismatch between coverslip glass, embedding medium, and the specimen, and light scattering are major determinants to consider. These are dependent on the imaging depth and the penetration depth into the sample, respectively. While the optical hardware is corrected for specific conditions (i.e., immersion medium, glass thickness and focal depth, temperature), the properties of the sample may easily deviate from these, which as a consequence reduces signal brightness or contrast and causes distortion artifacts. Thus, light scattering has to be minimized and the refractive properties of immersion oil and embedding medium have to be carefully matched to ensure optimal results with least distortions in the imaging depth of interest and with the given optical setup. This is particularly true for approaches that rely on mathematical post-processing and reconstruction steps and critically depend on optimal point spread function (PSF) behavior of the optical system as in 3D-SIM or wide-field deconvolution [53–56].

There is a range of practical measures to ensure optimal conditions. This starts by wiping the coverslip and objective properly using lint-free lens tissue and/or cotton swabs (pre-cleaning with 80 % ethanol, final cleaning with chloroform) to prevent light scattering by larger dust and dirt particles in the immersion oil. The coverslip thickness plays an important role, as the objective's optical pathway is corrected for a thickness of 0.17 mm and deviation thereof would lead to an optical mismatch. Hence, it is recommended to use borosilicate coverslips with minimal thickness deviation optimized for use with high NA objectives (e.g., Precision cover glasses #1.5H, thickness 0.170±0.005 mm, Marienfeld Superior).

The mounting medium typically has a greater refractive index than water ($n = 1.33$) hence reducing the mismatch between immersion oil and cover glass ($n > 1.51$) and the biological specimen ($n \approx 1.38$) and thereby ameliorating deterioration of image quality as the imaging depth increases. Moreover, it buffers and stabilizes the labeled specimen for long-term storage and reduces fluorophore bleaching by supplemented radical-scavenging reagents. Nonhardening glycerol-based formulations such as Vectashield (Vector Labs) or 10 mg/mL 1,4-diazabicyclo[2.2.2]octane (DABCO) in 90 % glycerol/PBS (pH 8.6) feature effective antifading properties for broad range of dyes [57] and are particularly suited to preserve cellular morphology, whereas hardening mounting media (e.g., Prolong Gold, Life Technologies) are ideal for long-term storage, but at the cost of flattening cells substantially.

Changes in the refractive index of the immersion oil can, to some extent, compensate for the above-mentioned sample-specific variations, and can be used as a most basic form of "adaptive optics" to minimize mismatch artifacts. In practical terms, the optimal refractive index should be determined empirically in advance for a given sample by using a set of graded immersion oils (e.g., Cargille Labs). Under optimal conditions, a distinct point-like fluorescent object (e.g., a synthetic bead, or a suitable labeling signal) in the depth of interest should have a symmetrical intensity distribution in both lateral and axial directions. This can be easily verified by creating orthogonal views of an acquired image stack [55].

The SI reconstruction algorithm uses the 3D optical transfer function (OTF) of the microscope in a given situation, mathematically described by the Fourier transform of the PSF. The PSF of the system needs to be carefully measured from synthetic fluorescent beads (e.g., 0.1 or 0.17 μm diameter PS-Speck microspheres, Life Technologies). Due to the fact that the light paths in a multichannel arrangement differ slightly, it is recommended to acquire individual OTFs for every channel separately. Importantly, the optical conditions when recording an OTF need to match the conditions in the sample itself. Consequently, for multicolor experiments the OTFs should be recorded with immersion oil of the same refractive index as used for the sample recording. The user must then choose which spectral range should have the most symmetrical PSF (this is often green emission but can vary with the application), choose an oil with the appropriate refractive index, and be prepared to accept asymmetric PSFs, and therefore suboptimal OTFs, in other spectral ranges.

4 Photobleaching and System Settings

When setting exposure times and laser powers for SI imaging a compromise between the dynamic ranges of the raw data and bleaching during acquisition must be reached. The minimal peak intensities required for good reconstruction critically depend on noise characteristics of the camera and the amount of blur (out-of-focus) background adjacent to the in-focus structure of interest and need to be determined for each specific case. In our experience this value may be as low as 1,000–2,000 counts (16-bit gray levels) for solitary structures such as small vesicles or cytoskeletal fibers, but must be more than 10,000 counts for large, densely folded biological structures, such as chromatin in mammalian cells. For cameras with poorer noise characteristics this threshold value can be substantially higher.

On the other hand, bleaching during the acquisition of hundreds of raw images should not exceed a critical level. As a rule of thumb, in order to prevent reconstruction artifacts, an intensity loss of no more than one-third between the initial (first angle) and the last (third angle) recorded images can be accepted as long as the last images do not fall below a critical signal-to-background level. Thus, in order to stay within the range of tolerable signal/bleaching range either the number of exposures must be reduced by restricting to smaller z-ranges, or alternative dyes and/or the use of enhancing reagents (Fig. 4) need to be considered to obtain brighter and more photostable signals allowing reduction of laser intensities and/or exposure times.

Besides bleaching, the local deposition of (light) energy during acquisition may also have a slight but measurable impact on the morphology of the observed object on a larger scale. Figure 5 exemplifies such morphological changes on samples caused by the absorbed light energy after several super-resolution acquisition cycles. This may not be of relevance for the standard acquisition routine with only one single round of acquisition, since the effect is only observed when the same position on the sample is repeatedly imaged. Nonetheless, this effect should be kept in mind when discussing and interpreting subtle localization differences, distortions, or long time series reconstructions.

Application of Three-Dimensional Structured Illumination Microscopy in Cell Biology... 179

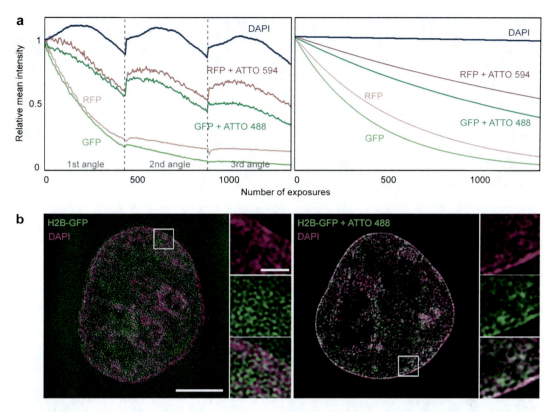

Fig. 4 Excessive bleaching of fluorescent proteins deteriorates reconstruction quality. (**a**) HeLa cells stably expressing GFP- or mRFP-tagged histone H2B were imaged with 3D-SIM. The chart displays the relative loss of nuclear fluorescence signal during acquisition of z-stacks with >10 μm height (average of three measurements). Post-detection with GFP- and mRFP-binding nanobodies (GFP-/RFP-booster; ChromoTek) coupled to organic ATTO-488 or ATTO-594 dyes, respectively, significantly enhances overall brightness and photostability. DAPI counterstaining signal shows only minor bleaching. *Fitted curves* are shown on the *right*. (**b**) Reconstructed 3D-SIM midsections of two representative H2B-GFP expressing cells from measurements shown in (**a**). While the reconstruction quality of GFP-labeled chromatin without post-labeling is very poor, as indicated by high-frequency noise pattern superimposing the underlying sample structure (*left*), the ATTO-488-amplified H2B-GFP shows decent reconstruction quality and structural resolution similar to the DAPI-stained chromatin (*right*). *Scale bar* represents 5 μm (*inset* 1 μm)

5 Labeling Density

Structural resolution is not only determined by the optical hardware. As one goes down in scale, the absolute number as well as the density of the fluorophores attached to specific structures or epitopes becomes equally important. Even though a fluorophore, e.g., Rhodamine, occupies only a very small volume of 1–2 nm in diameter, in most cases, more than one fluorophore needs to be attached to a confined site (i.e., an epitope), in order to provide an adequate signal-to-background ratio. In other words, structural resolution is limited not only by microscopic or technical properties but also by

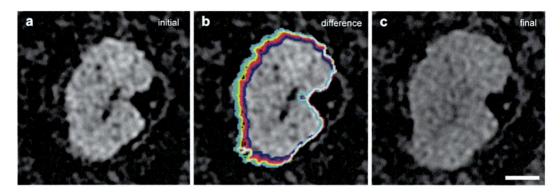

Fig. 5 Repeated 3D-SIM acquisitions can induce structural changes of the observed sample. Magnification of a DAPI-labeled chromocenter in a mouse C127 cell nucleus acquired with 3D-SIM. (**a**) Structure of the chromocenter after the first of total eight acquisitions. (**b**) The outline of this chromocenter has been labeled in different pseudo-colors, each representing one acquisition cycle, showing the progression of light-induced changes. (**c**) The structure of the same chromocenter has changed substantially after eight 3D-SIM acquisition cycles. Note that the DAPI dye still allows the generation of super-resolution images due to its photostability. *Scale bar* represents 0.5 μm

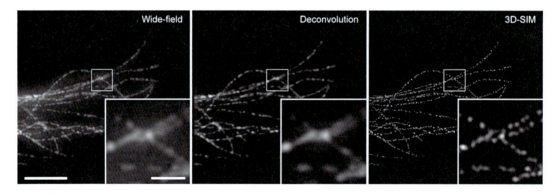

Fig. 6 Relationship between spatial resolution and labeling density. Immunofluorescence staining of a mouse macrophage cell with primary antibodies against α-tubulin and Alexa-488-conjugated secondary antibodies. Comparison of pseudo wide-field, wide-field deconvolution, and 3D-SIM z-projections of the same cell region illustrates the increasingly discontinuous appearance of the microtubule labeling with enhanced spatial resolution. *Scale bar* represents 5 μm (*inset* 1 μm)

the density of the labeling. Discontinuity of the labeling along a given structure, such as microtubules, becomes more and more evident with increasing resolution (Fig. 6). This fact is carried to the extreme in localization microscopy approaches that while featuring 10-nm-range localization precision (often mistakenly stated as resolution) commonly have a several-fold lower effective structural resolution due to their dependency on the labeling density [58–61]. Furthermore the distance of the labeling fluorophore to the actual target structure, particularly given that the size of typical primary/secondary IgG antibody complexes is 15–20 nm, can become of significance for distance measurements.

These observations illustrate a paradigm shift in super-resolution imaging. Previously, the major focus has been onto improving the optical instrumentation in order to further increase resolution. Now, it becomes clear that inevitably these improvements in theoretically achievable resolution have to go along with optimized labeling agents regarding size, affinity/avidity, photophysical properties, dye coupling, and versatility, in order to provide dense and high specific coverage of biological structures in fixed and living cells [28, 52, 62].

6 Channel Alignment

As mentioned above and described elsewhere in detail [25, 37, 38], the classical OMX configuration employs separate cameras for each color channel. This multi-camera configuration bears the advantage of higher acquisition speed granted by the time saved from switching or rotating filter sets when acquiring different channels and additional mechanical stability by the omission of moving parts, such as filter sets or mirrors. With often more than 1,000 raw images for one dataset, the resistance of the microscope against vibrations, stage drift, and other physical disturbances is a prerequisite to receive reliable super-resolution images. This implies a major drawback: The acquisition with several cameras makes an adequate alignment mandatory in order to interpret the resulting data correctly and to draw correct conclusions about the localization of the acquired signals. The signal alignment has to occur in two steps, first mechanically via precise adjustment of the cameras, and second virtually by post-processing (through coordinate transformation and applying a translational and rotational shift parameter) the reconstructed raw data to obtain the best overlap of all channels in all spatial dimensions.

The correct channel alignment is a general issue in multicolor imaging and typically achieved by acquiring multispectral polystyrene beads conjugated with different fluorochromes (e.g., 0.5 µm diameter Tetra-Speck microspheres, Life Technologies) and using a best-fit algorithm to determine the system-specific alignment parameters. However, it should be noted that such reference bead slides do not reflect the exact circumstances within a real biological sample. When imaging the 3D volume of an entire cell, light passing through the specimen gets scattered and distorted by cellular structures due to local variations in refractive index. As the level of heterogeneity is sample dependent, the induced optical aberrations may be more or less severe. Furthermore, despite the correction of the employed objectives, chromatic aberrations in the entire optical system are not fully compensated. The effect is wavelength dependent and becomes more prominent with increasing depth, but has usually been neglected due to the comparably poor axial resolution of conventional microscopy.

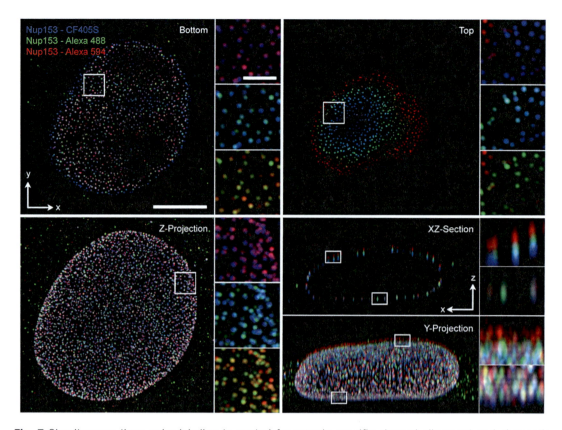

Fig. 7 Simultaneous three-color labeling to control for sample-specific channel alignment and chromatic aberration. *Upper panels* show optical sections from the bottom (apical) and the top (distal) end of a mouse C127 cell nucleus. The sample was immunostained with primary antibodies against Nup153, a component of the nuclear pore complex. These were simultaneously detected with three secondary antibody species, each coupled to different fluorochromes suitable for detection in the three color-cannels. *Lower left panel* shows a maximum z-projection of the same nucleus. Note that even on the apical side no perfect colocalization is observed, which may be caused by a different spatial orientation and/or binding competition of the differentially coupled antibody complexes. *Scale bar* represents 5 μm (*inset* 1 μm)

Using a slide with calibration beads attached on the coverslip surface, however, does not account for the three-dimensionality of this effect. It is therefore recommended to additionally prepare suitable biological calibration samples, which reflect more substantially the conditions within the sample to be analyzed. Figure 7 illustrates such an example of an immunofluorescence detection using a primary antibody against a well-defined cellular structure, in this case the nuclear pore complex. The primary antibody is simultaneously detected by three different secondary antibody species, each coupled to a different dye to result in distinct multicolor-labeled spots distributed within the entire volume of the biological specimen. Notably, after aligning the three channels to fit best at the lower cell surface level, a substantial misalignment can be observed with increasing z-depth. In specimen of significant depth (>5 μm) the alignment parameter should thus be modified to optimally fit on the z-level of the structure of interest or to the specimen center, respectively, to account for this chromatic mismatch and to avoid axial misalignment at the upper cell surface.

7 Colocalization and Optical Resolution

Even with best possible adjustment of the channel-specific optical beam paths and post-processing alignment, the signal overlap of differentially labeled antibodies binding to the same target epitope is by no means perfect when observed with increased level of resolution (Fig. 8). Immunofluorescence staining usually employs a complex of several secondary antibodies that target species specific to the constant region of primary antibody IgGs and carry several fluorochromes. Every one of these secondary antibodies may then randomly occupy a variable orientation in space. With the length of a typical IgG_1 antibody of about 7–10 nm [63, 64], the distance between desired antigen and fluorophore may sum up to about 15–20 nm, which is potentially doubled to 30–40 nm between two differentially colored antibody complexes detecting the same epitope. This clearly reaches a distance range that is visually detectable, as two differentially colored spots do not perfectly merge to a uniformly mixed color spot anymore, which is the common criterion for two signals being assigned as "colocalizing."

This issue becomes relevant when considering multicolor fluorescence applications to support or exclude possible biological interaction of two given molecules in vivo. The common way to quantify such relations is by colocalization analyses [65–67]. These determine the degree of overlap between two pixel intensity values in each color channel. A simple and in many cases sufficient approach is given by the Pearson correlation coefficient (PC), which is obtained by dividing the covariance of the two variables by the product of their standard deviations. PC values range from −1 (perfect anticorrelation) to +1 (perfect correlation), with 0 being not correlated. A drawback of PC is, however, the need for comparable signal levels in both channels to extract reasonable results and its strong dependency on background noise and, thus, thresholding. Alternative approaches, such as the Manders' coefficient, have been introduced to account for these variables as well as for negative effects of out-of-focus blur and detector noise [65–67], but still, substantial ambiguities remain associated with these pixel-based methods.

So far, such analyses were at least to some extent comparable as the optical resolution of the analyzed microscopic images was typically within the same range diffraction-limited. However, the picture changes drastically with the much-improved optical resolution provided by super-resolution techniques. Here any two given

184 Daniel Smeets et al.

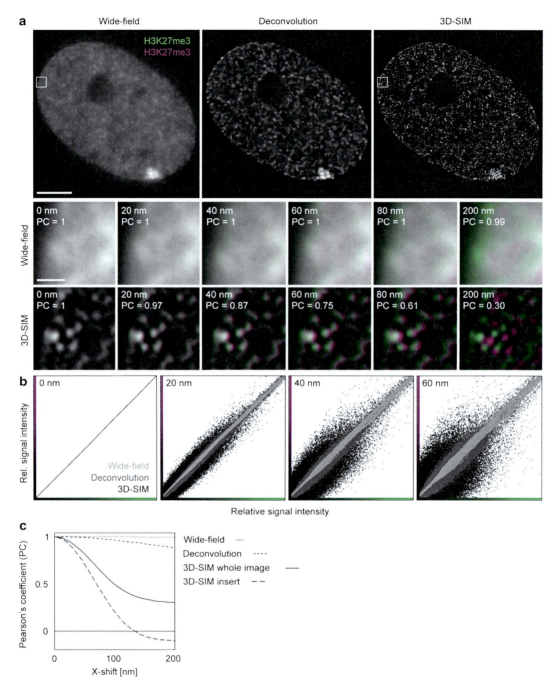

Fig. 8 Quantitative measurements of colocalization depend on the spatial resolution. (**a**) To simulate perfect colocalization as well as defined sub-pixel deviations thereof following procedure was performed: an exemplary image set (mouse cell nucleus immunostained with antibodies against histone H3 tri-methylated at lysine 27; H3K27me3 in pseudo wide-field, deconvolution, and 3D-SIM) was first magnified by bicubic transformation to obtain images with pixel sizes of 5 nm. The gray scale information was copied into two channels and false colored *green* and *magenta* to generate an image of perfectly "colocalizing" white signals. *Lower panel* shows the magnified view of the boxed regions in the *upper panel* with the *magenta* channel being successively shifted in *x*-direction in steps of 20 nm relative to the *green channel*. In each case the Pearson's

macromolecular-sized objects need to be much more closely spaced to generate correlating pixel intensity values. In other words, the lower the resolution, the higher the amount of correlation, irrespective of the real spatial distance of two fluorescing objects (Fig. 7). This exemplifies the fact that common pixel-based approaches to quantify colocalization are becoming increasingly inadequate with improving resolution, in particular when comparing correlation coefficients from image data obtained at different resolution levels. Moreover, the PC strongly depends on the labeling density, which adds another source of inaccuracy when comparing these values. Thus, alternative approaches that are less resolution-dependent, such as nearest neighbor distribution analyses, should be considered to quantitatively determine spatial relationships between two populations of differentially labeled targets [65, 68–70].

This chapter aimed to raise awareness that considerably higher standards have to be met in all experimental steps, from the sample preparation to the quantitative evaluation, in order to generate meaningful biological data from 3D-SIM in particular and super-resolution microscopy in general. As these technologies become more and more accessible to a broader research community we will undoubtedly see an enormous impact on cell biological research with exciting discoveries lying ahead. Among a variety of complementing techniques, 3D-SIM with its specific features will prove its value to address a wide range of fundamental biological questions.

Acknowledgments

The authors wish to thank Heinrich Leonhardt and the BioImaging Network Munich for generous support, Andreas Maiser for technical help, Fabio Spada and Markus Rehberg for providing samples, and Ian Dobbie, Justin Demmerle, and Yolanda Markaki for valuable comments on the manuscript. This work was supported by grants from the Center of Integrated Protein Science Munich and the DFG (TR5 and SCHE1596/2-1). The authors declare no conflict of interests.

Fig. 8 (continued) coefficient (PC) was measured as indicated. *Scale bars* represent 5 and 0.5 μm (*inset*). (**b**) Relative pixel intensity plots of images shown in (**a**) with channels shifted between 0 and 60 nm. Perfect colocalizing signals have the same pixel intensity in both channels and thus lie exactly on the normal of the plot (*left plot*, no shift). Larger distances from the normal indicate lower degree of colocalization. For each shift applied, a broadened distribution of plotted measure points can be noted with increasing image resolution. (**c**) Pearson's coefficient plotted as a function of the applied channel shifts. Note that at the 3D-SIM level of resolution already small distance shifts of below 100 nm lead to a significant decrease in the corresponding Pearson's coefficient, while almost no changes are observed at the resolution level of conventional imaging

References

1. Conchello J-A, Lichtman JW (2005) Optical sectioning microscopy. Nat Methods 2:920–931
2. Lichtman JW, Conchello J-A (2005) Fluorescence microscopy. Nat Methods 2:910–919
3. Fernández-Suárez M, Ting AY (2008) Fluorescent probes for super-resolution imaging in living cells. Nat Rev Mol Cell Biol 9:929–943
4. Cremer M, Grasser F, Lanctôt C, Müller S, Neusser M, Zinner R, Solovei I, Cremer T (2008) Multicolor 3D fluorescence in situ hybridization for imaging interphase chromosomes. Methods Mol Biol 463:205–239
5. Giepmans BNG, Adams SR, Ellisman MH, Tsien RY (2006) The fluorescent toolbox for assessing protein location and function. Science 312:217–224
6. Romer T, Leonhardt H, Rothbauer U (2011) Engineering antibodies and proteins for molecular in vivo imaging. Curr Opin Biotechnol 22:882–887
7. Betzig E, Patterson GH, Sougrat R, Lindwasser OW, Olenych S, Bonifacino JS, Davidson MW, Lippincott-Schwartz J, Hess HF (2006) Imaging intracellular fluorescent proteins at nanometer resolution. Science 313:1642–1645
8. Hess ST, Girirajan TPK, Mason MD (2006) Ultra-high resolution imaging by fluorescence photoactivation localization microscopy. Biophys J 91:4258–4272
9. Rust MJ, Bates M, Zhuang X (2006) Sub-diffraction-limit imaging by stochastic optical reconstruction microscopy (STORM). Nat Methods 3:793–795
10. Klar TA, Jakobs S, Dyba M, Egner A, Hell SW (2000) Fluorescence microscopy with diffraction resolution barrier broken by stimulated emission. Proc Natl Acad Sci U S A 97:8206–8210
11. Hell SW, Wichmann J (1994) Breaking the diffraction resolution limit by stimulated emission: stimulated-emission-depletion fluorescence microscopy. Opt Lett 19:780–782
12. Dyba M, Jakobs S, Hell SW (2003) Immunofluorescence stimulated emission depletion microscopy. Nat Biotechnol 21:1303–1304
13. Gustafsson MG (2000) Surpassing the lateral resolution limit by a factor of two using structured illumination microscopy. J Microsc 198:82–87
14. Gustafsson MGL (2005) Nonlinear structured-illumination microscopy: wide-field fluorescence imaging with theoretically unlimited resolution. Proc Natl Acad Sci U S A 102:13081–13086
15. Gustafsson MGL, Shao L, Carlton PM, Wang CJR, Golubovskaya IN, Cande WZ, Agard DA, Sedat JW (2008) Three-dimensional resolution doubling in wide-field fluorescence microscopy by structured illumination. Biophys J 94:4957–4970
16. Toomre D, Bewersdorf J (2010) A new wave of cellular imaging. Annu Rev Cell Dev Biol 26:285–314
17. Patterson G, Davidson M, Manley S, Lippincott-Schwartz J (2010) Superresolution imaging using single-molecule localization. Annu Rev Phys Chem 61:345–367
18. Hell SW (2007) Far-field optical nanoscopy. Science 316:1153–1158
19. Huang B, Babcock H, Zhuang X (2010) Breaking the diffraction barrier: super-resolution imaging of cells. Cell 143:1047–1058
20. Schermelleh L, Heintzmann R, Leonhardt H (2010) A guide to super-resolution fluorescence microscopy. J Cell Biol 190:165–175
21. Hirvonen LM, Wicker K, Mandula O, Heintzmann R (2009) Structured illumination microscopy of a living cell. Eur Biophys J 38:807–812
22. Shao L, Kner P, Rego EH, Gustafsson MGL (2011) Super-resolution 3D microscopy of live whole cells using structured illumination. Nat Methods 8(12):1044–1046
23. Kner P, Chhun BB, Griffis ER, Winoto L, Gustafsson MGL (2009) Super-resolution video microscopy of live cells by structured illumination. Nat Methods 6:339–342
24. Fiolka R, Shao L, Rego EH, Davidson MW, Gustafsson MGL (2012) Time-lapse two-color 3D imaging of live cells with doubled resolution using structured illumination. Proc Natl Acad Sci 109:5311–5315
25. Schermelleh L, Carlton PM, Haase S, Shao L, Winoto L, Kner P, Burke B, Cardoso MC, Agard DA, Gustafsson MGL, Leonhardt H, Sedat JW (2008) Subdiffraction multicolor imaging of the nuclear periphery with 3D structured illumination microscopy. Science 320:1332–1336
26. Baddeley D, Chagin VO, Schermelleh L, Martin S, Pombo A, Carlton PM, Gahl A, Domaing P, Birk U, Leonhardt H, Cremer C, Cardoso MC (2010) Measurement of replication structures at the nanometer scale using super-resolution light microscopy. Nucleic Acids Res 38:e8
27. Brown ACN, Oddos S, Dobbie IM, Alakoskela J-M, Parton RM, Eissmann P, Neil MAA, Dunsby C, French PMW, Davis I, Davis DM

(2011) Remodelling of cortical actin where lytic granules dock at natural killer cell immune synapses revealed by super-resolution microscopy. PLoS Biol 9:e1001152

28. Guizetti J, Schermelleh L, Mantler J, Maar S, Poser I, Leonhardt H, Muller-Reichert T, Gerlich DW (2011) Cortical constriction during abscission involves helices of ESCRT-III-dependent filaments. Science 331:1616–1620

29. Markaki Y, Gunkel M, Schermelleh L, Beichmanis S, Neumann J, Heidemann M, Leonhardt H, Eick D, Cremer C, Cremer T (2010) Functional nuclear organization of transcription and DNA replication: a topographical marriage between chromatin domains and the interchromatin compartment. Cold Spring Harb Symp Quant Biol 75:475–492

30. Elia N, Sougrat R, Spurlin TA, Hurley JH, Lippincott-Schwartz J (2011) Dynamics of endosomal sorting complex required for transport (ESCRT) machinery during cytokinesis and its role in abscission. Proc Natl Acad Sci 108:4846–4851

31. Green LC, Kalitsis P, Chang TM, Cipetic M, Kim JH, Marshall O, Turnbull L, Whitchurch CB, Vagnarelli P, Samejima K, Earnshaw WC, Choo KHA, Hudson DF (2012) Contrasting roles of condensin I and II in mitotic chromosome formation. J Cell Sci 125(pt 6):1591–1604

32. Mikeladze-Dvali T, von Tobel L, Strnad P, Knott G, Leonhardt H, Schermelleh L, Gönczy P (2012) Analysis of centriole elimination during C. elegans oogenesis. Development 139:1670–1679

33. Weil TT, Parton RM, Herpers B, Soetaert J, Veenendaal T, Xanthakis D, Dobbie IM, Halstead JM, Hayashi R, Rabouille C, Davis I (2010) Drosophila patterning is established by differential association of mRNAs with P bodies. Nat Cell Biol 14:1305–1313

34. Strauss MP, Liew ATF, Turnbull L, Whitchurch CB, Monahan LG, Harry EJ (2012) 3D-SIM super resolution microscopy reveals a bead-like arrangement for FtsZ and the division machinery: implications for triggering cytokinesis. PLoS Biol 10:e1001389

35. Lawo S, Hasegan M, Gupta GD, Pelletier L (2012) Subdiffraction imaging of centrosomes reveals higher-order organizational features of pericentriolar material. Nat Cell Biol 14:1148–1158

36. Sonnen KF, Schermelleh L, Leonhardt H, Nigg EA (2012) 3D-structured illumination microscopy provides novel insight into architecture of human centrosomes. Biol Open 1:965–976

37. Carlton PM, Boulanger J, Kervrann C, Sibarita J-B, Salamero J, Gordon-Messer S, Bressan D, Haber JE, Haase S, Shao L, Winoto L, Matsuda A, Kner P, Uzawa S, Gustafsson M, Kam Z, Agard DA, Sedat JW (2010) Fast live simultaneous multiwavelength four-dimensional optical microscopy. Proc Natl Acad Sci 107:16016–16022

38. Dobbie IM, King E, Parton RM, Carlton PM, Sedat JW, Swedlow JR, Davis I (2011) OMX: a new platform for multimodal, multichannel wide-field imaging. Cold Spring Harb Protoc 2011:899–909

39. Thompson RE, Larson DR, Webb WW (2002) Precise nanometer localization analysis for individual fluorescent probes. Biophys J 82:2775–2783

40. Bennett BT, Bewersdorf J, Knight KL (2009) Immunofluorescence imaging of DNA damage response proteins: optimizing protocols for super-resolution microscopy. Methods 48:63–71

41. Markaki Y, Smeets D, Cremer M, Schermelleh L (2013) Fluorescence in situ hybridization applications for super-resolution 3D-structured illumination microscopy. Methods Mol Biol 950:43–64

42. Jones SA, Shim S-H, He J, Zhuang X (2011) Fast, three-dimensional super-resolution imaging of live cells. Nat Methods 8:499–508

43. Wombacher R, Heidbreder M, van de Linde S, Sheetz MP, Heilemann M, Cornish VW, Sauer M (2010) Live-cell super-resolution imaging with trimethoprim conjugates. Nat Methods 7:717–719

44. Pellett PA, Sun X, Gould TJ, Rothman JE, Xu M-Q, Corrêa IR, Bewersdorf J (2011) Two-color STED microscopy in living cells. Biomed Opt Express 2:2364–2371

45. Manley S, Gillette JM, Patterson GH, Shroff H, Hess HF, Betzig E, Lippincott-Schwartz J (2008) High-density mapping of single-molecule trajectories with photoactivated localization microscopy. Nat Methods 5:155–157

46. Sadoni N, Sullivan KF, Weinzierl P, Stelzer EH, Zink D (2001) Large-scale chromatin fibers of living cells display a discontinuous functional organization. Chromosoma 110:39–51

47. Solovei I, Cavallo A, Schermelleh L, Jaunin F, Scasselati C, Cmarko D, Cremer C, Fakan S, Cremer T (2002) Spatial preservation of nuclear chromatin architecture during three-dimensional fluorescence in situ hybridization (3D-FISH). Exp Cell Res 276:10–23

48. Daneshtalab N, Doré JJE, Smeda JS (2010) Troubleshooting tissue specificity and antibody selection: procedures in immunohistochemical studies. J Pharmacol Toxicol Methods 61:127–135

49. Salic A, Mitchison TJ (2008) A chemical method for fast and sensitive detection of DNA synthesis in vivo. Proc Natl Acad Sci 105:2415–2420

50. Buck SB, Bradford J, Gee KR, Agnew BJ, Clarke ST, Salic A (2008) Detection of S-phase cell cycle progression using 5-ethynyl-2′-deoxyuridine incorporation with click chemistry, an alternative to using 5-bromo-2′-deoxyuridine antibodies. Biotechniques 44:927–929

51. Hinner MJ, Johnsson K (2010) How to obtain labeled proteins and what to do with them. Curr Opin Biotechnol 21:766–776

52. Dempsey GT, Vaughan JC, Chen KH, Bates M, Zhuang X (2011) Evaluation of fluorophores for optimal performance in localization-based super-resolution imaging. Nat Methods 8: 1027–1036

53. Wallace W, Schaefer LH, Swedlow JR (2001) A workingperson's guide to deconvolution in light microscopy. Biotechniques 31:1076–1078, 1080, 1082 passim

54. Parton RM, Davis I (2006) Lifting the fog: image restoration by deconvolution. In: Celis JE (ed) Cell biology: a laboratory handbook, 3rd edn. Academic, New York, pp 187–200

55. Gibson SF, Lanni F (1992) Experimental test of an analytical model of aberration in an oil-immersion objective lens used in three-dimensional light microscopy. J Opt Soc Am A 9:154–166

56. Arigovindan M, Sedat JW, Agard DA (2012) Effect of depth dependent spherical aberrations in 3D structured illumination microscopy. Opt Express 20:6527–6541

57. Cordes T, Maiser A, Steinhauer C, Schermelleh L, Tinnefeld P (2011) Mechanisms and advancement of antifading agents for fluorescence microscopy and single-molecule spectroscopy. Phys Chem Chem Phys 13:6699–6709

58. Cremer C, Kaufmann R, Gunkel M, Pres S, Weiland Y, Müller P, Ruckelshausen T, Lemmer P, Geiger F, Degenhard S, Wege C, Lemmermann NAW, Holtappels R, Strickfaden H, Hausmann M (2011) Superresolution imaging of biological nanostructures by spectral precision distance microscopy. Biotechnol J 6:1037–1051

59. Huang B, Bates M, Zhuang X (2009) Super-resolution fluorescence microscopy. Annu Rev Biochem 78:993–1016

60. Gould TJ, Verkhusha VV, Hess ST (2009) Imaging biological structures with fluorescence photoactivation localization microscopy. Nat Protoc 4:291–308

61. Baddeley D, Cannell MB, Soeller C (2010) Visualization of localization microscopy data. Microsc Microanal 16:64–72

62. Grunwald C, Schulze K, Giannone G, Cognet L, Lounis B, Choquet D, Tampé R (2011) Quantum-yield-optimized fluorophores for site-specific labeling and super-resolution imaging. J Am Chem Soc 133:8090–8093

63. Silverton EW, Navia MA, Davies DR (1977) Three-dimensional structure of an intact human immunoglobulin. Proc Natl Acad Sci U S A 74:5140–5144

64. Dong Y, Shannon C (2000) Heterogeneous immunosensing using antigen and antibody monolayers on gold surfaces with electrochemical and scanning probe detection. Anal Chem 72:2371–2376

65. Bolte S, Cordelières FP (2006) A guided tour into subcellular colocalization analysis in light microscopy. J Microsc 224:213–232

66. Waters JC (2009) Accuracy and precision in quantitative fluorescence microscopy. J Cell Biol 185:1135–1148

67. Zinchuk V, Zinchuk O, Okada T (2007) Quantitative colocalization analysis of multicolor confocal immunofluorescence microscopy images: pushing pixels to explore biological phenomena. Acta Histochem Cytochem 40: 101–111

68. Jaskolski F, Mulle C, Manzoni OJ (2005) An automated method to quantify and visualize colocalized fluorescent signals. J Neurosci Methods 146:42–49

69. Wörz S, Sander P, Pfannmöller M, Rieker RJ, Joos S, Mechtersheimer G, Boukamp P, Lichter P, Rohr K (2010) 3D geometry-based quantification of colocalizations in multichannel 3D microscopy images of human soft tissue tumors. IEEE Trans Med Imaging 29: 1474–1484

70. Lachmanovich E, Shvartsman DE, Malka Y, Botvin C, Henis YI, Weiss AM (2003) Co-localization analysis of complex formation among membrane proteins by computerized fluorescence microscopy: application to immunofluorescence co-patching studies. J Microsc 212:122–131

Chapter 9

Scanning Near-Field Optical Microscopy for Investigations of Bio-Matter

Christiane Höppener

Abstract

Optical near-fields can be employed for a wide range of applications, e.g., light localization, light scattering, and field enhancement. In this chapter the principles of near-field scanning optical microscopy (NSOM) will be outlined. The basic idea of this technique is the extension of the bandwidth of accessible spatial frequencies beyond the limits of conventional light microscopy. This strategy has been implemented in different ways. By now this technique covers a broad spectrum of optical contrasts. Here, special attention is turned on the high-resolution spectroscopic imaging of biological samples.

Key words Scanning near-field optical microscopy (SNOM, NSOM), Fluorescence microscopy, Raman spectroscopy, Membrane proteins

1 Introduction

The idea for breaking the diffraction limit of light reaches back to the first half of the last century. Encouraged by *Albert Einstein*, *Edward Hutchinson Synge* proposed in 1928 a Gedankenexperiment for establishing a microscope with submicroscopic resolution [1]. The strategy of making use of a tiny hole in an opaque metal screen as a submicroscopic light source as well as using small scattering particles as secondary light sources forms the basis of what is known today as aperture and apertureless near-field scanning optical microscopy (NSOM). These techniques aim at the confinement of the photon flux between a probe and an object by guiding the probe in a distance much smaller than the wavelength of the irradiating field across a sample in order to ensure a near-field interaction. In the near-field regime the optical resolution is no longer dependent on the wavelength of the applied irradiation field. Instead the resolution is determined by the size of the probe. By reciprocity, the probe can be employed as either a nanoscopic light source or detector.

Being ahead of time the practical implementation of Synge's idea in a light microscope was impossible at that time due to the lack of technology to fabricate submicroscopic structures and to manipulate these with nanometer accuracy in space. Unfortunately, with progressing time research had lost sight of Synge's pioneering ideas. Until the first realization of an optical near-field microscope slightly modified ideas were proposed by different groups [2–6]. In a first instance the near-field concept was experimentally proven in 1956 with acoustic waves [3] and two decades later for the first time with electromagnetic waves in the nonoptical frequency domain [7]. The invention of what is known today as aperture-type scanning near-field optical microscopy dates back to the year 1982, when *Dieter Pohl* and coworkers at the IBM research laboratories in Rüschlikon developed the first experimental realization of an aperture-probe near-field microscope [8]. At the same time the group of *Aaron Lewis* at Cornell University published a similar approach toward sub-diffraction-limited microscopy [9]. Shortly after this breakthrough and stimulated by the developments in the field of surface enhanced Raman spectroscopy (SERS) *John Wessels* published the idea of apertureless near-field optical microscopy [6]. However, it took almost a decade to accomplish the first experimental realization of an apertureless scanning near-field microscope [10–15].

Today, both approaches have been widely employed for high-resolution spectroscopic imaging [16–21]. Several probe concepts have been introduced in order to further strengthen the applicability of this technique. Whereas aperture probes mainly utilize the effect of light localization associated with the existing near-field, apertureless techniques may in addition take advantage of the field enhancement effect associated with pointed metal structures. Latest trends in this field consider different probe designs in terms of optical antennas [22–24] and merge the established concepts of light localization and field enhancement into new probe designs with improved efficiency.

Already *E.H. Synge* proposed the near-field approach in the context of biological imaging. Since the first demonstration of sub-diffraction limited optical resolution and single molecule sensitivity [25], the potential ability of these near-field techniques for investigations of cellular components and proteins has been widely discussed [16, 26–32]. By now, many of the technical challenging hurdles have been mastered to employ this technique for biologically relevant studies. In particular within the last decade various examples have been published exploiting the exceptional capabilities of near-field optical techniques for biological studies [33–41].

In this chapter, the fundamental principles and major concepts in the field of near-field optical (NFO) microscopy in connection with its ability to be employed for biomolecular and cellular imaging on the nanometer scale are discussed and a few of the recent applications of this technique for studying biological sample systems are exemplified.

2 Theoretical Background

The restriction of the optical resolution in conventional microscopy to roughly half of the wavelength of the light, as stated by the *Abbe* or *Rayleigh* criterion [42, 43], has its origin in the loss of non-propagating field components of a source field upon propagation in space. A mathematical tool to analyze the propagation and focussing of optical fields is the angular spectrum representation. Hence, it can be used to obtain a fundamental understanding of the imposed resolution limit in optical microscopy. The angular spectrum representation enables the decomposition of electromagnetic fields in homogenous propagating waves and inhomogeneous nonpropagating (evanescent) waves. The evanescent field components are strictly associated with inhomogeneities in space, e.g., plane interfaces or submicroscopic nanoobjects. In the angular spectrum representation a field $\mathbf{E}(x,y,z)$ in an arbitrary collector's plane z can be evolved by a superposition of plane waves of the form $\exp(\mathbf{kr})\exp(-i\omega t)$ as

$$\mathbf{E}(x,y,z) = \iint_{-\infty}^{+\infty} \hat{\mathbf{E}}(k_x,k_y,0)e^{\pm ik_z z}e^{i(k_x x + k_y y)} \tag{1}$$

Here, $\hat{\mathbf{E}}(k_x,k_y,0)$ denotes the spectral amplitude of the harmonic waves, which arise from the two-dimensional Fourier transformation of the source field $\mathbf{E}(x,y,0)$. The propagation direction is given by the wavevector $\mathbf{k}$, which is defined by $k = \sqrt{k_x^2 + k_y^2 + k_z^2} = n\omega/c = 2\pi n/\lambda$, with n being the index of refraction. According to (1) the propagation of the source field is characterized by an optical transfer function, which is given by the term $\exp(\pm ik_z z)$. This function transforms the corresponding waves either in propagating waves for real k_z, i.e., $k_x^2 + k_y^2 \leq k^2$, or in evanescent waves in case that k_z becomes imaginary, i.e., $k_x^2 + k_y^2 > k^2$. Hence, the field can be separated in far-field components associated with homogeneous solutions, i.e., propagating plane waves, and near-field components. The corresponding evanescent modes are characterized by an exponential decay. Thus, evanescent waves are strictly bound to their source and cannot radiate to the far-field zone ($z \gg \lambda/2\pi$). The decomposition of a source field in its near- and far-field contributions provides a descriptive way to analyze the influence of the near-field interactions on the image formation in a light microscope.

Figure 1a shows schematically the reciprocal relation between the optical resolution Δx, defined as the full width at the half maximum (FWHM) of the point spread function, and the bandwidth of the corresponding wavevector components. The propagation of an electromagnetic field radiating from a submicroscopic light source in homogeneous space involves a narrowing of the bandwidth of

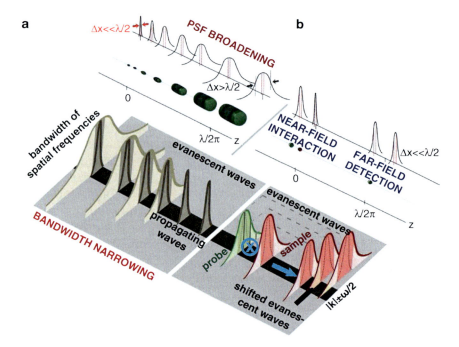

Fig. 1 (**a**) Schematic representation of the propagation of the electromagnetic field of a submicroscopic, radiating light source located at $z=0$ in a homogeneous space and the characteristic alteration of the one-dimensional point spread function and angular spectrum upon propagation in z-direction. (**b**) Schematic representation of the conversion and shifting of evanescent field components into propagating waves by scattering at a nanoscopic object

the corresponding spatial frequencies. Whilst the source field is characterized by a broad spectrum of spatial frequencies, i.e., the spectrum is dominated by wavevector components $|k_z| > \pm\omega/c$, at a distance $z \geq \lambda/2\pi$ these are restricted to $|k_z| \leq \pm\omega/c$ since the propagation of the corresponding waves is subject to dispersion. Thus, the image formation is connected to a loss of information, which is encountered in standard light microscopy by the progressive broadening of the point spread function to $\sim \lambda/2$. Information on a submicroscopic length scale of a sample are encoded in the scattered evanescent waves, i.e., in the high spatial frequencies. The higher these spatial frequencies are, the faster the corresponding evanescent waves decay. Therefore, their inclusion in the image formation becomes more difficult.

The structure of an arbitrary small object can be only exactly represented in an image plane, if the full bandwidth of spatial frequencies can be recovered. This can be achieved by conversion of the evanescent field components in propagating plane waves. The near-field interaction of a source with a sample and the transformation of evanescent waves in propagating modes can also be understood by analyzing the electromagnetic field in the source, sample, and detector's plane in the angular spectrum representation

[44, 45]. Disregarding effects related to the strong coupling of source and sample in the near-field regime, the field propagated to the detector can be analyzed similarly to the field propagation in a homogeneous medium. The field arriving after interaction with the sample at the detector is defined by the spatial spectrum of the sample field, which derives from the convolution of the irradiating field and a function describing the optical properties of the sample. In this simplified picture, the high spatial frequencies are shifted to the far-field detection window $|k_z| \leq \pm \omega/c$ (c.f. Fig. 1b). Thus, nanoscale information of the sample are preserved in the waves propagating to the far-field zone.

3 Experimental Realization

The central concept of near-field optical techniques is the extension of the bandwidth of spatial frequencies, and thus, near-field scanning optical microscopy (NSOM) is mandatorily connected to the implementation of a nanoscopic probe. This probe is characterized by a large fraction of high spatial frequencies and can act as either a nanoscopic light source or photon collector. Interaction of the corresponding evanescent modes requires this probe to be positioned in the near-field zone of the sample at a distance of $\ll \lambda/2\pi$. Whilst the development of various probe designs is related to different types of near-field microscopy, the controlled three-dimensional positioning of the nanoscopic probe follows a unique principle. The probe-sample distance is regulated by means of an electronic feedback loop, which utilizes a certain distance-dependent short-range interaction between the probe and the sample. Figure 2 depicts schematically a typical NSOM setup. Commonly, the basis of a near-field microscope is formed by an inverted microscope comprising a high NA objective. The sample is mounted on an xy(z) piezoelectric scanner, which enables the manipulation of the sample independently from the external illumination and the detection system. The illumination of the probe-sample region can be accomplished externally by tightly focussing the laser light on the probe, or by launching the light directly into a probe. For the detection of the induced spectroscopic responses sensitive photon detectors, such as Avalanche photodiodes, CCD cameras or spectrometers are used. The key element of any near-field optical microscope is the scan head, which hosts the near-field probe and often implements piezoelectric actuators for its lateral as well as axial manipulation relative to the optical axis of the inverted microscope. Instead of using the xy(z)-scanner for raster-scanning the sample underneath the probe, the scan head actuators can be utilized to raster-scan the probe across the sample surface. In addition, the scan head can host the preamplifiers for the detection of the feedback signals and optical components for the illumination

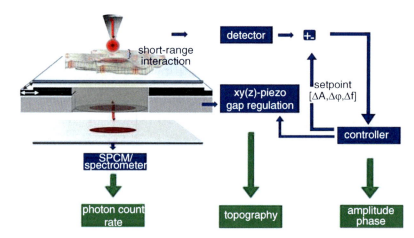

Fig. 2 Schematic representation of a near-field scanning optical microscope (NSOM): An optical probe is kept in close distance to the sample surface by means of an external distance control unit utilizing a short-range interaction, e.g., the interaction force. The probe-sample gap is adjusted by a piezoelectric actuator which is controlled by an electronic feedback loop. For a force-distance controlled system information on the sample are gained by the simultaneously recorded optical signature and topography. In addition information can be extracted from the amplitude, phase of the recorded signals

of the probe or the collection of the optical signals. Modification of this generalized setup can be implemented for different NSOM methods. The principles and the properties of different probe concepts will be addressed in the Sects. 4 and 5. In the following I will discuss the implementation of distance control schemes for measurements of biological matter in liquid environments.

3.1 Distance Control

One of the major problems, which hindered an earlier implementation of the near-field concept in the field of optical microscopy, and in particular challenging measurements in liquid environments, was related to the development of a sufficient probe-sample distance control. The theoretical considerations outlined in Sect. 2 demonstrate that the obtainable resolution of a near-field optical microscope is directly correlated with the capability to collect the nonpropagating wave components of a nanoscale object, and thus critically depends on the ability to precisely probe and control the tip-sample distance. As pointed out in Fig. 1a the distance needs to be clearly smaller than the FWHM of the SPF of a confocal light microscope, and thus, falls on a scale of several nanometer to tens of nanometer. Stimulated by the progress in the development of other scanning probe techniques an NFO-microscope adopts the same general concepts for the control and adjustment of the probe-sample gap [46, 47]. An electronic feedback loop enables to reliably maintain the probe-sample distance on the nanometer scale with sub-nanometer accuracy by the detection of a monotonically altering interaction between probe and sample, e.g., the interaction force or alternatively the tunneling current between probe and sample (see Fig. 2). The measured electrical signal is transformed into a corresponding z-displacement by means of a piezoelectric actuator. The combination of near-field optical microscopy with other scanning probe techniques to regulate the probe-sample separation enables to probe the optical properties of a sample and simultaneously correlate these with further sample properties, e.g., the topology, the sample elasticity, surface friction, conductivity, etc. Nevertheless, the external control of the probe-sample separation bears the risk of inducing topography artifacts in the NSOM images [48], and thus, NFO imaging requires careful data analysis.

Most commonly, the probe-sample distance in NSOM is controlled by piezoelectric sensing of the normal or lateral interaction forces between the probe and the sample [49–54]. Piezoelectric quartz tuning forks, which can provide a high quality factor are often employed as sensitive force sensors [51]. In common, the tuning fork is mechanically excited at its resonance frequency by an external dither piezo. Alterations in the amplitude, phase, or frequency from the free oscillating system upon probe–sample interaction can be utilized as a sensitive distance-dependent signal. A tuning fork can be treated as a harmonic oscillator. Within this simplified model the measurable dampening forces can be approximated to $\lesssim 50$ pN [51, 54].

The distance control units based on a piezoelectric quartz tuning fork can be mainly divided into two methods. In shear-force microscopy the orientation of the tuning fork is perpendicular to the sample, and the probe is attached along one of the prongs (c.f. Fig. 3a). Thus, the probe is excited to a lateral oscillation at the resonance of the sensor, and thus, it is sensitive to the shear- or friction forces. Alternatively, the tuning fork can also be rotated by 90°, such that its prongs are aligned parallel to the sample plane. As schematically displayed in Fig. 3b a short probe can be attached to the lower prong of the tuning fork, and oscillates perpendicularly to the sample surface, similarly to the well-known tapping-mode in conventional atomic force microscopy. The major difference to standard tapping-mode schemes in conventional atomic force microscopy is the significantly smaller oscillation amplitude. In contrast to shear-force feedback circuits the tuning fork's sensitivity is critically affected by the attached mass of the probe, which restricts the overall length of the probe to several millimeters. Piezoelectrical detection methods are the most commonly employed techniques for the application of heavy aperture or field enhancement probes. Although several cantilever-based probe concepts have been introduced over the last decades [55–61], piezoelectric tuning fork-based detection of shear-forces is most widely spread in this field to control the probe-sample distance. Other techniques such as scattering-type NSOM mostly rely on cantilever-based probes such as standard AFM probes [62, 63].

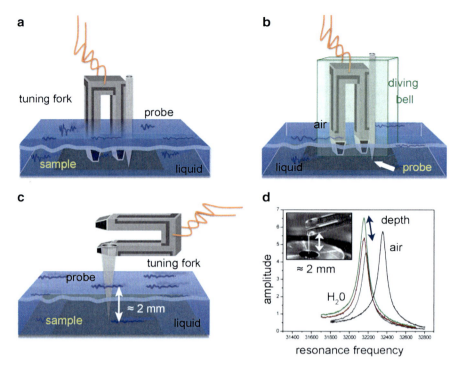

Fig. 3 Schematic comparison of different force-distance control methods applied in NSOM for measurements in liquid environments. (**a**) Common shear-force distance control based on a piezoelectric quartz tuning fork: Measurements in liquid environments require an immersion of the tuning fork. (**b**) Diving bell concept combined with a shear-forced feedback: A sealed chamber keeps the piezoelectric tuning fork in air upon immersion in liquid. (**c**) Tapping-mode like force distance control: An excitation of an oscillation along the tip axis enables the probe to be mounted such that a millimeter-long part protrudes beyond the lower prong of the tuning fork. Only the tip is immersed in the liquid. (**d**) Typical resonance curve for a tuning fork with attached near-field probe acquired in air and upon immersion of the probe in water for different immersion depths. Inset: Photograph of the probe-tuning fork region in a tapping-mode-based NSOM system

The shear-force and the tapping-mode distance control schemes are widely employed in the field of NSOM. Both techniques turn out to be comparable with respect to their stability and sensitivity under ambient conditions [34, 54, 64]. Major differences occur only for measurements in liquid environments. In order to prevent high losses in the force sensitivity the tuning fork sensor has to be kept entirely out of the liquid environment [34, 54, 65], since a contact of the liquid with the tuning fork prongs causes a strong dampening of the oscillation. The differences for shear-force and tapping-mode measurements in liquid environments find their origin in the limitation of the protruding length of an attached optical probe. In shear-force controlled systems the protruding length of the probe beyond the foremost part of the prong is restricted to several 100 μm in order to avoid a direct excitation of the probe's eigenmodes. This leads inevitably to a contact between the liquid and the tuning fork, if no additional precautions are taken into account. Originally this problem has been approached by working in externally controlled μm-thin liquid layers [66] or by a partial immersion of a protected tuning fork [65]. A successful implementation of shear-force-based NSOM measurements of cellular membranes in liquid environments was not achieved until *Koopman* et al. introduced a diving bell concept in order to prevent a liquid-tuning fork contact. This approach utilizes the effect of liquid exclusion in a sealed chamber upon immersion in a liquid reservoir. Thus, the tuning fork is kept in air, and only a short part of the attached probe, which leaves the entirely immersed chamber through a tiny hole at its bottom, is exposed to the liquid.

A simple way to keep the tuning fork out of the liquid is provided by the tapping-mode arrangement. Due to the length of the protruding probe an operation in liquid environments with high sensitivity is possible even for reservoirs of several millimeter height without an independent protection of the sensor. Therefore, the dampening of the oscillation arises only from the liquid layer acting on the walls of the probe, i.e., the dampening strength is correlated with the total probe's surface area immersed in the liquid. Hence, the provided force sensitivity is critically affected by the tip shape. For imaging of soft biological matter under physiological conditions the detection of normal forces seems to be preferable since lateral forces can likely damage or destroy soft matter.

4 Aperture-Type Near-Field Microscopy

Aperture probes have been the initial and so far the most widely used concept in near-field fluorescence microscopy. In this section I will introduce and compare different types of aperture probes with respect to their light confinement and light transmission. These properties will be discussed in the direction of biological applications. Finally, their capability to study biological matter is pointed out by several examples.

4.1 Circular Apertures

Tiny holes in an opaque metal film have attracted a lot of interest in recent years due to their unique nano-optical effects, e.g., extraordinary light transmission and light confinement [67–71]. The diffraction of electromagnetic waves at a tiny hole has been addressed theoretically by *H.A. Bethe* and *C.J. Bouwkamp* [72, 73]. They derived the electromagnetic field for a circular hole in an infinitely thin, perfectly conducting metal film with a diameter much smaller than the wavelength of the irradiation field. The electromagnetic field propagated to the far-field zone is well represented by a superposition of the fields of a perpendicularly oriented electric and an in-plane magnetic dipole located in the center of the aperture. Experimental data obtained by mapping the vectorial field distribution of aperture probes with single molecules show a qualitatively good agreement with the theoretical predictions in the *Bethe–Bouwkamp* model [25, 74].

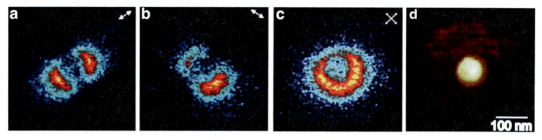

Fig. 4 Scalar field distribution of a circular aperture probe addressed by imaging of a 20 nm polystyrene bead doped with fluorescent molecules with an aperture probe. (**a**) Intensity distribution of the fluorescence signal imaged with a circular aperture probe with a diameter of ~90 nm illuminated by a linearly polarized plane wave. The polarization direction is indicated by the arrow in the upper right corner of the image. (**b**) Intensity distribution of a polarization direction perpendicular to the one in (**a**). (**c**) Merged image of (**a**) and (**b**) showing the field enhancement located at the metallic rim of the probe. (**d**) Topography image of the aperture probe as imaged by a 20 nm bead

Figure 4 reflects the characteristic scalar field distribution of a circular aperture obtained by mapping the lateral intensity distribution of a fluorescent polystyrene bead scanned through the near-field of an illuminated aluminum-coated aperture fiber probe. Here, the aperture probe was fabricated by successive squeezing of an entirely metal-coated tip against a smooth substrate [75]. Therefore, the glass core protrudes from the flat metallic plateau at the foremost end of the probe (c.f. topography image of Fig. 4d). Polystyrene beads doped with fluorescent molecules are well suited for mapping the scalar field distribution of an aperture probe because of their large brightness and their isotropic optical properties. Thus, the in-plane and out-of-aperture plane field components are recorded simultaneously. For linearly polarized plane wave excitation (indicated by the arrows in the corresponding images of Fig. 4b, c) a double-lobe pattern is observed, which is aligned along the polarization direction. The field distribution at a circular aperture is characterized by strong field components with an orientation perpendicular to the aperture plane located at the glass metal boundary, whilst the in-plane field components are strongest in the center of the aperture. The merged image of perpendicular polarization directions demonstrates an enhanced field located at the metallic rim of the aperture (see Fig. 4d). The strong field located at the metal-dielectric boundary is connected to a geometrically imposed field line crowding in this region. This observation gave first evidence for a field enhancement effect associated with pointed metal structures, which forms the basis for the apertureless near-field technique discussed in Sect. 5. Note that the intensity distribution in the center of the probe does not drop to zero, but represents the in-plane field components, which are naturally weaker than the enhanced field at the metallic rim. The lower intensity in the center of the aperture is also conditioned by an increased aperture-bead distance caused by the protruding glass tip. In agreement with the theoretically calculated field distribution of an aperture probe the electric field partly penetrates into the metal coating, which causes a slight broadening of the effective aperture size [19]. In the context of utilizing the near-field interaction for breaking the diffraction limit of light, the collimation of the aperture field at a certain distance from the aperture plane is essential. The collimation of the aperture field remains constant for a distance, which corresponds to the size of the aperture [76]. In this near-field region an image of the aperture would be almost exact, and the confinement depends only on the aperture diameter. Beyond this regime the confined field rapidly diverges. In order to separate the in-plane and axial field components single molecules with a fixed orientation of the transmission dipole moment are used [74, 77].

In addition to their ability to confine fields the applicability of aperture probes for near-field optical methods is determined by their overall optical transmission, e.g., the achievable brightness or the detection efficiencies of the nanoscopic probe. The transmission through a subwavelength hole in a metallic screen scales for an ideal circular aperture with the 6th power of its radius [72]. This is in contradiction to the effort to minimize the aperture size in order to achieve a strong light confinement, and hence, to obtain super-resolution. The light transmission in aperture probes is also critically affected by their conical angle. Essential is the propagation length of the existing waveguide modes in the metal-coated structure. In a metal-coated tapered dielectric waveguide all existing modes will be even stronger than exponentially damped if the core diameter decreases below a certain length [78]. For example, in an aluminum-coated tapered fiber even the fundamental HE_{10}-mode will propagate only until the diameter of the probe is reduced to 150 nm. Beyond this cutoff, the modes become evanescent. Partly the launched radiation is reflected and the energy is dissipated in the metal coating. This leads to a local heating of the metal coating, which imposes an upper limit for the damage threshold or the total power of the light coupled into the probe without damaging the metal coating.

Although several cantilever-based aperture probes have been introduced to the NSOM field [55–59], primarily, aperture probes formed from optical fibers are used, nowadays. The fabrication process can be separated in the taper formation in the dielectric core, the subsequent metallization and the aperture formation [79, 80]. Pointed fiber probes can be fabricated by thermally assisted pulling [79] or chemical etching processes [66, 81–84]. Also etching of the glass core improves the tip geometry, the major drawback of standard glass fiber aperture probes remains. Their conical angle is restricted to 20–40° [85]. Thus, at best the light transmission of these probes yields 10^{-7} for aperture diameters of ~100 nm [17, 86]. Attempts to further increase the light transmission for fiber probes aim at the development of selective etching methods to induce double tapered structures, which further minimize the distance between the cutoff diameter and the aperture plane [87–90]. In addition, the conditions for the metal coating have to be chosen carefully and have been explored in detail [79, 91]. Small imperfections as well as an increased surface roughness of the metal film can affect the electromagnetic field distribution significantly. The formation of the aperture can be obtained by rotational shadow evaporation [17], by subsequent tip pounding procedures [8, 34, 92] as well as subsequent focussed ion beam milling of entirely metal-coated pointed fibers [93, 94]. The latter procedure has been proven to result in aperture probes with well-defined optical properties due to the removal of protruding metal grains, e.g., a highly symmetric field distribution, well-defined polarization behavior, and improved light transmission.

The demand for aperture probes which combine a high optical resolution with a good light transmission has led to the implementation of new aperture geometries. Noncircular aperture probes can exhibit better light confinement properties. This concept goes well beyond simply increasing the conical angle in order to increase the light transmission. On the basis of triangular-shaped aperture probes this strategy will be outlined in the following section [95–97].

4.2 Triangular Apertures

The high symmetry of circular apertures results in a distribution of the strongest field components along the metallic ring. Thus, the light confinement is characterized by a double-lobe pattern, with a FWHM of each lobe being significantly smaller than the aperture diameter. Unambiguously, a confinement to a single spot cannot be achieved with this type of probes. Consequently, the optical resolution is determined by the aperture diameter and not by the FWHM of the well-separated lobes. A stronger light confinement can be only achieved if the geometry of the aperture is altered.

Tetrahedral glass fragments have been considered as bodies for both apertureless and aperture near-field optical probes [95, 98, 99]. The tetrahedrally shaped glass fragments are formed by cleavage of a coverslip such that the resultant edges enclose a large taper angle of ~90° [99]. The cleaving process can lead to very small radii of curvature for the foremost end of this probes, and thus, in principle to aperture sizes as small as tens of nanometers. As in the case of circular aperture fiber probes the glass body is entirely metal coated and the aperture can be subsequently opened by a squeezing process or by focussed ion beam milling. As shown by the SEM of Fig. 5a this leads to the formation of an aperture with a triangular shape [95, 96]. The large taper angle reduces the cutoff length for the supported modes in this structure significantly. Thus, triangular aperture probes provide an optical transmission, which is at

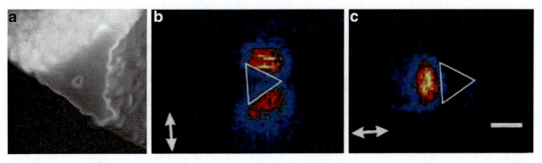

Fig. 5 Light confinement in triangular aperture probes. (**a**) Scanning electron microscopy image of a triangular aperture probe fabricated by focussed ion beam milling. Reprinted from [96] with permission of the Optical Society of America, copyright 2005. (**b** and **c**) Scalar intensity map of a fluorescently doped polystyrene sphere illuminated by a triangular aperture probe for perpendicular polarization directions of the irradiating field, i.e., parallel (**b**) and perpendicular (**c**) to the left edge of the triangle. A perpendicular polarized plane wave illumination results in a high degree of confinement of the incident field to a single spot close to the corresponding side of the triangular aperture. Reprinted from [95] with permission of the American Physical Society, copyright 2002

least three orders of magnitude larger than for circular aperture probes providing a comparable resolution [95], and also holds for a significantly higher damage threshold.

Owing to the equilateral triangular shape of the aperture these near-field optical probes exhibit advanced polarization-dependent light confinement properties [95]. Initial studies addressed this ability by mapping the scalar intensity distribution of fluorescent nanobeads illuminated by means of a triangular aperture probe. Whilst a suppression of the double-lobe pattern for highly symmetric circular apertures is impossible, a triangular-shaped aperture illuminated by a plane wave polarized perpendicularly to one of the triangle sides leads to a confinement of the light field in a single spot located at this edge of the triangular aperture [95, 96] (c.f. Fig. 5b). Similar to the light confinement in circular apertures, the field is strongest close to the metal-dielectric transition and scales laterally with roughly half the side length of the triangle. For a polarization of the excitation light parallel to one of the equilateral sides of the triangle the field is split into two spots located close to the opposite triangle corner (c.f. Fig. 5c). Mapping of the vectorial field components at a triangular aperture as well as numerical simulations of the electric field distribution demonstrate that the field component along the direction of polarization is always strongest, and the corresponding perpendicular in-plane field component is negligible. The field component perpendicular to the aperture plane is significantly smaller than the strongest field component [96].

4.3 Biomolecular Studies

The optical as well as the geometrical properties of aperture probes have evidently strong influence on their capability to exploit bio-matter. Geometrical effects concern not only their light confinement and transmission but also the sensitivity of the force distance control. Whilst standard fiber probes can be employed in liquid environments with high sensitivity and reliability, probes formed by a tetrahedral glass body experience several disadvantages. The geometry of the glass fragment requires a tilting of the tuning fork of ~45°. Consequently, lateral as well as normal force components determine the distance control. Moreover, the large mass of the tetrahedron causes a significant asymmetry for the two tuning fork prongs and thus, results in strongly decreased Q-factors. Although this effect can be in principle counterbalanced by attaching an additional mass to the other prong, the tetrahedral shape of the probe leads to a significantly larger interaction force of the liquid molecules acting on the walls of the tetrahedron. Therefore, it is more difficult to implement an operation in liquid environments with this type of probe, and requires special working schemes to enable nondestructive imaging of bio-matter. Nonetheless, the higher brightness of these probes and thus, the significantly larger signal-to-noise ratio opens up new prospects for biomolecular studies on the nanometer scale. Although the distance control of simple circular fiber probes is less critical, their low light throughput limits their applicability in terms of the required size and brightness of the structures under investigations.

The application of aperture probes marks the early stage of high-resolution near-field biomolecular investigations. Initially only employed under ambient conditions [33, 100–102], within the last decade the transition to study biological system under physiological conditions has been achieved [34–36]. Initially focussing on resolving individual membrane structures, such as macromolecular complexes and nanodomains of membrane proteins, today investigations in this field address the distribution of individual membrane proteins, such as membrane receptors, ion channels etc., and their arrangement in nanocluster and nanodomains, as well as the formation of microscopic domains [37, 38, 103–106]. The interaction of specific membrane building blocks, e.g., observed in diffusion or transport phenomena, has been also addressed by means of aperture-based nano-optical imaging combined with fluorescence correlation analysis [40, 107–109].

Similar to the field of confocal microscopy, biological issues are approached by the aperture near-field technique most frequently in terms of immunofluorescence studies. Although NFO-imaging is not restricted to the fluorescence contrast [20, 21, 110, 111], investigations of bio-matter with aperture-type probes take prevailingly advantage of the specific staining protocols, which are well established in connection with standard fluorescence microscopy. Recently, *Chen et al.* used nanoscale immunofluorescence imaging combined with quantum dot labeling of antigen-specific T-cell receptors (TCRs) to address changes in the distribution of this receptor in non-activated and stimulated V2γ2Vδ2 T cells [112].

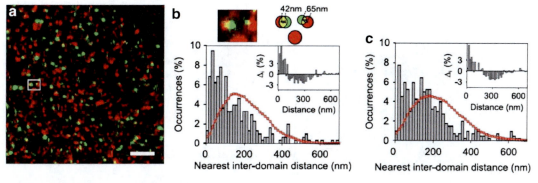

Fig. 6 NSOM investigations of the organization of the raftophilic protein CD55 with respect to raft ganglioside GM1 crosslinked to its ligand cholera-toxin CTxB-GM1. The latter form small nanodomains in the cell membrane of monocytes. (**a**) Nanoscale colocalization of CD55 (*green*) and CTxB-GM1 (*red*) nanodomains by means of dual color excitation/detection NSOM. Scale bar: 1 μm. (**b** and **c**) Nearest interdomain distance of CD55 and LFA-1 to the closest CTxB-GM1 nanodomain. *Red curve* shows simulations of random spatial distribution of proteins and CTxB-GM1 nanodomains. Reprinted form [41] with permission of PNAS, copyright 2010

These investigations reveal clear differences in the TCR aggregation for $\alpha\beta$- and $\gamma\delta$-TCRs and demonstrate that in-vivo clonal expansion of V2γ2Vδ2 T cells is associated with a redistribution of TCR receptors, i.e., the formation of nanoclusters, nanodomains, and microdomains. Quantum dot-based nanoscale immunofluorescence studies take advantage of the higher brightness and photostability of semiconductor quantum dots compared to organic fluorophores. This enabled imaging of TCRs with an optical resolution of ~50 nm.

Several groups have demonstrated similar studies on other biological systems and issues. For example, *Garcia-Parajo* and coworkers have addressed questions related to the organization of cellular membranes in small nanodomains such as lipid rafts [35, 38, 41]. Single-molecule aperture near-field microscopy has been employed in connection with the diving bell concept for regulation of the aperture-sample distance to map the nanoscale organization of raft ganglioside GM1 after its tightening by the cholera toxin ligand. The data support the hypothesis of the existence of raft-based interconnectivity mediated by cholesterol at the resting state in the presence of an intact cytoskeleton. Figure 6 shows colocalization data of CTxB-GM1 nanodomains and the raftophilic proteins CD55 and LFA-1 acquired by dual color excitation/detection near-field microscopy providing a localization accuracy of ~2–5 nm. These investigations demonstrate that CTxB-GM1 nanodomains do not colocalize with these proteins. Interdomain-nearest neighbor distance (i-NND) analysis revealed that CD55 and LFA-1 not only show clearly shorter i-NNDs than expected for a totally random distribution (c.f. Fig. 6b, c) but also indicate a preferential spatial proximity to the CTxB-GM1 nanodomains. In contrast the i-NND-distribution of the non-raft marker protein CD71 reflects a random organization with respect to the CTxB-GM1 nanodomains in these studies.

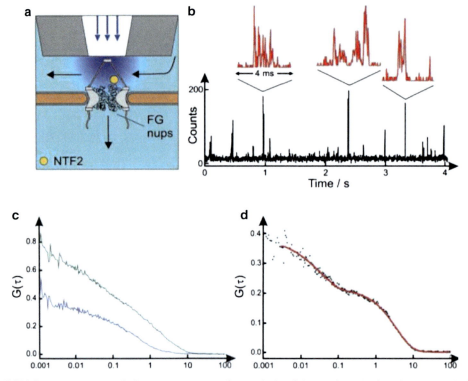

Fig. 7 NSOM fluorescence correlations spectroscopy for analysis of the nucleocytoplasmic transport of NTF-2 through a single nuclear pore complex (NPC). (**a**) Schematic representation of the experiment, which employs a triangular-shaped aperture probe. Fluorescence time traces of labeled NTF-2 diffusion are recorded by centering the near-field probe above a single NPC in an unsupported membrane patch of a Xenopus laevis nuclear envelope. (**b**) Typical fluorescence time trajectory of the NTF-2 transport through a NPC, which reveals strong fluorescence burst of varying fluorescence intensity. (**c**) Autocorrelation function (ACF) of a time trace of labeled NTF-2 measured above an NPC (*green*) and ACF of a time trace of the background signal (*blue*) recorded next to an NPC. (**d**) Background-corrected ACF of the NTF-2 diffusion and least square fit (*red*). Reprinted from [40] with permission of the American Chemical Society, copyright 2009

The high 3D-light confinement associated with metal nanostructures has also been exploited for high sensitivity fluorescence correlation spectroscopy (FCS) [40, 107–109]. FCS is a powerful tool to study diffusion and translocation processes [113]. *Herrmann et al.* studied the nucleocytoplasmic transport of fluorescence-labeled NTF2 by means of single molecule tracking with a triangular aperture probe (see Fig. 7). Initial studies on nuclear pore complexes (NPCs) aimed at their identification using circular aperture probes and immunofluorescence labeling against specific NPC proteins (Nups) [36]. In order to understand the translocation of a substrate between the cytoplasm and the nucleus in cells, the interaction of the substrate with the Nups has to be analyzed. To track the transport of a fluorescent substrate through the NPC and to identify intermolecular interactions, FCS studies on unsupported membranes were carried out. For this purpose, a triangular aperture probe was precisely positioned on top of a single nuclear pore complex and the count rate of the fluorescence signal of a single NTF2 molecule is recorded as a function of time (c.f. Fig. 7a). The recorded fluorescence time trajectories reveal fluorescence bursts on the order of 1–5 ms. These signals arise from binding of the substrate to the nuclear pore complex. On a μs-time scale these bursts show clearly a substructure (c.f. Fig. 7b). These fluctuations in the intensity of the signal have been interpreted in terms of a relocation of the bound substrate within the transport channel. Hence, the fluorescent substrate molecule frequently enters and exits the region corresponding to the strongest evanescent field, indicating a high mobility of the translocated substrate in the direction of the channel axis. The fluctuating signal has been further analyzed by means of the autocorrelation function (c.f. Fig. 7c, d). From this the diffusion constant at the channel entrance was determined to 17 μm/s. The residence time of NTF2 is on the order of 3.6 ms and corresponds to the known length of the transport channel of NPCs of ~85 nm. The autocorrelation function obtained from the experimental data support the assumption of a facilitated diffusion of NTF with the Brownian motion being the driving force, and the hypothesis of the interaction of translocated substrates with phenylalanine-glycine-(FG)-rich nucleopore proteins in the transport channel.

In general, it is well accepted today that apertures can at best reach optical resolutions of ~50 nm, due to the finite skin depth of ~10 nm for metals at optical frequencies and their limited light transmission (see Sect. 4.1). Although other aperture geometries may slightly suppress this limit, the finite skin depth sets a lower limit. Hence, different near-field optical probe concepts are mandatory. In the following section I will discuss apertureless approaches, which have been already shown to be able to overcome the limit of aperture probes and can reach the regime of true protein resolution of ~10 nm [60, 98, 99, 114–119].

5 Apertureless Scanning Near-Field Microscopy

Optical near-field interaction with matter has been established in several ways. As pointed out before, aperture probes can be used as direct nanoscopic illumination sources. Alternatively, methods which rely on apertureless probes, e.g., solid metal tips or finite metal particles, take advantage of light scattering at the probe [19]. In the latter approach one has to distinguish between different scattering regimes. Dependent on the material properties of the probe and the sample different scattering interactions can dominate the optical contrast formation. Externally illuminated strong scattering metal probes can act as secondary light sources based on the field enhancement effect. Under certain conditions, the light–sample interaction can be much stronger than the scattering of the incident light at the pointed-probe. In this situation, the probe is employed as a passive element, which scatters the near-field components of the externally illuminated sample [63, 120]. Although biological applications can be found for both techniques [121–125], so far tip-enhanced near-field microscopy is more widespread, and thus will be discussed in the following.

5.1 Optical Antennas

Coupling of an external laser field to nanoscale metal structures is widely discussed in the context of optical antennas [23]. Already *J. Wessel* drew the analogy of a sharply pointed metal tip externally illuminated by a laser field and electromagnetic antennas of the radiofrequency and microwave regime [6]. In a generalized form, an antenna is a device, which efficiently converts free propagating radiation into localized energy, and vice versa. In this context an antenna is a tool, which efficiently mediates between far-field and near-field interaction of radiation. In the strong interaction regime, the field scattered by the near-field probe can be resonant with the incident irradiating field, and thus, the scattered field interacting with the sample in the near-field regime can be strongly enhanced with respect to the incident electromagnetic field. The electromagnetic field enhancement present in a nanoscale structure is associated with different effects, e.g the lightning rod effect, and plasmonic effects such as surface, edge, and particle plasmons. Just as the brightness of an aperture probe is determined by the light transmission properties, the brightness of active scattering probes is strongly correlated with the achievable field enhancement at the metallic structure. Most commonly, etched solid metal tips, metallized dielectric or semiconductor tips, such as standard AFM cantilever tips, finite metal nanoparticles, and FIB milled finite metal structures have been employed for fluorescence microscopy, Raman-Scattering, absorption measurements, and nonlinear signal generation [15, 60, 110, 111, 116, 118, 121, 126–128]. Figure 8a schematically reflects the principle of this technique.

5.1.1 Sharply Pointed Solid Metal Tips

Sharply pointed solid metal tips are characterized by a strong electromagnetic field enhancement [14]. This effect is fueled by the crowding of the field lines at sharp tips and edges of metallic structures, similarly to the principle of a lightning rod. This quasi-electrostatic effect is largely non-resonant, and thus can be utilized for a large range of spectroscopy techniques. Under certain conditions the incident light illuminating the metal tip can excite a surface charge oscillation, which drives the free electrons in the metal structure to the foremost end of the metal structure, leading to a strong charge accumulation in the confined region of the extended metal structure. The field associated with this charge density oscillation can be significantly stronger than the illuminating field. A crowding of the field lines can occur for structures much smaller

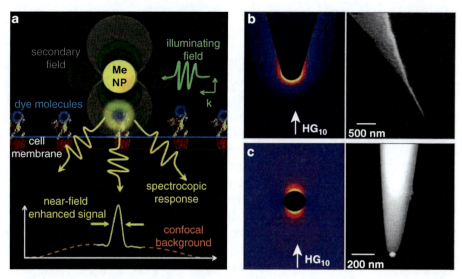

Fig. 8 Tip-enhanced near-field optical microscopy (TENOM) with pointed probes. (**a**) Schematic outline of the principle of TENOM with metal nanostructures and of the signal composition, i.e., near-field and far-field contributions. (**b**) Sharply pointed metal tips fabricated by electrochemical etching from a Au wire. Calculated electric field distribution (E^2) at the tip apex of a solid metal tip for an illumination of the tip with an on-axis focused Hermite-Gaussian (1, 0) laser mode, which provides a strong longitudinal polarization component (*left*). The field distribution has been scaled by a factor of 0.02. Scanning electron micrograph of an etched solid Au tip (*right*). (**c**) Calculated electric field distribution (E^2) of a single spherical Au nanoparticle illuminated under the same conditions as in (**b**) (*left*). Scanning electron micrograph of a typical 60 nm-Au nanoparticle antenna (*right*). *Left images* of (**b**) and (**c**) are reprinted from [19] with permission of Annual Reviews, copyright 2006

than the wavelength of light and the corresponding field enhancement is expected to be significantly stronger for decreasing sizes of these structures. Consequently, apertureless probes provide a new concept to high-resolution imaging, which is in principle non-diffraction-limited. This perspective was the driving force to introduce this concept to near-field optics since it necessarily will overcome the geometrical and material-imposed limitations of the aperture concept. The experimentally required conditions to obtain sufficient field enhancement factors have been approached in multitude studies [14, 91, 129–132]. Figure 8b shows a typical electron micrograph of an electrochemically etched gold tip, and the calculated electric field distribution at the tip for an illumination with a Hermite–Gaussian HG_{10} mode. The obtainable field enhancement is known to be correlated with the tip material, geometry, roughness, crystallinity of the material and also the efficiency of light coupling to the structure, e.g., the polarization of light [14, 132, 133].

At first the TENOM-concept has been successfully demonstrated by utilizing nonlinear optical interaction, e.g., two-photon excitation, Raman-spectroscopy, or Coherent anti-Stokes Raman (CARS) microscopy [15, 111, 115]. In conjunction with higher-order nonlinear processes the signal-to-background ratio is less critically affected by the confocal background. The demand for high-resolution biological imaging is still naturally interrelated to the fluorescence contrast. The weak extinction coefficients of biological materials and molecules as well as weak Raman- and Rayleigh-scattering cross-sections often prevent their label-free detection. Exemptions are rarely occurring naturally fluorescent proteins and complexes, such as tryptophan-rich proteins, bacteriorhodopsin, or light harvesting complexes. In this context, tip-enhancement techniques gain new interest, since this technique not only fuels high-resolution spectroscopic imaging but also benefits from the signal enhancement. Thus, TENOM provides the perspective to further increase the excitation and detection sensitivity in high-resolution imaging. Hence, it is expected that TENOM is most sufficiently employed to study weakly fluorescent entities, e.g., naturally fluorescent biomolecules and samples with low photostability [134]. Although the initial TENOM studies drew a lot of interest and fueled its frequent implementation for studies in the fields of solid state physics and material science, applications of solid metal tips for high-resolution fluorescence imaging failed. Therefore, TENOM was considered to be less applicable to biological studies for a long time. Mainly, this limitation has its origin in the modification of the local density of states of the environment of the quantum emitter. Thus, the close proximity of the metal structure opens up new non-radiative decay channels, i.e., the corresponding energy is dissipated in the metal [135–139]. Hence, the aim is to accomplish pointed structures which provide a strong electric field enhancement, and at the same time are characterized by low radiation dampening. In this context, finite metal nanoparticles and tetrahedral geometries have been proposed [99, 130].

5.1.2 Nanoparticle Antennas

Nanoparticles have attracted huge interest in the field of nanoscale biosensing and bioimaging due to their unique plasmonic properties. Similarly to surface charge oscillations supported at dielectric-metal interfaces, i.e., surface plasmons, finite metal nanostructures with different geometries are characterized by localized plasmon modes. Here, the induced surface charge oscillations are restricted to the size of the metal structure, and the existing modes depend strongly on the geometry and material of the structure, which enables a precise adjustment of the resonance over a broad spectral range [140–142]. The simplest geometry supporting localized plasmon modes, and thus forming a particle plasmon probe, consists of a single spherical metal particle (c.f. Fig. 8c). Although spherical metal nanoparticles are not the most efficient optical antennas in terms of the provided field enhancement, they form an ideal model system, which can be analyzed theoretically and experimentally to gain a detailed understanding of its influence on the fluorescence emission of quantum emitters [118, 127, 143–145].

In the *Drude-Sommerfeld*-model the electrons are treated as a free electron gas. These electrons are displaced by an external field with respect to the stationary lattice. Thus, the incident field E_0 induces a dipole moment $\mu = \alpha\, E_0$ in the particle, which depends on the polarizability α of the particle. In the quasi-static limit, the resultant oscillation of the electron gas for a spherical particle corresponds to a dipolar mode. Disregarding any interband transitions in the metal, the plasmon resonance occurs at $\omega_{sp} = \omega_p (1 + 2\epsilon_d)^{-1/2}$, with ω_p denoting the plasma frequency and ϵ_d is the dielectric function of the particle's environment. At the plasmon resonance the light scattering at the particle can be extremely large and the corresponding modes are associated with enhanced electromagnetic fields. In this context metal nanoparticles are often discussed also as alternative spectroscopic labels. However, strong scattering is prevailingly supported by large particles, since the scattering cross-section scales with the sixth power of the particle's radius, which requires their size to be on the order of 10–100 nm. The use of the particle plasmon resonances as probes for near-field interactions in the optical regime was initially approached in an experiment by *Fischer and Pohl* in 1989. They illuminated a Au-coated polystyrene sphere absorbed on a glass supported Au film in a *Kretschmann* configuration and measured the surface plasmon scattering as a function of the distance of a dielectric substrate [146].

Figure 9 shows high-resolution images of dye samples with varying molecule densities imaged with a single spherical gold nanoparticle externally illuminated by a radially polarized laser beam. The diameter of the nanoparticle is ~50 nm yielding an optical resolution of ~40 nm. These images demonstrate the capability of this technique for biomolecular studies with single-molecule sensitivity at biologically relevant concentrations. Apart from spherical geometries or nanoshells, the plasmonic properties can be tuned across the entire visible spectrum by selecting different materials and non-spherical geometries, e.g., elongated particles such as ellipsoids and rods, which will be characterized by a red-shifted plasmon resonance [130, 141]. In addition, the plasmon-resonance critically depends on the dielectric properties of the environment, and their interparticle separation, i.e., the spatial coupling of nanoparticles and the associated formation of gap plasmons [147].

As the plasmon-mediated field is strongly bound to the metal nanoparticle the resultant enhancement effects are only ascertainable in the near-field regime. Hence, the application of an optical antenna takes place in the strong coupling regime [23]. Consequently, the physical properties of an antenna are critically interrelated to the sample properties, i.e., its performance depends on the considered optical transitions. Vice versa, an optical antenna

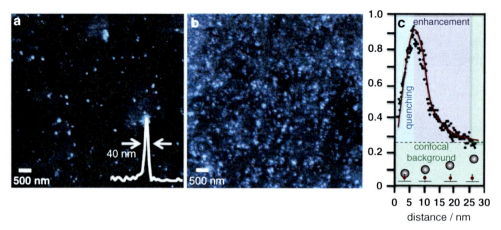

Fig. 9 Antenna-assisted single molecule fluorescence investigations. (**a** and **b**) High-resolution single molecule images of Alexa633 dye samples with different molecule densities recorded by means of a ~50 nm spherical Au particle antenna. Individual molecules are clearly identified (**a**). The image contrast is affected by the larger confocal background signal for the higher molecule density in (**b**), which reflects the expected density of abundant membrane-proteins in cellular membranes. (**c**) Typical distance versus fluorescence intensity curve recorded with a single spherical nanoparticle antenna on top of a single molecule. The approach curve clearly demonstrates the influence of the antenna on the emission properties of the dye molecule underneath the antenna. The strongest fluorescence enhancement is obtained at a distance of ~5 nm. Smaller antenna-molecule separations lead to strong quenching of the molecule's fluorescence

in close proximity to a quantum emitter can modify its optical properties. This influence of a metal structure on the fluorescence emission of a quantum emitter is known as the *Purcell* effect and has been theoretically and experimentally approached in multiple studies [118, 127, 135, 137–139, 148]. Changing the local environment of a quantum emitter, e.g., by means of a metallic structure, alters the local density of states of the environment. Consequently, the excitation process and the subsequent relaxation of the excited state for photoluminescent or fluorescent entities can be modified. In the situation of weak excitation, i.e., far from the saturation of the excited state, the emission rate of a quantum emitter can be separated in the processes of excitation from the ground state to the first excited state, and relaxation back to the ground state by emission of a fluorescence photon. Hence, the fluorescence emission rate is given by the product of the excitation rate and the probability to relax to the ground state via emission of a photon: $\gamma_{em} = \gamma_{exc} \cdot Q_i$. The latter is given by the intrinsic quantum yield $Q_i = \gamma_{rad} / (\gamma_{rad} + \gamma_{nrad})$. Both excitation rate and the radiative and non-radiative rate can be modified by the presence of an optical probe. Therefore, the fluorescence enhancement factor can be expressed as:

$$M_{fl} = \gamma_{exc}^{probe} / \gamma_{exc}^{0} \cdot Q_{probe} / Q_i = E_{probe}^{2} / E_{0}^{2} \cdot Q_{probe} / Q_i \quad (2)$$

Equation (2) shows that the fluorescence enhancement factor is determined by the enhancement of the incident electric field at the metal structure, but can also be clearly affected by influence of the probe on the radiative rate, i.e., by the possibility to modify the quantum yield of a molecule. Obviously, one has to aim at the introduction of metal structures which hold for a sufficient strong electric field enhancement and at same time pander the radiative relaxation of a molecule or minimize non-radiative relaxation processes, i.e., dissipation of the energy in the metal. In this context, finite metal structures have been found to be more suitable than solid metal tips, since the latter allow for a relaxation of the excited state by coupling to plasmon modes propagating along the tip shaft. The highest efficiency for antenna-assisted fluorescence imaging will be obtained for weak quantum emitters, since their low quantum yield can be significantly increased by affecting the radiative rate [149, 150]. In contrast, strong quantum emitters with quantum yields close to unity will experience strong quenching of the fluorescence emission. The resultant modification of the fluorescence emission rate and excited state lifetime have been analyzed both theoretically and experimentally in distance-dependent studies [118, 127, 144, 151, 152]. Figure 9 shows a typical approach curve recorded on a vertically aligned single dye molecule, which is characterized by a high quantum yield. The maximum signal enhancement is obtained at a distance of ~5 nm. At distances shorter than 5 nm the quenching rate exceeds the gain in the excitation rate provided by the enhanced electric field at the nanoparticle.

A major concern related to the application of tip-enhanced near-field probes for high-resolution fluorescence microscopy is the external illumination scheme. Compared to the largely background-free operation of aperture probes, measurements by means of solid metal structures and particle plasmon probes are characterized by a significant confocal background signal, which can obscure weaker near-field contributions [60, 153]. This is especially an issue for investigations of samples with high molecule densities and samples with strong variations in the entity density, like the ones found in many biological applications. Recently, it has been shown that the constant confocal background can be suppressed in the antenna-assisted images by means of a modulation of the distance-dependent near-field signal [153–156].

5.1.3 Tip on Aperture Probes

Although the background signal can be entirely rejected by a demodulation of periodically altered fluorescence signals, the employed excitation power is significantly reduced to the one used in standard confocal laser scanning microcopy (approximately $\lesssim 100$ nW) and the presence of the nanoparticle antenna is expected to lower the photobleaching rate of molecules in their direct environment [157], the external illumination scheme raises some concerns with respect to the photobleaching of labeled molecules during image acquisition. In this regard a combination of the

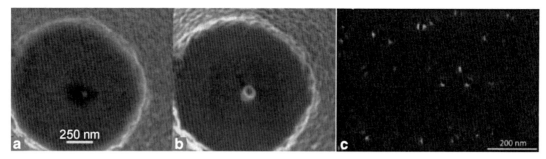

Fig. 10 Optical Antenna fabricated on top of an aperture fiber probe by means of e-beam epitaxy. (**a** and **b**) Electron micrograph of a typical tip-on-aperture (TOA) probe before and after metal deposition on the fabricated structure. Reprinted from [158] with permission of the American Physical Society. (**b**) Fluorescence Image of a single molecule sample imaged with a TOA-probe. Reprinted from [116] with permission of the American Physical Society

so far introduced probe concepts, i.e., aperture probes and enhancement-tips, provides the prospect of background-free measurements under illumination conditions, which will retain the majority of fluorescence labels in these investigations due to a significantly reduced illumination area and a shorter total exposure time.

Frey et al. introduced for the first time a new tip concept, which utilizes the illumination of a finite metal tip through a common aperture probe. In this approach the metal tip is fabricated on top of a fiber aperture probe by means of electron beam deposition followed by the subsequent metallization of the grown antenna (c.f. Fig. 10). The principle of these TOA-probes has been transferred to FIB fabrication schemes, e.g., to craft the antenna structure from entirely metal-coated aluminum fiber probes. By means of this approach *Taminiau et al.* have demonstrated that the formed aluminum tip can be fabricated such that it acts as an optical monopole antenna in analogy with its radiofrequency counterpart [119]. Although an aperture is used to couple the irradiating field to the optical antenna in the near-field regime, the aperture probe also can illuminate the sample, and thus can lead to a confined background signal. Nevertheless, this contribution is clearly reduced compared to an external illumination, and thus helps to improve the signal-to-background ratio and image contrast, respectively. Moreover, this approach decreases the total exposure time of a molecule significantly during the image acquisition.

5.2 TENOM Applications

5.2.1 Tip-Enhanced Raman Spectroscopy of Biomolecules

Raman spectroscopy has been considered for a long time as a powerful tool to investigate the chemical composition of materials. This information is encoded in the vibrational modes, and thus Raman spectroscopy can give a fingerprint of the matter under investigation. Being naturally a label-free optical technique, a lot of interest has been directed to the idea of biomolecular imaging by addressing specific Raman-bands of certain amino acids such as cysteine, tyrosine, phenylalanine, histidine and others, in proteins and lipids in order to study bacteria and cells, and to enable sequencing of DNA and RNA by identifying specific nucleotide bases. A major drawback of this optical contrast is the very low Raman cross-section associated with these vibrational transitions, which is ~14 orders of magnitude smaller than the one for strong fluorescent molecules. Cumulative effects on rough metal films can lead to a strong enhancement of the Raman-signal, known as surface-enhanced Raman spectroscopy (SERS). Enhanced Raman spectroscopy can overcome some of the limitations [159–161]. Nonetheless, broad variations in the obtainable signal-enhancement are observed, which are frequently discussed in the context of the existence of hot spots [162, 163]. Consequently, the idea of localized and high-resolution Raman microscopy stimulated by means of sharply pointed metal was pursued. Tip-enhanced Raman spectroscopy (TERS) benefits from similar effects as surface-enhanced Raman spectroscopy, but utilizes a single element nanostructure, which provides strong electromagnetic field enhancement or supports the formation of localized hot spots between a metal film and the tip [19, 21, 161, 164]. In the latter case, one takes advantage of the additional generation of gap-plasmons [165], which further increase the electric field enhancement. Today, the provided resolution can be on the sub-15 nanometer scale. The proof of single molecule sensitivity has been addressed in several experiments [166–168]. Recently, *Streitner et al.* have reported combined Raman- and STM measurements which proved strong evidence for a single-molecule detection sensitivity [168].

A major step toward the single-base sequencing of nucleic acids such as DNA and RNA was achieved by high sensitivity TENOM studies of adenine, thymine, cytosine, and guanine on Au(111) surfaces [169, 170]. Molecular concentrations in the picomolar range have been well distinguished in the background-corrected spectra. Figure 11 shows the characteristic tip-enhanced spectra of samples composed of one specific base. Further examples for biomolecular TERS investigations addressed entities such as bacteria, cytochrome c, the surfaces of the ommatidial lens, and others [171–176]. Recently, *Richter et al.* have demonstrated the potential of TERS for label-free simultaneous topographical and chemical characterization of cell membranes [175]. In their study they addressed the lateral distribution of proteins and lipids in human colon-cancer cells by taking advantage of automated data processing based on statistic methods. Protein-rich areas were assigned by means of the characteristic amide I and amide III Raman-bands and their spectral shifts were correlated with distinct structural conformations, whilst lipids can be identified by means of their symmetric and antisymmetric PO_2 stretching modes and other bands (CH_2–, CH_3–, and ester group). Clear differences in the local abundance, i.e., protein-to-lipid ratio, have been observed with nanometer-scale resolution.

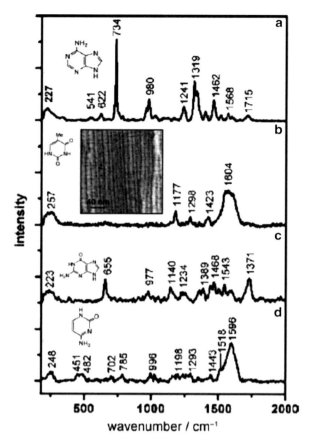

Fig. 11 Background-corrected Raman spectra of nucleotide bases (**a**) adenosine, (**b**) thymine, (**c**) guanine and (**d**) cytosine recorded by means of TENOM on Au(111) surfaces. TERS enables specific detection of the individual bases with extremely high sensitivity. The inset shows a typical scanning tunneling microscopy image of a self-assembled monolayer of thymine. Integration time: 1 s at a excitation power of ~2 MW. Reprinted form [169] with permission of the American Chemical Society. Copyright 2006

Although TERS has been employed already for nanoscale analysis of biological entities and, by this, has been proven to have great potential for biomolecular application, this technique is still in its infancy. Especially, further progress has to be made in terms of its applicability in cell science under biological relevant conditions. So far none of the reported TERS studies has been carried out under physiologically relevant conditions. Nonetheless, due to its ability to gain complementary information in a label-free detection scheme it can be expected that this field rapidly develops.

5.2.2 Antenna-Assisted Fluorescence Microscopy

A detection sensitivity on a single quantum-emitter level with nanoscale resolution has been reported multiple times with different types of apertureless probes [60, 116, 118, 119, 122, 127, 177]. *Gerton et al.* employed silicon cantilever probes, which are

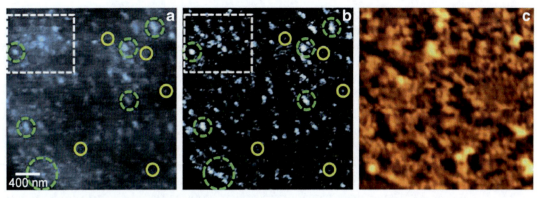

Fig. 12 Map of plasma membrane-bound Ca^{2+}-ATPases in red blood cells. (**a**) Regular antenna-assisted fluorescence image displays the lateral organization of PMCA1 and PMCA4 across the cell membrane. (**b**) Background-suppressed fluorescence image of the same area shown in (**a**). The improved image contrast enables the identification of individual membrane proteins and clusters of membrane proteins. (**c**) Simultaneously imaged topography of the cytoplasmic side of the RBC plasma membrane showing the actin and spectrin filaments lining the plasma membrane

characterized by weak fluorescence quenching at small tip-quantum emitter distances compared to gold nanostructures, but at the same time, these tips provide clearly lower electromagnetic field enhancement factors. They reported on imaging of single quantum dots with sub-10 nm resolution. Single-molecule sensitivity has been demonstrated for spherical plasmon probes and tip-on-aperture (TOA)-probes [116, 118, 119, 127]. Nanoscale investigations of the organization of cellular membranes have been addressed by means of nanoparticle plasmon probes and tip-on-aperture probes. Similarly to aperture probes these optical antennas have also been demonstrated to be applicable in liquid environments [41, 121, 122, 177].

For the first time, the application of non-aperture probes in liquid environments for single-protein mapping has been demonstrated by means of single spherical Au nanoparticle probes combined with a tapping-mode like distance control [121, 122]. Although the Ca^{2+}-pumps addressed in these experiments are of relatively low abundance, individual PMCAs cannot be identified in the plasma membrane of red blood cells by conventional confocal fluorescence microscopy, due to their organization in μm-size areas. An optical resolution of ~50 nm provided by the applied optical antenna enables the identification of individual proteins. Figure 12 displays an example for antenna-assisted fluorescence measurements of individual fluorescence-labeled PMCAs in the plasma membrane of red blood cells imaged in a buffered medium. On average each protein is labeled with 2–3 dye molecules. This fact and the large density of proteins arranged in microscopic domains lead to a significant confocal background, which certainly lowers the achievable image contrast. Consequently, the fluorescence signals arising from weakly labeled proteins might be obscured in regions of high protein density (marked square region in Fig. 12a).

Figure 12b displays a background-suppressed fluorescence image of the same area shown in Fig. 12a. An entire suppression of the background signal is achieved by introducing a modulation of the intensity of the optical signal from the highest enhancement to the background level. This is accomplished by a direct modulation of the feedback-loop [153]. The elimination of the constant confocal background signal by means of a demodulation of the optical signal at the modulation frequency clearly improves the image contrast, and consequently, more image details become visible. Individual PMCAs (yellow solid circles) as well as clusters of PMCAs (green dashed circles) can be found in the corresponding antenna-assisted image. The major advantage of antenna-assisted microscopy is the ability to simultaneously map protein distributions (c.f. Fig. 12b) and the membrane topology (c.f. Fig. 12c). This enables to address structural modifications and chemical redistributions involved in specific membrane-associated processes, e.g., the formation of vesicles in endocytosis and exocytosis. Figure 12c shows the corresponding membrane topology of the area imaged in Fig. 12a, b, which is characterized by actin and spectrin filaments lining the cytoplasmic side of the plasma membrane.

Recently the *Garcia-Parajo* and coworkers reported on the application of a TOA-type probe for studying the organization of transmembrane proteins in monocytes [41]. These studies demonstrate the high potential for the investigation of membrane nanodomains of proteins and lipids by means of optical antennas. They focused on mapping the distribution of the leukocyte-specific integrin LFA-1. The transmembrane receptors mediate the migration of immune cells. Previously, aperture-probe type near-field fluorescence studies on this system revealed the existence of nanoclusters of LFA-1 in the plasma membrane of monocytes prior to ligand-receptor binding [178]. In order to determine the real cluster size optical antennas are employed, which are similar to previously introduced monopole antennas fabricated by FIB milling on top of a circular aperture probe [119]. These antennas were tuned to resonance by carefully adjusting the length of the antenna. Figure 13 displays a confocal fluorescence large-scale image of an LFA-1 immunofluorescence labeled monocyte (a) and a magnified confocal image (b) of the marked area in (a), as well as probe-based antenna-assisted fluorescence images (c and d) of the marked areas in A. The significantly higher optical resolution provided by these antennas allowed for the analysis of the formed LFA-1 nanodomains. Probe-based antenna mapping of LFA-1 receptors confirmed the formation of nanoclusters. The corresponding fluorescence spots have a larger size than the provided lateral resolution of the applied probe. The determined average value of 72 ± 21 nm is clearly smaller than the one determined in previous aperture-probe-based studies.

These initial examples for the application of TENOM techniques to biological materials demonstrate its high potential for high-resolution

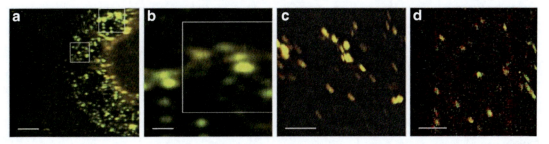

Fig. 13 Distribution of LFA-1 receptors in the plasma membrane of monocytes. (**a** and **b**) Confocal fluorescence large-scale and high magnification image of fluorescence-labeled LFA-1 transmembrane proteins. Scale bars: 5 μm and 1 μm. (**c** and **d**) LFA-1 maps of the marked regions in (**a**) images with a probe-based monopole antenna. Scale bar: 1 μm. The high-resolution fluorescence images clearly reveal small nanodomains of LFA-1 receptors

imaging. The versatility of this technique is not only reflected by its ability to approach biological studies by different optical contrast and by the potential to enhance spectroscopic signals, but it enables also the direct comparison of the chemical organization, i.e., protein distributions, with structural features.

6 Summary

Near-field optical microscopy has been successfully implemented in biomolecular and cellular science. The introduction of the optical antenna concept to the field of fluorescence imaging under relevant biological conditions marks a milestone in the field of near-field microscopy, since it is now serving the demand for studying biological matter on a single protein level and breaks the limitations of aperture probes. It can be expected that the highly dynamic field of optical antennas will further improve their optical properties, and that this progress will accomplish measurements with true-protein resolution, i.e., on the sub-10 nm scale. Additionally, the effect of the radiative decay rate engineering and the increased photostability of molecules in the presence of an antenna significantly broaden the spectrum of applications, and hold also for the perspective to study entities in biological membranes not addressable with other types of fluorescence microscopy techniques. Although NFO approaches are mandatorily surface sensitive methods, and thus cannot address inner-cellular structures in a noninvasive way, the highly exponential distance-dependent character of the optical near-field provides antenna-assisted microscopy as well as aperture-based NSOM with an axial resolution, which cannot be achieved by means of other techniques. This capability opens up the possibility for high sensitivity investigations of membrane-associated processes, which demand for three-dimensional nanoscale resolution and localization.

Acknowledgments

I would like to thank the "Ministerium für Wissenschaft, Forschung und Technologie" (MWFT) of the State North Rhine-Westphalia, Germany for their financial support within the initiative "Rückkehrer-Programm Nanotechnology".

References

1. Synge EH (1928) A suggested model for extending microscopic resolution into the ultramicroscopic region. Philos Mag 6:356–362
2. O'Keefe JA (1956) Resolving power of visible light. J Opt Soc Am 46:359
3. Baez AV (1956) Is resolving power independent of wavelength possible? An experiment with a sonic 'macroscope'. J Opt Soc Am 46:901
4. McCutchen CW (1967) Superresolution in microscopy and the abbe resolution limit. J Opt Soc Am 57:1190–1192
5. Massey GA (1984) Microscopy and pattern generation with scanned evanescent waves. Appl Opt 23:658–660
6. Wessel J (1985) Surface-enhanced optical microscopy. J Opt Soc Am B 2:1538–1540
7. Ash EA, Nicholls G (1972) Super-resolution aperture scanning microscope. Nature 237:510–513
8. Pohl DW, Denk W, Lanz M (1984) Optical stethoscopy: image recording with resolution $\lambda/20$. Appl Phys Lett 44:651–653
9. Lewis A, Isaacson M, Harootunian A, Muray A (1984) Development of a 500Å spatial resolution light microscope. Ultramicroscopy 13:227–231
10. Zenhausern F, O'Boyle MP, Wickramasinghe HK (1994) Apertureless near-field optical microscope. Appl Phys Lett 65:1623–1625
11. Bachelot R, Gleyzes P, Boccara AC (1994) Near-field optical microscopy by local perturbation of a diffraction spot. Microsc Microanal Microstruct 5(4–6):389–397
12. Kawata S, Inouye Y (1995) Scanning probe optical microscopy using a metallic probe tip. Ultramicroscopy 57:313–317
13. Keilmann F, van der Weide DW, Eickelkamp T, Merz R, Stockle D (1996) Extreme subwavelength resolution with a scanning radio-frequency transmission microscope. Opt Commun 129:15–18
14. Novotny L, Sánchez EJ, Xie XS (1998) Near-field optical imaging using metal tips illuminated by higher-order Hermite–Gaussian beams. Ultramicroscopy 71:21–29
15. Sanchez EJ, Novotny L, Xie XS (1999) Near-field fluorescence microscopy based on two-photon excitation with metal tips. Phys Rev Lett 82 (20):4014–4017
16. Dunn B (1999) Near-field scanning optical microscopy. Chem Rev 99:2891–2928
17. Hecht B et al (2000) Scanning near-field optical microscopy with aperture probes:fundamentals and applications. J Chem Phys 112:7761–7774
18. Hecht B (2004) Nano-optics with single quantum systems. Philos Trans R Soc Lond A 362:881–899
19. Novotny L, Stranick SJ (2006) Near-field optical microscopy and spectroscopy with pointed probes. Annu Rev Phys Chem 57:303–331
20. Hartschuh A, Qian H, Meixner A, Anderson N, Novotny L (2006) Tip-enhanced optical spectroscopy for surface analysis in biosciences. Surf Interface Anal 38:1472–1480
21. Kawata S, Inouye Y, Verma P (2009) Plasmonics for near-field nano-imaging and superlensing. Nat Photon 3:388–394
22. Pohl DW (2000) Near-field optics seen as an antenna problem. In: Zhu X, Ohtsu M (eds) Near-field optics, principles and applications. World Scientific, Singapore, p 9–21
23. Bharadwaj P, Deutsch B, Novotny L (2009) Optical antennas. Adv Opt Photon 1:438–483
24. Novotny L, van Hulst NF (2011) Antennas for light. Nat Photon 5:83–90
25. Betzig E, Chichester RJ (1993) Single molecules observed by near-field scanning optical microscopy. Science 262:1422–1425
26. Lewis A (1991) The optical near-field and cell biology. Semin Cell Biol 2:187–192
27. Betzig E, Chichester RJ, Lanni F, Taylor DL (1993) Near-field fluorescence imaging of cytoskeletal actin. Bioimaging 1:129–135
28. Gheber LA, Hwang J, Eddidin M (1998) Design and optimization of a near-field scanning optical microscope for imaging samples in liquid. Appl Opt 373574–373581
29. Subramaniam V, Kirsch AK, Jovin T (1998) Cell biological applications of scanning near-field optical microscopy. Cell Mol Biol 44:689

30. Edidin M (2001) Near-field scanning optical microscopy, a siren call to biology. Traffic 2:1797–803
31. de Lange F et al (2001) Cell biology beyond the diffraction limit: near-field scanning optical microscopy. J Cell Sci 114:4153–4160
32. Garcia-Parajo MF (2008) Optical antennas focus in on biology. Nat Photon 2:201–203
33. Hwang J, Gheber LA, Margolis L, Edidin M (1998) Domains in cell plasma membranes investigated by near-field scanning optical microscopy. Biophys J 74:2184–2190
34. Höppener C, Molenda D, Fuchs H, Naber A (2003) Scanning near-field optical microscopy of a cell membrane in liquid. J Microsc 210:288–293
35. Koopman M et al (2004) Near-field scanning optical microscopy in liquid for high resolution single molecule detection on dendritic cells. FEBS Lett 573:6–10
36. Höppener C, Siebrasse JP, Peters R, Kuitscheck U, Naber A High-resolution near-field optical imaging of single nuclear pore complexes under physiological conditions. Biophys J 88:3681–3688
37. Ianoul A et al (2005) Imaging nanometer domains of beta-adrenergic receptor complexes on the surface of cardiac myocytes. Nat Chem Biol 1:196–202
38. de Bakker BI et al (2007) Nanoscale organization of the pathogen receptor dc-sign mapped by single-molecule high-resolution fluorescence microscopy. Chem Phys Chem 8:1473–1480
39. Höppener C, Novotny L (2008) Antenna-based optical imaging of single Ca^{2+} transmembrane proteins in liquids. Nano Lett 8:642–646
40. Herrmann M et al (2009) Near field optical study of protein transport kinetics at a single nuclear pore. Nano Lett 9:3330–3336
41. van Zanten TS et al (2010) Direct mapping of nanoscale compositional connectivity on intact cell membranes. Proc Natl Acad Sci USA 107:15437–15442
42. Abbe E (1873) Beiträge zur Theorie des Mikroskops und der mikroskopischen Wahrnehmung. Archiv f Miroskop Anat 9:413
43. Rayleigh L (1896) On the theory of optical images with special reference to the microscope. Philos Mag 5:167–195
44. Vigoureux JM, Depasse F, Girard C (1992) Superresolution of near-field optical microscopy defined from properties of confined electromagnetic waves. Appl Opt 31:3036–3045
45. Novotny L, Hecht B (2006) Principles of nano-optics. Cambridge University Press, Cambridge
46. Binnig G, Rohrer H (1982) Scanning tunneling microscopy. Helv Phys Acta 55:726
47. Binnig G, Quate CF, Gerber C (1986) Atomic force microscope. Phys Rev Lett 56:930
48. Hecht B, Bielefeldt H, Inouye Y, Pohl DW, Novotny L (1997) Facts and artifacts in near-field optical microscopy. J Appl Phys 81:2492–2498
49. Betzig E, Finn PL, Weiner SJ (1992) Combined shear force and near-field scanning optical microscopy. Appl Phys Lett 60:2484–2486
50. Toledo-Crow R, Yang PC, Chen Y, Vaez-Iravani M (1992) Near-field differential scanning optical microscope with atomic force regulation. Appl Phys Lett 60:2957–2959
51. Karrai K, Grober RD (1995) Piezoelectric tip-sample distance control for near field optical microscopes. Appl Phys Lett 66:1842–1844
52. Brunner R, Bietsch A, Hollrichter O, Marti O (1997) Distance control in near-field optical microscopy with piezoelectric shear-force detection suitable for imaging in liquids. Rev Sci Instrum 68:1769
53. Tsai DP, Lu YY (1998) Tapping-mode tuning fork force sensing for near-field scanning optical microscopy. Appl Phys Lett 69:2724–2726
54. Naber A, Maas H-J, Razavi K, Fischer U (1999) Dynamic force distance control suited to various probes for scanning near-field optical microscopy. Rev Sci Instrum 70:3955–3961
55. Quate C (1994) Near-field scanning optical and force microscope including cantilever and optical waveguide. US patent 5,354,985
56. Danzebrink HU, Wilkening G, Ohlsson O (1995) Near–field optoelectronic detector probes based on standard scanning force cantilevers. Appl Phys Lett 67:1981
57. Eckert R et al (2000) Near-field fluorescence imaging with 32 nm resolution based on microfabricated cantilevered probes. Appl Phys Lett 77:3695Ű3697
58. Oesterschulze E, Georgiev G, Müller-Wiegand M, Vollkopf A, Rudow O (2000) Transmission line probe based on a bow–tie antenna. J Microsc 202:39–44
59. Heisig S, Rudow O, Oesterschulze E (2000) Scanning near-field optical microscopy in the near-infrared region using light emitting cantilever probes. Appl Phys Lett 77:1071–1073
60. Gerton JM, Wade LA, Lessard GA, Ma Z, Quake SR (2004) Tip-enhanced fluorescence microscopy at 10 nanometer resolution. Phys Rev Lett 93:180801–4
61. Farahani JN et al (2007) Bow-tie optical antenna probes for single-emitter scanning

near-field optical microscope. Nanotechnology 18:1255061–4
62. Knoll B, Keilmann F (1999) Near-field probing of vibrational absorption for chemical microscopy. Nature 399:134
63. Hillenbrand R, Keilmann F (2002) Material-specific mapping of metal/semiconductor/dielectric nanosystems at 10 nm resolution by backscattering near-field optical microscopy. Appl Phys Lett 80:25
64. Ruiter A, Veerman J, van der Werf K, van Hulst N (1997) Dynamic behavior of tuning fork shear-force feedback. Appl Phys Lett 71:28–30
65. Rensen WHJ, van Hulst NF (2000) Imaging soft samples in liquids with tuning fork based shear force microscopy. Appl Phys Lett 75:1557–1559
66. Lambelet P, Sayah A, Pfeffer M, Philipona C, Marquis-Weible F (1998) Chemically etched fiber tips for near-field optical microscopy: a new process for smoother tips. Appl Opt 37:7289
67. Ebbesen TW, Lezec HJ, Ghaemi HF, Thio T, Wolff PA (1998) Extraordinary optical transmission through sub-wavelength hole arrays. Nature 391:667–669
68. Lezec HJ et al Beaming light from a subwavelength aperture. Science 297:820–822
69. Fischer UC (1986) Submicrometer aperture in a thin metal film as a probe of its microenvironment through enhanced light scattering and fluorescence. J Opt Soc Am B 3:1239–1244
70. Kuhn H (1987) Self-organizing molecular electronic devices? In: Carter F (ed) Molecular electronic devices II. Dekker, New York, p 411–426
71. Betzig E, Trautman JK, Harris TD, Weiner JS, Kostelar RL (1991) Breaking the diffraction barrier: optical microscopy on a nanometric scale. Science 251:1468–1470
72. Bethe HA (1944) Theory of diffraction by small holes. Phys Rev 66:163–182
73. Bouwkamp CJ (1950) On Bethe's theory of diffraction by small holes. Philips Res Rep 5:321–332
74. Veerman J, García-Parajó M, Kuipers L, van Hulst N (1999) Single molecule mapping of the optical field distribution of probes for near-field microscopy. J Microsc 194:477
75. Höppener C, Molenda D, Fuchs H, Naber A (2003) Simultaneous topographical and optical characterization of near-field aperture probes by way of imaging fluorescent nanospheres. Appl Phys Lett 80:1331–1333
76. Leviatan Y (1986) Study of near-zone fields of a small aperture. J Appl Phys 60:1577–1583
77. Moers MHP (1995) Near-Field Optical Microscopy. ISBN 90-9008593-9, University of Twente, The Netherlands, CIP-Gegevens Koninklijke Bibliotheek, Den Haag
78. Novotny L, Hafner C (1994) Light propagation in a cylindrical waveguide with a complex, metallic, dielectric function. Phys Rev E 50:4094–4106
79. Valaskovic GA, Holton M, Morrison G (1995) Parameter control, characterization, and optimization in the fabrication of optical fiber near–field probes. Appl Opt 34:1215
80. Hollars CW, Dunn RC (1998) Evaluation of thermal evaporation conditions used in coating aluminum on near-field fiber-optic probes. Rev Sci Instrum 69:1747–1452
81. Turner D (1984) Etch procedure for optical fibers. US patent 4,469,554
82. Hoffmann P, Dutoit B, Salathé R-P (1995) Comparison of mechanically drawn and protection layer chemically etched optical fiber tips. Ultramicroscopy 61:165–170
83. Zeisel D, Dutoit B, Nettesheim S, Zenobi R (1996) Pulsed laser-induced desorption and optical imaging on a nanometer scale with scanning near-field microscopy using chemically etched fiber tips. Appl Phys Lett 68:2491
84. Stöckle R et al (1999) High-quality near-field optical probes by tube etching. Appl Phys Lett 75:160–162
85. Obermüller C, Karrai K, Kolb G, Abstreiter G (1995) Transmitted radiation through a subwavelength sized tapered optical fiber tip. Ultramicroscopy 61:171–178
86. Novotny L, Pohl DW (1995) Light propagation in scanning near-field optical microscopy. In: Marti O, Möller R (eds) Photons and local probes, NATO Advanced Study Institute, Series E. Kluwer Academic, Dordrecht, pp 21–33
87. Pangaribuan T, Yamada K, Jiang S, Ohsawa H, Ohtsu M (1992) Reproducible fabrication technique of nanometric tip diameter fiber probe for photon scanning tunneling microscope. Jpn J Appl Phys 31:L1302
88. Saiki T, Mononobe S, Ohtsu M, Saito N, Kusano J (1996) Tailoring a high-transmission fiber probe for photon scanning tunneling microscope. Appl Phys Lett 68:2612
89. Monobe S, Saiki T, Suzuki T, Koshihara S, Othsu M (1998) Fabrication of a triple tapered probe for near–field optical spectroscopy in uv region based on selective etching of a multistep index fiber. Opt Commun 146:45–48
90. Yatsui T, Kourogi M, Ohtsu M (1998) Increasing throughput of a near-field optical fiber probe over 1000 times by the use of a triple-tapered structure. Appl Phys Lett 73:2089–2091

91. Martin OJF, Paulus M (2002) Influence of the surface roughness on the near-field generated by an aperture / apertureless probe. J Microsc 205:147
92. Saiki T, Matsuda K (1999) Near-field optical fiber probe optimized for illuminationŬcollection hybrid mode operation. Appl Phys Lett 74:2773–2775
93. Muranishi M et al (1997) Control of aperture size of optical probes for scanning near-field optical microscopy using focused ion beam technology. Jap J Appl Phys 36:L942–L944
94. Veerman JA, Otter AM, Kuipers L, van Hulst NF (1998) High definition aperture probes for near-field optical microscopy fabricated by focused ion beam milling. Appl Phys Lett 72:3115–3117
95. Naber A et al (2002) Enhanced light confinement in a near-field optical probe with a triangular aperture. Phys Rev Lett 89:210801
96. Molenda D, des Francs GC, Fischer UC, Rau N, Naber A (2005) High-resolution mapping of the optical near-field components at a triangular nano-aperture. Opt Express 13:10688–10696
97. des Francs GC, Molenda D, Fischer UC, Naber A (2005) Enhanced light confinement in a triangular aperture: Experimental evidence and numerical calculations. Phys Rev B 72:165111–6
98. Fischer UC, Koglin J, Fuchs H (1994) The tetrahedal tip as a probe for scanning near–field optical microscopy at 30 nm resolution. J Microsc 176:231–237
99. Koglin J, Fischer U, Fuchs, H (1996) Scanning near–field optical microscopy with a tetrahedral tip at a resolution of 6 nm. J Biom Opt 1:75–78
100. Enderle T et al (1997) Membrane specific mapping and colocalization of malarial and host skeletal proteins in the plasmodium falciparum infected erythrocyte by dual-color near-field scanning optical microscopy. Proc Natl Acad Sci USA 94:520–525
101. Meixner AJ, Kneppe H (1995) Scanning near-field optical microscopy in cell biology and microbiology. Cell Mol Biol 44:673–688
102. Ianoul A et al (2004) Near-field scanning fluorescence microscopy study of ion channel clusters in cardiac myocyte membranes. Biophys J 87:3525–3535
103. Qiao W et al (2005) Imaging of p-glycoprotein of H69/VP small-cell lung cancer lines by scanning near-field optical microscopy and confocal laser microspectrofluorometer. Ultramicroscopy 105:330–335
104. Chen Y et al (2008) Excitation enhancement of cdse quantum dots by single metal nanoparticles. Appl Phys Lett 93:053106.
105. van Zanten TS et al (2009) Hotspots of GPI-anchored proteins and integrin nanoclusters function as nucleation sites for cell adhesion. Proc Natl Acad Sci USA 106:18557–18562
106. Abulrob A et al (2010) Nanoscale imaging of epidermal growth factor receptor clustering 285:3145–3156
107. Vobornik D et al (2008) Fluorescence correlation spectroscopy with sub-diffraction-limited resolution using near-field optical probes. Appl Phys Lett 93:1639041–4
108. Vobornik D et al (2009) Near-field optical probes provide subdiffractionlimited excitation areas for fluorescence correlation spectroscopy on membranes. Pure Appl Chem 81:1645–1653
109. Manzo C, van Zanten TS, Garcia-Parajo MF (2010) Nanoscale fluorescence correlation spectroscopy on intact living cell membranes with NSOM probes. Biophys J 100:L08–L10
110. Hartschuh A, Pedrosa HN, Novotny L, Krauss T (2003) Simultaneous fluorescence and raman scattering from single carbon nanotubes. Science 301:1354–1356
111. Ichimura T, Hashimoto M, Inouye Y, Kawata S (2004) Tip-enhanced coherent anti-stokes raman scattering for vibrational nanoimaging. Phys Rev Lett 92:220801–4
112. Chen Y et al (2010) A surface energy transfer nanoruler for measuring binding site distances on live cell surfaces. J Am Chem Soc 132:16559–16570
113. Schwille P, Haupts U, Maiti S, Webb W (1999) Molecular dynamics in living cells observed by fluorescence correlation spectroscopy with one- and two- photon excitation. Biophys J 77:2251–2265
114. Sanchez EJ, Novotny L, Holtom GR, Xie XS (1997) Fluorescence imaging of single molecules by two-photon excitation. J Phys Chem A 101:7019–7023
115. Hartschuh A, Sanchez E, Xie X, Novotny L (2004) High-resolution near-field Raman microscopy of single-walled carbon nanotubes. Phys Rev Lett 90:0955031–4
116. Frey HG, Witt S, Felderer K, Guckenberger R (2004) High-resolution imaging of single fluorescent molecules with the optical near-field of a metal tip. Phys Rev Lett 93:200801
117. Kalkbrenner T et al (2005) Optical microscopy via spectral modifications of a nanoantenna. Phys Rev Lett 95:2008011–2008014
118. Anger P, Bharadwaj P, Novotny L (2006) Enhancement and quenching of single molecule fluorescence. Phys Rev Lett 96:1130021–4
119. Taminiau TH, Moerland RJ, Segerink FB, Kuipers L, van Hulst NF (2007) Resonance

of an optical monopole antenna probed by single molecule fluorescence. Nano Lett 7:28
120. Keilmann F, Hillenbrand R (2004) Near-field microscopy by elastic light scattering from a tip. Philos Trans R Soc Lond A 362:787–797
121. Höppener C, Novotny L (2008) Antenna-based optical imaging of single Ca^{2+}-transmembrane proteins in liquids. Nano Lett 8:642–646
122. Höppener C, Novotny, L (2008) Imaging of membrane proteins using antenna-based optical microscopy. Nanotechnology 19:3840121–3840128
123. van Zanten TS, Lopez-Bosque MJ, Garcia-Parajo MF (2010) Imaging individual proteins and nanodomains on intact cell membranes with a probe-based optical antenna. Small 6:270–275
124. Brehm M, Taubner T, Hillenbrand R, Keilmann F (2006) Mapping of single nanoparticles and viruses at nanoscale resolution. Nano Lett 6:1307–1310
125. Paulite M et al (2011) Imaging secondary structure of individual amyloid fibrils of a ?2-microglobulin fragment using near-field infrared spectroscopy. J Am Chem Soc 133:7376–7383
126. Hartschuh A, Qian H, Meixner A, Anderson N, Novotny L Tip-enhanced optical spectroscopy of single-walled carbon nanotubes. In: Kawata S, Shalaev VM (eds) Photons and local probes. Advances in nano-optics and nano-photonics, vol 1, Tip Enhancements. Elsevier, Amsterdam, New York
127. Kühn S, Hakanson U, Rogobete L, Sandoghdar V (2006) Enhancement of single-molecule fluorescence using a gold nanoparticle as an optical nanoantenna. Phys Rev Lett 97:017402
128. Palomba S, Danckwerts M, Novotny L (2009) Nonlinear plasmonics with gold nanoparticle antennas. J Opt A 11:1140301-6
129. Martin YC, Hamann HF, Wickramasinghe HK (2001) Strength of the electric field in apertureless near-field optical microscopy. J Appl Phys 89:5774–5778
130. Sönnichsen C et al (2002) Drastic reduction of plasmon damping in gold nanorods. Phys Rev Lett 88:077402
131. Novotny L, Sanchez EJ, Xie XS (1999) Near-field optical spectroscopy based on the field enhancement at laser illuminated metal tips. Opt Photon News 10:24
132. Demming AL, Festy F, Richards D (2005) Plasmon resonances on metal tips: Understanding tip-enhanced Raman scattering. J Chem Phys 122:1847161–1847167
133. Ossikovski R, Nguyen Q, Picardi G (1956) Simple model for the polarization effects in tip-enhanced Raman spectroscopy. Phys Rev B 75:045412
134. Höppener C, Novotny L (2012) Exploiting the light-metal interaction for biomolecular sensing and imaging. Quart Rev Biophys 45:209–255
135. Purcell EM (1946) Spontaneous emission probabilities at radio frequencies. Phys Rev 69:681
136. Drexhage KH (1970) Influence of a dielectric interface on fluorescence decay time. J Lumin 1/2:693–701
137. Drexhage KH (1974) Interaction of light with monomolecular dye layers. In: Wolf E (ed) Progress in optics, vol 12. North Holland, Amsterdam, p 161–232
138. Bian RX, Dunn RC, Xie XS, Leung PT (1995) Single molecule emission characteristics in near-field microscopy. Phys Rev Lett 75:4772–4775
139. Novotny L (1996) Single molecule fluorescence in inhomogeneous environments. Appl Phys Lett 69:3806–3808
140. Kreibig U, Vollmer M (1995) Optical properties of metal clusters. Springer, Berlin
141. Kelly KL, Coronado E, Zhao LL, Schatz GC (2003) The optical properties of metal nanoparticles: The influence of size, shape, and dielectric environment. J Phys Chem B 107:638–677
142. Jain PK, Lee K-S, El-Sayed I H, El-Sayed MA (2006) Calculated absorption and scattering properties of gold nanoparticles of different size, shape, and composition: Applications in biological imaging and biomedicine. J Phys Chem B 110:7238–7248
143. Bharadwaj P, Anger P, Novotny L (2007) Nanoplasmonic enhancement of single-molecule fluorescence. Nanotechnology 18:044017
144. Rogobete L, Kaminski F, Agio M, Sandoghdar V (2007) Design of plasmonic nanoantennae for enhancing spontaneous emission. Opt Lett 32:1623–1625
145. Härtling T, Reichenbach P, Eng LM (2007) Near-field coupling of a single fluorescent molecule and a spherical gold nanoparticle. Opt Express 15:12806
146. Fischer UC, Pohl DW (1989) Observation on single-particle plasmons by near-field optical microscopy. Phys Rev Lett 62:458–461
147. Li K., Stockman MI, Bergman DJ (2003) Self-similar chain of metal nanospheres as an efficient nanolens. Phys Rev Lett 91:227402
148. Wokaun A, Lutz HP, King AP, Wild UP, Ernst RR (1983) Energy transfer in surface enhanced luminescence. J Chem Phys 79:509–514

149. Bardhan R, Grady NK, Cole JR, Joshi A, Halas NJ (2009) Fluorescence enhancement by Au nanostructures: Nanoshells and nanorods. Nano 3:744–752
150. Bharadwaj P, Novotny L (2010) Plasmon-enhanced photoemission from a single Y3N@C80 fullerene. J Phys Chem C 210:7444–7447
151. Bharadwaj P, Novotny L (2007) Spectral dependence of single molecule fluorescence enhancement. Opt Express 15:14266–14274
152. Seelig J et al (2007) Nanoparticle-induced fluorescence lifetime modification as nanoscopic ruler: Demonstration at the single molecule level. Nano Lett 7:685–689
153. Höppener C, Novotny L (2009) Background suppression in near-field optical imaging. Nano Lett 9:903–908
154. Xie C, Mu C, Cox JR, Gerton JM (2006) Tip-enhanced fluorescence microscopy of high-density samples. Appl Phys Lett 89:143117
155. Yano T et al (2007) Confinement of enhanced field investigated by tip-sample gap regulation in tapping-mode tip-enhanced raman microscopy. Appl Phys Lett 91:121101l
156. Mangum BD, Mu C, Gerton JM (2008) Resolving single fluorophores within dense ensembles: contrast limits of tip-enhanced fluorescence microscopy. Opt Express 16:6183–6193
157. Kuhn H (1970) Classical aspects of energy transfer in molecular systems. J Chem Phys 53:101–108
158. Frey H, Keilmann F, Kriele A, Guckenberger R (2002) Enhancing the resolution of scanning near-field optical microscopy by a metal tip grown on an aperture probe. Appl Phys Lett 81:5530–5532
159. Kerker M, Wang D-S, Chew H (1980) Surface enhanced Raman scattering (SERS) by molecules adsorbed at spherical particles. Appl Opt 19:4159–4174
160. Moskovits M (1985) Surface-enhanced Raman spectroscopy. Rev Mod Phys 57:783–826
161. Pettinger B (2010) Single-molecule surface- and tip-enhanced Raman spectroscopy. Mol Phys 108:2039–2059
162. Markel VA et al (1999) Near-field optical spectroscopy of individual surface-plasmon modes in colloid clusters. Phys Rev B 59:10903
163. Moskovits M (2005) Surface-enhanced Raman spectroscopy: a brief retrospective. J Raman Spectrosc 36:485–496
164. Hartschuh A (2008) Tip-enhanced near-field optical microscopy. Angew Chem Int Ed 47:8178–8191
165. Pettinger B, Domke KF, Zhang D, Picardi G, Schuster R (2009) Tip-enhanced raman scattering: influence of the tip-surface geometry on optical resonance and enhancement. Surf Sci 603:11335–1341
166. Neacsu CC, Dreyer J, Behr N, Raschke MB (2006) Scanning probe raman spectroscopy with single molecule sensitivity. Phys Rev B 73:19234061–4
167. Zhang Y, Tan YW, Stormer HL, Kim P (2005) Single molecule tip-enhanced raman spectroscopy with silver tips. J Phys C 111:1733–1738
168. Steidtner J, Pettinger B (2008) Tip-enhanced raman spectroscopy and microscopy on single dye molecules with 15 nm resolution. Phys Rev Lett 100:361011–4
169. Domke KF, Zhang D, Pettinger B (2007) Tip-enhanced raman spectra of picomole quantities of dna nucleobases at au(111). J Am Chem Soc 129:6708–6709
170. Bailo E, Deckert V (2008) Tip-enhanced raman spectroscopy of single rna strands: Towards a novel direct-sequencing method. Angew Chem 47:1658–1661
171. Watanabe H, Ishida Y, Hayazawa N, Inouye Y, Kawata S (2004) Tip-enhanced near-field raman analysis of tip-pressurized adenine molecule. Phys Rev B 69:5155418
172. Anderson MS, Gaimari SD (2003) Raman-atomic force microscopy of the ommatidial surfaces of dipteran compound eyes. J Struct Biol 142:364–368
173. Neugebauer U et al (2006) Resonances of individual metal nanowires in the infrared. Chem Phys Chem 7:1428–1430
174. Yeo BS, Madler S, Schmid T, Zhang W, Zenobi R (2008) Tip-enhanced raman spectroscopy can see more: The case of cytochrome c. JPCC 112:4867–4873
175. Richter M, Hedegaard M, Deckert-Gaudig T, Lampen P, Deckert V (2011) Laterally resolved and direct spectroscopic evidence of nanometer-sized lipid and protein domains on a single cell. Small 7:209–214
176. Wood BR et al (2011) Tip-enhanced raman scattering (ters) from hemozoin crystals within a sectioned erythrocyte. Nano Lett 11:1868–1873
177. Frey HG, Paskarbeit J, Anselmetti D (2009) Tip-enhanced single molecule fluorescence near-field microscopy in aqueous environment. Appl Phys Lett 94:2411161–2411163
178. Cambi A et al (2006) Organization of the integrin LFA-1 in nanoclusters regulates its activity. Mol Biol Cells 17:4270

Chapter 10

Atomic Force Microscopy of Living Cells

David Alsteens and Yves F. Dufrêne

Abstract

Atomic force microscopy (AFM) is a powerful technique for analyzing the structure, properties, and interactions of living cells down to molecular resolution. Rather than using an incident beam as in optical and electron microscopies, AFM measures the tiny forces acting between a sharp tip and the sample surface. While AFM imaging provides information about the nanoscale surface architecture of living cells in real time, single-molecule force spectroscopy analyzes the localization, mechanics, and interactions of the individual cell surface constituents, thereby contributing to elucidate the molecular bases of cellular events like cell adhesion and mechanosensing. In this chapter, we describe the principles of AFM and explain relevant experimental procedures, we survey recent progress made in applying AFM to microbial cells, and we discuss two recent case studies carried out in our laboratory in which the technique could unravel the mechanical and clustering behavior of cell surface sensors and adhesion proteins.

Key words Atomic force microscopy, AFM, Adhesion, Cell adhesion proteins, Cell wall, Cell surface, Clustering, Mechanics, Live cell imaging, Pathogens, Sensors, Single-molecule manipulation, Yeast

1 Introduction

During the past 40 years, the importance of the microbial cell surface in biology, medicine, industry, and ecology has been increasingly recognized [1, 2]. Because they constitute the frontier between the cells and their environment [3, 4], microbial cell walls play several key functions: supporting the internal turgor pressure of the cell [4], protecting the cytoplasm from the outer environment, imparting shape to the organism, acting as a molecular sieve, sensing environmental stresses/stimuli [5–8], and controlling interfacial processes (molecular recognition, cell adhesion, and aggregation) [9–13]. These processes have major consequences, which can be either beneficial, such as in biotechnology (wastewater treatment, bioremediation, and immobilized cells in reactors), or detrimental in industrial systems (biofouling and contamination) and in medicine (interactions of pathogens with animal host tissues and accumulation on implants). Hence, there is considerable interest in improving our understanding of the functions of microbial cell surfaces.

Knowledge of the functions of microbial cell surfaces requires determination of their structural and biophysical properties. Light microscopy has long been a favorite tool of microbiologists, enabling them to count and identify microbial cells, and to observe general structural details. Fluorescence techniques provide valuable information on the cell wall organization, assembly and dynamics, but the resolution remains limited to the wavelength of the light source [14, 15]. Recently, there has been progress in breaking the diffraction limit barrier with the advent of super-resolution microscopy, also named far-field optical nanoscopy, but this super-resolution method is not yet well established in microbiology [16, 17]. Our current perception of microbial cell walls owes much to the development of cryoelectron microscopy techniques, which have allowed researchers to obtain high-resolution views of microbial structures in conditions close to their native state [18]. Notably, cryoelectron tomography—or three-dimensional electron microscopy—provides images of whole bacterial cells, at resolutions that are one to two orders of magnitude higher than those currently attained with light microscopy [19]. Yet, cryoelectron microscopy methods are demanding and require vacuum conditions during the analysis, meaning that live cells cannot be imaged in aqueous solution.

In biophysics, several methods are available to probe the interactions and mechanical properties of single molecules, including flow chamber experiments (using laminar flow, single intermolecular forces are measured between the coated walls of a chamber and coated spheres [20]), microneedles (a bendable glass microneedle, coated with biomolecules, is brought in contact with a functionalized surface and the force generated on the microneedle is measured by beam deflection), the biomembrane force probe (BFP; single-molecule forces are measured using a force transducer made of a phospholipid vesicle maintained by a glass micropipette; the spring of the membrane is tuned by the controlled aspiration pressure applied by the micropipette), optical and magnetic tweezers (force probes are microscopic beads manipulated by an external field, i.e., photon or magnetic fields), and atomic force microscopy (AFM) [21–27]. These force assays cover a wide range of force strengths and length scales that are relevant to biology, going from weak intermolecular interactions to strong covalent bonds. There are basically two ways to exert force on molecules: either via mechanical force transducers which directly apply or sense forces as in microneedles, BFP and AFM, or via external fields (hydrodynamic, magnetic, or photon fields) acting from a distance, as in flow field, magnetic and optical tweezers. Optical and magnetic tweezers enable to noninvasively manipulate biomolecules in solution, thus including within living cells.

By contrast, BFP is a surface technique which offers the advantage that the transducer sensitivity can be tuned in operation to measure a wide range of forces at various biological interfaces including on cell surfaces. Of these methods, only AFM can both localize and force-probe single molecules on live cells. As we discuss below, this key benefit is being increasingly applied to microbial cell walls to gain insight into the organization, interactions, and nanomechanics of their individual constituents [26, 28].

2 Atomic Force Microscopy

2.1 General Principle

A new family of microscopes was born in the early 1980s, the scanning probe microscopes (SPMs). The SPM story started in 1981 with the birth of the scanning tunnelling microscope (STM), which uses tunnelling currents to image conducting and semiconducting flat surfaces down to the atomic level [29]. In parallel, the scanning near-field optical microscope (SNOM) was invented, allowing microscopy to break the resolution limit by exploiting the properties of evanescent waves [30, 31]. Few years later, the AFM was invented by Binnig et al. [32]. The principle is based on monitoring the interaction forces between an ultrasharp tip and a sample surface to generate a topographical surface image. In addition, force measurements make it possible to quantitatively measure forces between interacting moieties.

The main parts of the AFM are the cantilever, the tip, the sample stage, and the optical deflection system consisting of a laser diode and a photodetector (Fig. 1) [33]. The sample is mounted on a piezoelectric scanner which ensures three-dimensional positioning with high precision and accuracy. By applying appropriate voltages, piezoelectric materials expand in contact in some directions in a defined way. This is the basic idea behind the piezoelectric scanner (Fig. 1). Commercial scanners offer scan ranges up to 100 μm in the x, y direction and a few μm in the z range. One problem associated with the piezoscanner is that the expansion (i.e., the response) is not exactly linear to the applied voltage and shows hysteresis when comparing increasing and decreasing applied voltages. For large displacements, this nonlinearity can be important and be a source of inaccuracies.

Most instruments today use an optical method to measure the cantilever deflection with high resolution. The deflection of the cantilever is measured by the shift of a laser beam on the backside of the cantilever onto a split position-sensitive photodetector (Fig. 1). As the cantilever bends, the position of the laser beam on the detector shifts from its initial central point. The resulting voltage difference between the segments of the photodiode gives a direct and linear measure for the deflection of the cantilever.

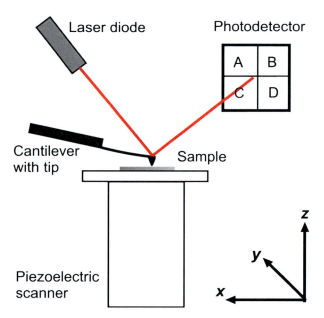

Fig. 1 AFM setup. The sample is mounted on a three-axis piezoelectric scanner. The topography of the sample is monitored by the deflection of the cantilever measured using a laser beam reflected of its back. The voltage difference between the upper and lower half of the photodetector (A + B − (C + D)) is a measure for the vertical deflection of the cantilever

Typically, AFM tips are made of silicon (Si) or silicon nitride (Si_3N_4) using microfabrication techniques. Often the backside of the cantilever is covered with a thin gold layer to enhance its reflectivity. The probe shape has a conical or pyramidal apex with a 2–50 nm curvature radius. For biological specimens, best results are generally obtained with cantilevers exhibiting small spring constants, i.e., in the range of 0.01–0.10 N/m. Note that actual spring constants may substantially differ from values quoted by the manufacturer. Therefore, when accurate knowledge of the force is required, such as in quantitative force measurements, spring constants must be determined experimentally [25].

AFMs can operate in air, vacuum, or liquid [34] and at various temperatures. Because AFM can be operated in aqueous solution, microscopists can use it to visualize in real time how cell surfaces interact with external agents or change during cell growth and division [35].

2.2 Imaging Modes

Various AFM imaging modes are available, which are referred to as contact or dynamic modes. The most widely used in biology is the contact mode, in which sample topography can be measured in different ways. In the constant force contact mode, the sample height is adjusted to keep the deflection of the cantilever (and hence the applied force) constant using a feedback loop (Fig. 2a) [33].

Atomic Force Microscopy of Living Cells   229

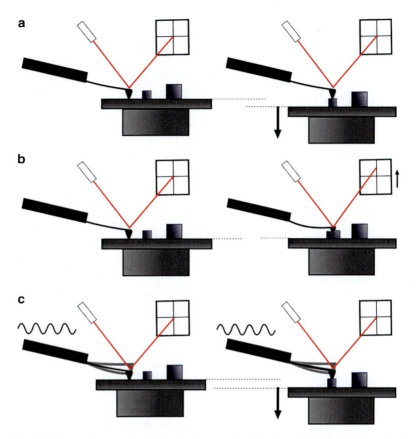

Fig. 2 AFM imaging modes. In contact mode, the idea is to keep the tip in contact with the surface to reveal its topography, either by maintaining a constant force (**a**) or a constant height (**b**). In tapping mode (**c**), the cantilever is oscillating and the amplitude is kept at a constant value

The feedback output is used to display a true height image. Small errors in the regulation of the force generally occur because the feedback loop can only react as the cantilever bends. The error signal is used in error signal imaging (called deflection imaging) and provides an image related to the first derivate of the topography in the scan direction, thus emphasizing edges and asperities of the surface. This mode minimizes large deflections of the cantilever and thus maintains the applied force at low values, which is often necessary to prevent sample damage. Tip–sample force can be adjusted from gently touching to deeply carving the surface. Typical contact forces during imaging of biological samples in liquid range from 100 to 250 pN. Higher forces can deform soft structures and lead to lower resolution due to increased contact area. In the constant-height mode, one simply records the cantilever deflection while the sample is scanned horizontally, i.e., at constant height (Fig. 2b) and the variation of the force is monitored. This mode can be used only for extremely flat surfaces in order to avoid damage to the sample.

Tapping mode is a dynamic AFM imaging mode that has been developed to reduce sample deformation by minimizing the contact time [36]. The cantilever is oscillating near its resonance frequency. During oscillation, the tip interacts with the sample and the amplitude tends to change. It is kept constant using a feedback loop, which is used to generate an image (Fig. 2c). Due to dissipative tip–sample interactions, phase shifts in the cantilever oscillation can be observed and exploited to gain additional information about the physical properties of the scanned surface [36].

2.3 Force Spectroscopy

Force spectroscopy relies on measuring the forces between the tip and the sample with piconewton (pN) sensitivity [25]. The sample is moved up and down by the piezoelectric scanner while measuring the cantilever deflection. This AFM mode can address crucial questions about inter- and intramolecular interactions of cell surfaces at the single-molecule level [37]. Molecular forces that can be measured range from very weak van der Waals (VDW) forces to covalent bonds. Typical interactions occurring at biological surfaces include hydrophobic, hydrophilic, electrostatic, VDW, and hydrogen bonds, and the force measured for a specific biological interaction reflects the sum of all contributions. The measurement of such small forces requires (1) the use of very soft cantilevers, i.e., small spring constants (in the range of 0.01–0.1 N/m), (2) a tip with appropriate nature and size to sense the local force, and (3) a piezoelectric system to precisely control the relative position of the tip.

2.3.1 Force–Distance Measurements

AFM force curves are obtained at a given (x, y) location by monitoring the cantilever deflection as a function of the vertical displacement of the piezoelectric scanner (Z_p) [38, 39]. Raw data are presented as a "voltage–displacement" curve (Fig. 3a) and can be converted into a "force–distance" curve (Fig. 3b) in two steps. First, the voltage is converted into a deflection of the cantilever (Z_c) using the sensitivity of the AFM detector. Secondly, the cantilever deflection is converted into a force using Hooke's law ($F = k_c Z_c$), where k_c is the cantilever spring constant. To obtain the distance, the deflection of the cantilever is subtracted from the scanner displacement.

A typical force distance curve is shown in Fig. 3b. At large tip–sample separation, there is no interaction (Fig. 3, inset 1). Consequently, the deflection of the free cantilever recorded far away from the surface defines the zero-force baseline. When the sample and the tip are close enough, surface forces cause the cantilever to bend. A jump to contact can be observed. Then, the cantilever and the surface come into hard contact, yielding a vertical line in the force–distance curve, in the case of a nondeformable surface (Fig. 3, inset 2). On a soft sample, however, sample indentation may occur. Using appropriate models, information on sample elasticity can be extracted. Hence, the approach curve (blue line)

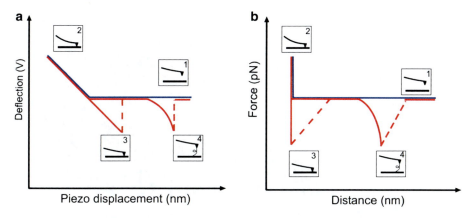

Fig. 3 AFM force experiments. (**a**) Voltage–displacement curve and (**b**) force–distance curve. The approach curve is in *blue* and the retraction in *red*. The cantilever is initially away from the surface (*inset 1*) and comes into contact with the surface (*inset 2*). During retraction, rupture events may occur at different distances due to nonspecific (*inset 3*) or specific (*inset 4*) interactions between the tip and the sample

can be used to measure nanomechanical properties as well as surface forces, including VDW, electrostatic, solvation, hydration, and steric forces [40].

The retraction curve (red line) often exhibits hysteresis due to adhesion. This results in a drop of the deflection below the baseline before the tip–sample contact is disrupted (Fig. 3, inset 3). This type of interaction at small separation distance can be related to the surface energy or the binding forces between rigid surfaces or small molecules. In the case of long, flexible molecules, an attractive force (or elongation force) develop up to large distances until rupture of the tip–molecule bond (Fig. 3, inset 4).

2.3.2 Spatially Resolved Force Spectroscopy

Because most biological samples show lateral heterogeneities at the nano- or microscale, it is very challenging to map the spatial distribution of properties and interactions on such heterogeneous samples. This can be achieved by recording an array of force curves over the surface and assembling these into a force volume (FV) (Fig. 4) [41]. The investigated properties can include charge density [42], adhesion [43], and stiffness [44, 45]. Two important, interconnected issues in FV imaging are the lateral and temporal resolutions. In general, the lateral resolution is related to various factors such as the tip radius, the drift, and the distance dependence of the interaction. In fact, the amount of curves collected, thus the size of FV files, are limited by the data acquisition time. Since the time required for recording a single force curve is between 0.1 and 10 s, the time to acquire a FV of 32×32 curves is on the order of 5 min to 5 h.

In the context of microbial surface analysis, adhesion force mapping offers avenues to explore the distribution of binding sites with nanoscale resolution. The approach to map interaction forces

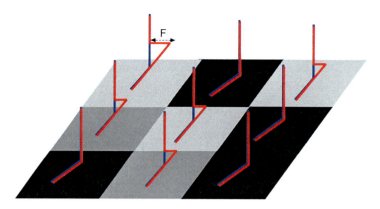

Fig. 4 Spatially resolved adhesion force mapping. Arrays of force curves are recorded in the *x, y* plane using functionalized tips. The unbinding force value (see label "F") is displayed as *grey pixels*, each bright pixel reflecting the detection of a specific adhesion event

across the surface is to record arrays of force curves in the *x, y* plane using functionalized tips (Fig. 4). From each curve, the adhesion force can be extracted and displayed as a grey pixel. The brightness of this pixel reflects the magnitude of the unbinding force (Fig. 4). In this way, direct visualization of the distribution of specific proteins have been reported on live microbial cells [46].

2.4 Experimental Procedure

2.4.1 Tip Functionalization

Applying force spectroscopy to measure chemical or molecular recognition forces and to localize specific binding sites requires functionalization of the tip with relevant chemical groups or biomolecules.

In so-called "chemical force microscopy" (CFM) [47], chemically functionalized tips are obtained by coating the AFM tips with a thin adhesive layer of chromium followed by another thin layer of gold. Gold tips are modified with self-assembled monolayers (SAMs) terminated by specific functional groups [47, 48].

In single-molecule force spectroscopy (SMFS), different strategies have been developed to attach bioligands on tips (Fig. 5). Although there is no "ideal" approach, coupling of nearly any ligand is possible by choosing a proper scheme. Note that these approaches are also valid for preparing flat supports. Several issues must be considered:

1. The force that attaches the molecules should be stronger than the intra- or intermolecular force to be studied.

2. The biomolecules should have enough mobility to freely interact with the receptor.

3. The density of ligands should be reduced to ensure single-molecule detection.

4. Oriented strategies should be preferred to maximize the probability of specific interaction.

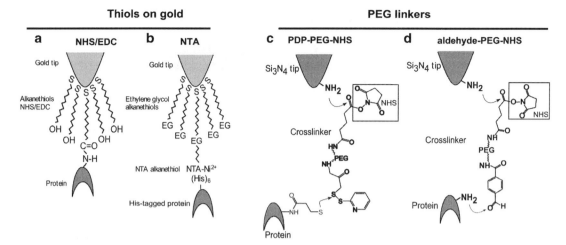

Fig. 5 Strategy to functionalize AFM tips with bioligands. (**a, b**) Alkanethiols on gold using either NHS/EDC (**a**) or NTA chemistry (**b**). (**c, d**) Covalent coupling using aminofunctionalization of silicon nitride tips and PDP-(PEG)$_{18}$-NHS linkers (**c**) or aldehyde-(PEG)$_{18}$-NHS linkers (**d**)

An elegant strategy relies on the strong binding of thiols on gold-coated tips. While some biomolecules that bear thiol groups can directly attached on gold surfaces [49, 50], others can be covalently attached onto carboxyl-terminated SAMs using 1-ethyl-3-(3-dimethylaminopropyl)carbodiimide (EDC) and *N*-hydroxysuccinimide (NHS) (Fig. 5a). It is also possible to orient all biomolecules in the same way by attaching recombinant histidine-tagged protein (His tag) onto AFM tips coated with nitrilotriacetate (NTA)-terminated alkanethiols [51, 52] (Fig. 5b). The disadvantage of this coupling method is the relatively low strength of the NTA-His bond.

A third approach is to covalently anchor biomolecules on silicon tips using various aminosilanization procedures [53–55] (using aminophenyltrimethoxysilane [APhS] or 3-aminopropyltriethoxysilane [APTES]) or amino-functionalization with ethanolamine-hydrochloride [55, 56]. After amino-functionalization of the silicon tips, a poly(ethylene glycol) (PEG) chain is usually attached via its reactive terminus (usually NHS ester). Sometimes, the PEG linker already carries the ligand at its second end [57, 58]. In other cases, heterobifunctional PEG linkers with one NHS ester function and another coupling function are used. The most prominent cross-linkers for amino-functionalized tips are:

1. Pyridyldithiopropionyl (PDP)-PEG-NHS (Fig. 5c); this is the most frequently used PEG linker in force spectroscopy [59–61]; the PDP function can couple a thiolated protein via disulfide bound formation in phosphate buffer; unfortunately, most ligands do not contain a free thiol group.

2. Aldehyde-PEG-NHS [56] (Fig. 5d); although this linker has two amino-reactive sites, the NHS ester is much more reactive

and will bind preferentially to the amine groups on the tip surface when chloroform is used as solvent; the other end remains active and can be further conjugated to amino groups of the lysine residues of the proteins; reaction of the protein with aldehyde results in the formation of a Schiff base, which is subsequently fixed by reduction with NaCNBH$_3$; the tendency for loop formation on the tip surface can be largely suppressed by applying PEG linkers at a concentration of 6.6 mg/mL rather than 1 mg/mL as with other linkers.

3. Acetal-PEG-NHS; recently, a new method developed by the Gruber team (Univ. Linz) was introduced, in which acetal linker replace aldehyde in the starting reagent; this linker cannot form loops on the tip and therefore low concentration is sufficient; an additional step to convert the acetal function into aldehyde function is needed.

Lastly, in single-cell force spectroscopy [62], single cells are attached on cantilevers or on tips in order to quantify cell–cell and cell–surface interactions. In microbiology, some groups have developed procedures for coating AFM cantilevers with bacteria [63, 64] and yeasts [65]. Microbial cells can be anchored by attaching single cells with a small amount of glue [65] or by using glutaraldehyde treatments to create covalent cross-linking between cells and tips [63, 64]. By combining this with the use of well-characterized isogenic strains, Razatos et al. [63] succeeded in discerning differences in the force of interaction among *Escherichia coli* strains with only small variations in the composition and length of core lipopolysaccharides. However, these treatments are likely to affect the structure and properties of cell surfaces. A second approach is based on attractive electrostatic interactions. In a first step, the AFM probe is coated with polyethyleneimine (PEI). In the second step, negatively charged cells are attached to the positively charged PEI-coated probes [66]. Another elegant method is to attach individual cells via lectins such as concanavalin A (Con A) which then allows specific adhesion forces to be measured between cells [67].

2.4.2 Sample Preparation

A key issue when using AFM to investigate biological processes in real time and under physiological conditions [68] is to attach the sample properly while avoiding its alteration or denaturation. An efficient method to immobilize adherent cells is to take advantage of their natural ability to spread or/and to adhere to surfaces [69, 70]. Animal cells, with their natural tendency to spread, can be imaged on solid surfaces such as glass [71, 72]. Coating the surface with proteins [73] or poly-L-lysine [74] may be used to enhance immobilization. Some prokaryotes can also be easily imaged without any chemical treatment. For example, mycobacteria were found to adhere on polycarbonate surfaces owing to hydrophobic interactions between the outer hydrophobic layer and the polymer surface [52].

As most bacteria do not spontaneously adhere to surfaces, other approaches have been developed to achieve bacterial immobilization. The choice depends on the cell type and on research objectives. Rod-shaped bacteria are often immobilized on a support pretreated with polycations (poly-L-lysine, PEI, etc.) [66, 75–77] or retained by covalent anchorage to the surface [78]. Recently, Colville et al. suggested avoiding the use of poly-L-lysine coating for bacterial attachment due to its bactericidal activity [79]. Furthermore, these treatments may induce rearrangement and denaturation of the cell wall, and contamination of the tip.

To avoid these issues, another way of immobilizing microbes is to trap them mechanically. The method consists in filtering a cell suspension through a Millipore filter with a pore size slightly smaller than the diameter of the cells [80]. This technique provides a strong fixation of the cell (which is a key point for obtaining good resolution) and circumvents problems associated with the large height of the sample (e.g., few µm for yeast cells). The main disadvantages are the access to only part of the cell surface and the risk of selecting a cell population based on the pore size. This technique is easily extended to a large variety of microbes owing to a large range of pore size commercially available. Additionally, fine tuning of pore size can be achieved by chemical treatment [81].

2.5 Recent Progress in Microbial AFM

Since the late 1980s, AFM has been increasingly used for analyzing biological samples and it is now established as an outstanding multifunctional nanotool (for reviews, see [24, 28, 37, 82–85]) offering many advantages over traditional methods: experiments on single live cells rather than solubilized or fragmented cell membranes and cell walls, information at the single-molecule level instead of averaged information on large ensembles of molecules, probing the biomolecular machinery of the cell in its native environment, and approaching the relationship between structure and function [24]. Here below, we survey recent advances made in applying AFM to address microbiological questions.

2.5.1 AFM as an Imaging Tool

There has been rapid progress in revealing structural details of purified cell walls and membranes. The first impressive images of purified membranes of bacteriorhodopsin in buffer solution were obtained in 1990 [86] with a resolution down to 1.1 nm. Similar resolutions were obtained on a large variety of other crystalline membrane proteins (purple membrane of *Halobacterium* [87], porins of *E. coli* [88, 89], S-layers from *Corynebacterium glutamicum* [90], from *Deinococcus radiodurans* [91], and from *Bacillus sphaericus* [92]). In these studies, AFM is not only used as an imaging tool but also serves to better understand the mechanism underlying function. For instance, Müller et al. [91] demonstrated the ability of AFM to image HPI S-layers of *D. radiodurans* and to directly monitor conformational changes of the individual molecules. In the same way, Müller and Engel [93] were able to observe voltage and pH-dependant conformational changes of extracellular loops of OmpF. The development of high-speed AFMs [94–96] (HS-AFM) now permits to reduce the image acquisition time to less than 20 ms. The application of HS-AFM enabled Casuso et al. [97] to record high-resolution movies acquired at a 100-ms frame acquisition time. In regions between different bacteriorhodopsin arrays, motions of the trimers were observed. Dynamic changes in protein conformation in response to illumination were visualized in bacteriorhodopsin using HS-AFM [98]. These findings provide novel perspectives for analyzing diffusion process of membrane proteins. A new AFM mode enables to control the force applied by the AFM probe on the sample. In this mode, the setpoint is continuously and automatically adjusted during the experiment, which enables to obtain high-resolution image of purple membranes at molecular resolution and of cells at high signal-to-noise ratio, for hours without intrinsic force drift [99].

Rapid advances have also been made in imaging live cells. The first images of live microbial cells in native conditions were obtained in 1995 by Kasas and Ikai [80]. Rapidly, many microbes were observed, revealing their native surface morphology (yeasts [80, 100], fungi [101], diatoms [102, 103], bacteria [104, 105], etc.). So far, the best resolution obtained on a live microbial cell in buffer condition is on the order of 10 nm. Such resolution was obtained by visualizing protein rodlets on *Aspergillus* fungal spores [106] and the S-layer of *C. glutamicum* [107]. Imaging the S-layer directly on live bacteria permitted to observe an underlying regular pattern of nanogrooves supposed to function as a molecular template promoting the 2D assembly of the S-layer.

AFM appears to be an ideal tool to image dynamic cellular processes. A beautiful example is the germination of *A. fumigatus* conidia [108], during which progressive disruption of the rodlet layer revealed the underlying inner cell wall structures. Similarly, Kailas et al. followed the process of cell division in *Staphylococcus aureus* [109, 110]. Detailed images of the surface of dividing cells showed ring-like and honeycomb structures at 20-nm resolution [81]. Structural analysis of *Bacillus atrophaeus* spores revealed previously unrecognized germination-induced alterations in spore coat architecture [111]. The nascent structure of the emerging vegetative cell showed a porous network of peptidoglycan, consistent with a honeycomb model structure. Insights into the nanoscale organization of cell wall peptidoglycan were recently revealed by Andre et al. [112]. Using mutant strains defective in cell wall polysaccharides, AFM images revealed that peptidoglycan forms periodic bands running parallel to the short axis. Such structures were missing on purified sacculi showing the importance to image directly on live cells and to avoid aggressive treatments.

Real-time AFM imaging also enables to investigate the activity of drugs on microbial cell walls [113–116]. Using HS-AFM, Fantner et al. [117] were able to investigate the kinetics of antimicrobial peptide activity. The action of the antimicrobial peptide CM15 on *E. coli* occurred in two steps: a first step or time variable incubation phase (which takes seconds to minutes to complete) was followed by a more rapid execution phase.

2.5.2 AFM as a Chemical Probe

AFM force spectroscopy with chemical tips, also called CFM, is a powerful tool to probe the physicochemical properties of cell walls. As a pioneering example, the surface hydrophobicity of *Phanerochaete chrysosporium* spores was mapped by recording multiple force–distance curves using OH (hydrophilic) and CH_3 (hydrophobic) terminated probes [101]. The curves always showed no adhesion, indicating that the spore surface was uniformly hydrophilic. Presumably, the nonsticky character of the fungal spore surface must play a role in determining its biological functions, namely protection and dispersion. More recently, Dague et al. [106] demonstrated with hydrophobic tips the feasibility to quantify and map hydrophobic heterogeneities on live cells. Similarly, the application of CFM with hydrophobic tips enabled to map chemical surface heterogeneities on two bacterial species, *Acinetobacter venetianus* and *Rhodococcus erythropolis* [118].

AFM probes functionalized with ionizable groups could also be employed to characterize electrostatic interactions and cell surface charge on yeast cells [119]. Hence, CFM appears as a complementary tool to classical experimental methods used to assess microbial surface hydrophobicity and surface charge. CFM circumvents the limitations of macroscopic assays and offers a way to map surface properties with a high lateral resolution.

2.5.3 AFM as a Nanomechanical Tool

AFM can also be used to probe the nanomechanical properties of microbial cell walls [120]. In pioneering work, Touhami et al. [45] showed that the bud scar of yeast cells is stiffer than the rest of the cell wall yielding Young's modulus (E_s) values of ~6 MPa and ~1 MPa for the bud scar and surrounding cell surface, respectively. Pelling et al. [121] reported for the first time the AFM Young's modulus measurements on a living bacterium in aqueous conditions. Studies using "bacteria threads" have estimated E_s values on "wet" bacteria 100 times higher that values obtained in aqueous conditions (≈ 200 kPa).

da Silva and Teschke [122] investigated the effect of an antimicrobial peptide on *E. coli*. The stiffness was monitored by AFM before and after injection of the peptide, revealing a decrease in the stiffness of the bacterial cell wall. Gaboriaud et al. [123]

probed the influence of environmental parameters, such as pH, on the mechanical surface properties of bacteria at a nanoscale level. The variations of E_s were associated with the swelling of a polymer fringe when varying the pH.

More recently, spatially resolved maps of elasticity enabled Francius et al. to quantitatively map the Young Modulus of diatoms [103] and to correlate the increase in stiffness to the incorporation of silica in the cell wall. Such nanomechanical measurements were also used to discriminate between dead or alive Gram-negative bacteria [124]. Cells with a damaged membrane presented a higher Young modulus than healthy cells. Possible explanations for this phenomenon include the collapsing of the polysaccharide layer, the stiffening of the Braun lipoproteins, and an increase of turgor pressure after heating (20 min, 45 °C).

2.5.4 AFM as a Single-Molecule Manipulation Tool

SMFS is a valuable approach to detect, localize, and characterize the molecular interactions that occur between complementary biomolecules, including streptavidin–biotin [125], DNA nucleotides [49], and antibody–antigen [53]. In addition, SMFS can measure intramolecular interactions, thereby providing insight into elasticity of single macromolecules, including proteins, polysaccharides, and DNA [49, 126–129].

During the past decade, there has been progress in applying these SMFS experiments to living microbial cells. Advances in such in vivo analyses have been rather slow worldwide because they are very delicate, complex, and require strong expertise in acquiring and interpreting the data. Dupres et al. [52] mapped the nanoscale distribution of heparin-binding hemagglutinin adhesins (HBHA) on live bacteria. Surprisingly, HBHA appeared to be organized into nanodomains. Gilbert et al. [130] used SMFS with antibiotic probes to explore the vancomycin binding sites on Gram-positive bacteria. Francius et al. [131] mapped and analyzed the conformation of individual polysaccharides directly on *Lactobacillus rhamnosus* GG (LGG) and on a mutant defective in cell wall polysaccharide production.

Notably, AFM can be used to stretch single polymers chains allowing determination of their nanomechanical properties. Performing SMFS on germinating spores showed elongation forces reflecting the macromolecular stretching of polysaccharides [132]. The elongation was consistent with the elastic deformation of dextran and amylose polysaccharides. Stretching the surface polysaccharides of *Pseudomonas putida* cells [133, 134] revealed the heterogeneity in surface biopolymers. Modeling of extension curves confirmed that the heterogeneity is more than a matter of differences in molecular weights, since a range of stiffnesses was also observed. More recently, lectin-functionalized tips permitted to discriminate two cell wall polysaccharides on the LGG [131] surface.

3 Case Studies

3.1 Measuring the Mechanical Behavior of Membrane Sensors

In *S. cerevisiae*, surface stresses acting on the cell wall or plasma membrane are detected by a group of five membrane sensors or wall stress component (Wsc): Wsc1, Wsc2, Wsc3, Mid2, and Mtl2 [8]. The sensors carry a cysteine motif near the amino terminal end, followed by a serine/threonine-rich (STR), highly mannosylated extracellular region, a single transmembrane domain (TMD), and a relatively short cytoplasmic tail. Mannosylation of the STR region is crucial for the sensor function and has been proposed to provide them with a rod-like structure. Yet, direct evidence for their biophysical properties was not available until now. We applied SMFS to investigate the mechanical properties of single Wsc1 sensors on living cells. We showed that Wsc1 behaves like a linear nanospring resisting to mechanical force and responding to cell surface stress [135].

3.1.1 Detection of Wsc1 Sensors on Live Cells

As the native Wsc1 sensor is too short to reach the outermost cell surface, the extracellular part of the Mid2 protein with a His tag was added to render the sensor specifically accessible to an Ni^{2+}–NTA tip (Fig. 6a). This method enabled us to detect and force probe individual His-tagged sensors on living yeast cells (Fig. 6). *S. cerevisiae* cells expressing the elongated Wsc1 derivative were immobilized on a polycarbonate membrane (Fig. 6b). Force–distance curves recorded between the yeast surface and an NTA-derivatized tip showed 162 ± 38 pN adhesion forces (Fig. 6c). A series of independent control experiments showed that these forces reflected the detection of individual His-tagged Wsc1 molecules.

3.1.2 Wsc1 Behaves Like a Nanospring

Force–extension curves obtained for proteins are usually well described by the worm-like chain (WLC) model. Notably, the

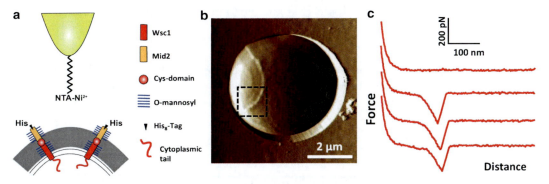

Fig. 6 AFM detects single Wsc1 sensors on live cells. (**a**) Force–distance curves were recorded on yeast cells expressing Wsc1 derivatives with an extended His tag, using AFM tips functionalized with Ni^{2+}-NTA groups. (**b**) AFM deflection image in buffer solution of a yeast cell expressing the elongated sensors and (**c**) representative force curves obtained with an Ni^{2+}–NTA tip. Adapted with permission from [135]

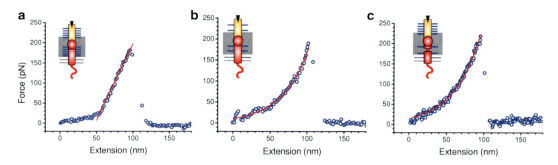

Fig. 7 Wsc1 sensors behave as linear springs while mutants altered in the extracellular STR region do not. (**a**) Representative force–extension curve obtained upon stretching a single Wsc1 molecule. Clearly visible are two extension regimes, reflecting elongation at nearly zero force, followed by a Hookean spring behavior. (**b**) Representative force–extension curves obtained upon stretching single Wsc1 sensors in a mutant in which mannosylation of the extracellular STR region is substantially reduced, and (**c**) in a mutant in which the STR region is interrupted by a run of glycines. Adapted with permission from [135]

force–extension curves recorded for Wsc1 (Fig. 7a) could not be fitted with the WLC model. Instead, the curves showed a first nearly constant force extension, corresponding to the straightening of the extracellular polypeptide chain. Following this first regime, a linear region where the force is directly proportional to the extension was observed. This behavior is characteristic of a Hookean spring and may be interpreted as follows [136]. Using the slope (s) of the linear portion of the curves, the sensor spring constant (k_s) can be calculated assuming that the deflection of the cantilever is only due to the elastic deformation of the sensor and knowing the cantilever spring constant (k_c). Because both the sensor and the cantilever can be modeled as springs (Fig. 8), both the tip and the sensor can move during force experiment. The distance that the piezo has moved, Z_p, is defined as zero where the two springs first make contact and become positive as the springs extend.

The force F, at any piezo position, can be calculated as [136]:

$$F(Z_p) = k_c Z_c = k_s i \qquad (1)$$

For two linear springs in contact, the magnitude of the piezo movement is equal to the sum of the compression of the two springs, or:

$$Z_p = Z_c + i \qquad (2)$$

Since the value of the linear slope is the ratio of Z_c to Z_p, combining Eqs. 1 and 2 gives:

$$k_s = \frac{k_c s}{1 - s} \qquad (3)$$

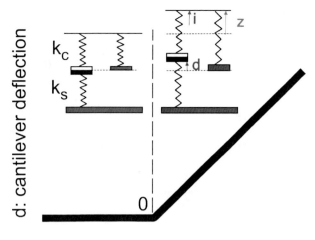

Fig. 8 Schematic diagram representing the Hookean behavior of two springs in series. On the *left side* of the figure, no extension has occurred and the piezo position is defined as zero. As the piezo moves a distance z, the cantilever spring (k_c) extends by a distance i and the sensor spring (k_s) extends by Z_p

Using Eq. 3, the spring constant of Wsc1 sensor is found to be 4.6 ± 0.4 pN/nm, a value that is very close to the behavior of ankyrin repeats [137]. We suggest that the sensor spring behavior reflects the straightening of the stiff, coil-like shape Wsc1 molecule, and possibly of the whole protein–cell wall complex as well. Interestingly, we observed a substantial reduction of the sensor spring constant when lowering the salt concentration or increasing the temperature. This indicates that the mechanical behavior of Wsc1 is influenced by cell surface stress.

3.1.3 Relative Contribution of the STR and Mid2 Regions to the Measured Spring Behavior

Glycosylation of the extracellular STR region of Wsc1 is believed to be responsible for its stiff, extended conformation [138]. Therefore, the relationship between the structure of Wsc1 and its mechanical properties was examined through mutagenesis. Two mutants affected in the extracellular STR-region were investigated. The first mutant (Fig. 7b) lacking one of the protein O-mannosyltransferases resulted in an underglycosylation of the sensor. The second mutant (Fig. 7c) contained an insertion of ten glycines residues in the center of the STR region. As can be observed in Fig. 7b, c, both mutants no longer exhibited a spring behavior, indicating that glycosylation of Wsc1 substantially contributes to its linear spring properties.

Could the measured spring properties be altered by the Mid2 elongation? To answer this question, we investigated a strain expressing His-tagged Wsc1 sensors lacking the Mid2 elongation. As expected, these native sensors could only be localized on the bud scar where the cell wall is thinner (Fig. 9a).

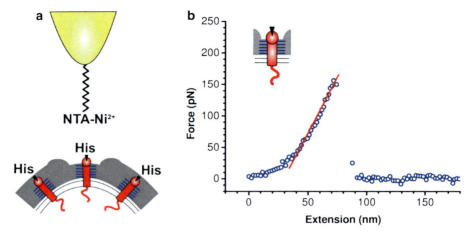

Fig. 9 Short, native Wsc1 sensors also behave as nanosprings. (**a**) The NTA tips reveal local accumulation of short native Wsc1 molecules in the bud scar region. (**b**) Representative force–extension curve revealing linear spring properties. Adapted with permission from [135]

Importantly, Fig. 9b shows that native sensors behaved like linear nanosprings with a spring constant of 4.8 ± 0.2 pN/nm, very close to the elongated sensors.

These experiments demonstrate that AFM can be combined with advanced protein design by yeast genetics to study the function and the mechanical properties of yeast sensors in living cells down to the single-molecule level. Specifically, SMFS is a valuable tool to force probe not only surface-associated proteins but also proteins that are embedded within the yeast cell wall.

3.2 Nanomechanics and Clustering of the Als5p Cell Adhesion Protein from *Candida albicans*

Studying the cellular "unfoldome" [139], i.e., cellular proteins that can be unfolded as part of their physiological function, has recently emerged as an important issue in cell biology. In animal cells, force-induced conformational changes in cell adhesion proteins like integrins are known to increase the protein binding strength [139–141]. It is also increasingly recognized that the early stage of adhesion involves the formation of adhesion domains composed of aggregated adhesion proteins [140, 142]. Such adhesion domains have been shown to grow and strengthen under force [143, 144]. Whether such force-induced unfolding and clustering occur with microbial cell adhesion proteins ("adhesins") remains essentially unknown.

We investigated the nanomechanical [145] and clustering [146] properties of cell adhesion proteins from the pathogen *Candida albicans*. Adhesion of *C. albicans* yeast cells to host tissues is a key factor in the maintenance of commensal and pathogenic states, which is mediated by a family of cell–surface proteins known as the agglutinin-like sequence (Als) proteins [9]. Of the eight different Als proteins, Als5p is one of the most extensively characterized. Als proteins possess four functional regions (Fig. 10a), i.e., an N-terminal

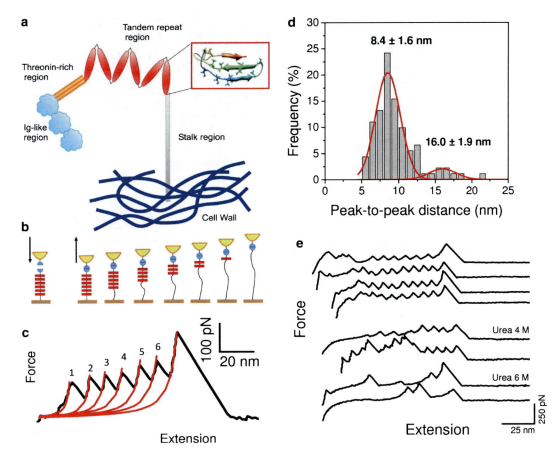

Fig. 10 Unfolding isolated Als5p proteins. (**a**) Representation of the Als5p from *C. albicans*. The tandem repeat (TR) region is made of multiple glycosylated 36-amino acid repeats that are arranged in antiparallel β-sheets (*inset*). (**b**) Principle of the SMFS experiments. Ig-T-TR$_6$ fragments were attached on a gold surface and stretched via their Ig domains using an Ig–T tip. (**c**) Force–extension curves obtained by stretching single Ig-T-TR$_6$ showed periodic features reflecting the sequential unfolding of the TR domains. Force peaks were well described by the WLC model (*red lines*). (**d**) Histogram of the peak-to-peak distances for the first six peaks. (**e**) Addition of urea dramatically altered the unfolding peaks due to hydrogen bond disruption (lower traces). Adapted with permission from [145]

immunoglobulin (Ig)-like region, which initiates cell adhesion, followed by a threonine-rich region (T), a tandem repeat (TR) region that participates in cell–cell aggregation, and a stalk region projecting the molecule away from the cell surface.

3.2.1 Unfolding Isolated Als5p Proteins

To probe the mechanical unfolding of isolated Als5p proteins, soluble Ig-T-TR$_6$ fragments terminated with a His tag were grafted onto gold surfaces modified with NTA groups and stretched by their Ig domain using AFM tips modified with Ig-T fragments (Fig. 10b). Force–extension curves showed a sawtooth pattern with well-defined peaks (Fig. 10c). The first six peaks showed ascending forces ranging from 160 to 240 pN corresponding to the

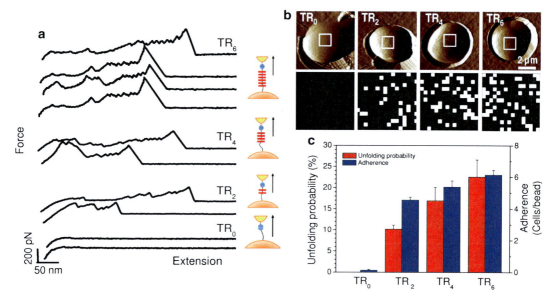

Fig. 11 Unfolding Als5p proteins on live cells. (**a**) Force–extension curves on *S. cerevisiae* expressing Als5p with six, four, two, and no TR. (**b**) Mapping unfolding forces on yeast cells expressing different numbers of TR. (**c**) Histogram showing the unfolding probability and the level of yeast adherence as a function of the number of TR. Adapted with permission from [145]

force-induced unfolding of the secondary structures of individual TR domains and could be fitted by the WLC model (Fig. 10c, red lines). Our measured forces (160–240 pN) are in the range of unfolding forces reported for β-fold domains, such as Ig and fibronectin domains in titin [126] and tenascin [127], and larger than the forces needed to unfold α-helical domains [128]. The change in contour length between the first consecutive six peaks was constant, $\Delta L_c = 8.4 \pm 1.6$ nm, and very close to the peak-to-peak distance (Fig. 10d) revealing that the chain is fully extended at this force. This length corresponds to the expected 36 amino acids of a single TR supporting our interpretation of the sawtooth pattern. The seventh peak showed a much higher force and larger spacing and could be attributed to the stretching of Ig domains, leading to the rupture of the Ig–Ig complex. Urea altered the shape of the unfolding peaks (Fig. 10e, lower traces), reflecting a loss of the mechanical stability of the TR domains due to hydrogen bond disruption. This observation correlated with cellular behavior since Als5p-mediated adhesion is known to be reversibly inhibited by urea [147].

3.2.2 Unfolding Als5p Proteins In Vivo

Notably, single Als proteins were localized and unfolded directly on live cells. Force–extension curves were recorded across the surface of yeast cells expressing Als5p with different numbers of TR (Fig. 11a). The curves obtained on cells expressing six repeats (TR$_6$) were similar to those found on isolated proteins. Cells expressing a smaller number of repeats showed patterns with fewer force peaks. The distribution of unfolding forces over the cell surface (Fig. 11b) revealed an increasing number of exposed proteins with the number of TR.

As shown in Fig. 11c, cells expressing a smaller number of repeats showed lower unfolding probability. The level of yeast adherence (number of cells per fibronectin-coated bead) was correlated with the unfolding probability (Fig. 11c), suggesting that the mechanical properties of the TR region are important for yeast adherence. These SMFS experiments demonstrate that stretching of Als5p leads to the unfolding of their individual TR domains. Presumably, unfolding of the domains leads to the exposure of hydrophobic groups that would promote fungal adhesion.

3.2.3 Imaging Force-Induced Als5p Nanodomains on Live Cell

Dual Detection of Als5p Proteins on Live Cell

Understanding how surface-associated proteins assemble to form nanodomains is an important issue [142]. To investigate whether Als5p is able to form such domains, we probed the surface distribution of Als5p on live cells using SMFS with tips bearing antibodies directed specifically towards the protein. To this end, a V5 epitope tag was inserted at the N-terminal end (Fig. 12a) and AFM tips terminated with anti-V5 antibodies (Fig. 12b) were used to detect and map individual Als5p by recording spatially resolved force curves on 1 μm^2 areas. As shown in Fig. 12c, 16 % of the force curves featured adhesion force peaks. The corresponding adhesion force histogram displayed four maxima centered at 45 ± 15, 98 ± 32, 180 ± 12, and 220 ± 20 pN. The two first peaks were attributed to the relatively weak binding of the anti-V5 antibody to the V5 epitope. Strikingly, force curves with large adhesion peaks (~180 and ~220 pN) showed profiles characteristic of Als5p binding to protein ligands (Fig. 12d, red curves). These curves showed sawtooth patterns with multiple force peaks that we reported earlier as the sequential unfolding of the Als5p TR domains. These observations lead us to conclude that the anti-V5 tip is capable of dual detection, i.e., molecular recognition of Als5p via the V5 epitope or strong multipoint attachment to ligands on the antibody (Fig. 12d, insets).

Mechanical Force Triggers Als5p Nanodomains

We next mapped the spatial arrangement of Als5p proteins. Adhesion force maps on 1 μm^2 obtained from wild-type cells (Fig. 13a, first column) on which force was never applied revealed that the proteins were homogeneously distributed (Fig. 13b, first column). Recording a second map on the same area led to an increase of the total density from 172 ± 16 to 268 ± 13 proteins/μm^2. In addition, proteins in the second force map formed clusters of 100–500 nm (Fig. 13c, first column). As shown in Fig. 13e, the fraction of clustered molecules increased from 8 to 61 %. This dramatic change in spatial organization was reproducibly observed upon recording additional maps.

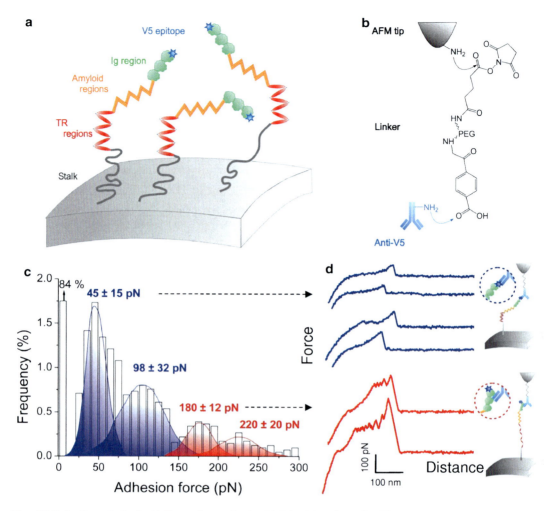

Fig. 12 Detection of single Als5p on live cells. (**a**, **b**) Principle of the SMFS experiment. *S. cerevisiae* cells expressing Als5p proteins tagged with a V5 epitope (**a**) were probed using AFM tips terminated with an anti-V5 antibody (**b**). Adhesion force histogram (**c**) and representative force curves (**d**) obtained by recording spatially resolved force curves over the cell surface using anti-V5 tips. The anti-V5 tip is capable of dual detection: the *blue curves* show single weak adhesion peaks reflecting recognition of the V5 tag, and the *red curves* feature sawtooth patterns with multiple force peaks documenting the unfolding of the entire protein via Ig binding. Reproduced with permission from [146]

Fig. 13 (continued) *squares* in **a**). *Blue* and *red pixels* correspond to forces smaller and larger than 150 pN, respectively, thus to V5-tagged Als5p recognition and unfolding. (**c**) Second adhesion force maps recorded on the same target area (maps #1′). (**d**) Adhesion force maps recorded on remote areas (maps #2) localized several hundred nanometers away (see *dashed squares* in **a**). (**e**) Surface density histograms showing the number of proteins per square micrometers measured for WT, WT$_k$, and for V326N cells in three different conditions. Reproduced with permission from [146]

Atomic Force Microscopy of Living Cells   247

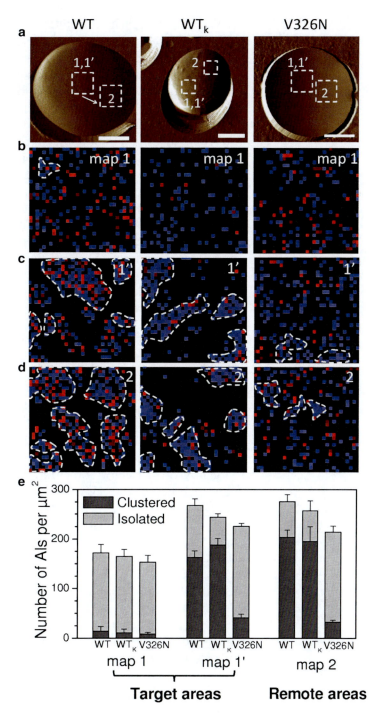

Fig. 13 Formation and propagation of Als5p nanodomains. (**a**) AFM topographic images of wild-type (*first column*), heat-killed (*second column*), and V326N mutant (*third column*) cells expressing V5-tagged proteins. (**b**) Adhesion force maps recorded with an anti-V5 tip on a given target area of the native cells, i.e., cells that were never subjected to force (maps #1; recorded on the *dashed*

We observed that heat-killed cells (WT$_k$) showed the same behavior as live cells (Fig. 13a–c, second column), i.e., increased Als5p density (Fig. 13e) and nanodomains formation. These results show that the force-induced clustering of Als5p is independent of cellular metabolic activity. We also found that clustering is abolished with a single-site mutation in the conserved amyloid-forming sequence. A V326N mutation in the Als5p sequence reduces TANGO β-aggregation potential of the amyloid region from 93 to 4 % [148]. To test whether the amyloid-forming sequence is critical to nanodomains formation, the V326N mutation was incorporated into full-length Als5p. Remarkably, Als5p clustering properties were almost completely abolished in the V326N mutant (Fig. 13a–c, third column). Since mutant cells also showed reduced aggregation, our data suggest that (1) amyloid interactions represent the driving force underlying Als5p clustering and (2) the resulting adhesion domains play a pivotal role in cellular aggregation.

Finally, we asked whether the force-induced nanodomains are localized or whether they can propagate across the surface. To this end, force curves were recorded on remote areas, i.e., areas localized several micrometers away from the first maps (Fig. 13d). Strikingly, nanodomains were observed on remote areas that were very similar to those observed on target areas for both WT and WT$_k$. These results indicate that delivering piconewton forces on a target area induces the formation of Als5p clusters across the entire cell surface. The extended length of Als proteins can be up to 500 nm, giving a radius of gyration of almost 1 μm across the cell surface. Owing to this radius of gyration, Als adhesins are able to reorganize on a scale of 100–500 nm even though they are covalently anchored to the cell wall.

Domain Formation and Propagation Are Slow, Time-Dependent Processes

Because clusters were essentially never seen in primary maps (see map 1 in Fig. 13), we postulated that nanodomains formation and propagation are rather slow processes. To further analyze this time dependency, cells were preactivated by force probing during increasing periods of time (Fig. 14), i.e., 12.5 min in Fig. 14b, 25 min in Fig. 14f and 37.5 min in Fig. 14j. These preactivation steps were followed by two classical maps. From these experiments, we learned that (1) 25 min of preactivation induces formation, but not propagation of the clusters and (2) propagation occurs in a time scale of above 35 min at a speed of ~20 nm/min. Hence, nanodomains formation and propagation are two slow, time-dependent processes.

4 Conclusion

With its ability to observe single cells at nanometer resolution, to monitor structural dynamics in response to environmental changes or drugs, and to detect and manipulate single-cell surface constituents,

Atomic Force Microscopy of Living Cells   249

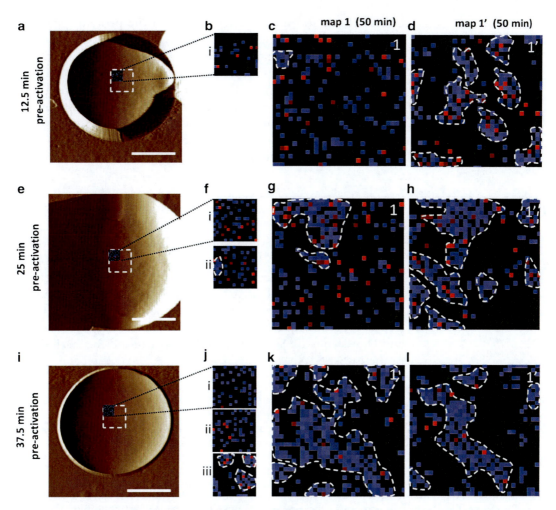

Fig. 14 Timescale of the Als5p reorganization. (**a**, **e**, **i**) AFM topographic images (scale bars: 1.5 μm), in buffer, showing three different cells expressing V5-tagged Als5p proteins. (**b**, **f**, **j**) Small preactivation maps recorded on 500 nm × 500 nm areas with an anti-V5 tip on the three native cells (maps #i, ii, and iii recorded in the *upper left corner* of the *dashed squares* shown in **a**, **e**, **i**). (**c**, **d**, **g**, **h**, **k**, **l**) Two consecutive adhesion force maps of 1 μm × 1 μm (maps #1 and #1′) recorded on the same preactivated areas as shown in **b**, **f**, **j**. Reproduced with permission from [146]

AFM provides new insight into the structure–function relationships of cell surfaces. We used AFM and genetic manipulation to measure the mechanical behavior of single Wsc1 sensors in living yeast cells, revealing they behave as nanosprings. This method offers exciting prospects for investigating how proteins respond to forces in living cells—including mammalian cells—and how mechanosensing events proceed in vivo. In the cell adhesion context, we applied AFM to measure the forces required to sequentially unfold Als5p adhesion proteins from the pathogen *C. albicans* both in vitro (on isolated proteins) and in vivo (directly on live cells). The data suggest that the modular and flexible nature of Als conveys both strength and toughness to the protein, providing high resistance to mechanical force and making it ideally suited to function as an adhesion molecule. We also used time- and spatially resolved AFM to demonstrate the force-induced formation and propagation of Als adhesion nanodomains on living cells. The results indicate that the force-induced clustering of adhesins may be a general mechanism for activating cell adhesion in microbial pathogens and open up exciting prospects for fighting fungal infections using antiadhesion molecules.

Acknowledgments

Work in our team was supported by the National Foundation for Scientific Research (FNRS), the Université catholique de Louvain (Fonds Spéciaux de Recherche), the Federal Office for Scientific, Technical and Cultural Affairs (Interuniversity Poles of Attraction Programme), and the Research Department of the Communauté française de Belgique (Concerted Research Action). Y.F. Dufrêne and D. Alsteens are Senior Research Associate and Research Fellow of the FRS-FNRS.

References

1. Beveridge TJ, Graham LL (1991) Surface layers of bacteria. Microbiol Rev 55(4):684–705
2. Mozes N et al (1991) Microbial cell surface analysis. Structural and physico-chemical methods. VCH Publishers, New York
3. Beveridge TJ (1981) Ultrastructure, chemistry and function of the bacterial cell wall. Int Rev Cytol 72:229–317
4. Wessels JGH (1993) Wall growth, protein excretion and morphogenesis in fungi. New Phytol 123(3):397–413
5. García-Rodríguez LJ et al (2005) Cell integrity signaling activation in response to hyperosmotic shock in yeast. FEBS Lett 579(27):6186–6190
6. Levin DE (2005) Cell wall integrity signaling in *Saccharomyces cerevisiae*. Microbiol Mol Biol Rev 69(2):262–291
7. Klis FM et al (2002) Dynamics of cell wall structure in *Saccharomyces cerevisiae*. FEMS Microbiol Rev 26(3):239–256
8. Rodicio R, Heinisch JJ (2010) Together we are strong—cell wall integrity sensors in yeasts. Yeast 27(8):531–540
9. Dranginis AM et al (2007) A biochemical guide to yeast adhesins: glycoproteins for social and antisocial occasions. Microbiol Mol Biol Rev 71(2):282–294
10. Lewin R (1984) Microbial adhesion is a sticky problem. Science 224:375–377
11. Florin EL, Moy VT, Gaub HE (1994) Adhesion forces between individual ligand receptor pairs. Science 264(5157):415–417
12. Van der Mei HC, Van de Belt-Grotter B, Busscher HJ (1995) Implications of microbial adhesion to hydrocarbons for evaluating cell surface hydrophobicity 2. Adhesion mechanisms. Colloids Surf B Biointerfaces 5:117–126
13. Sundstrom P (2002) Adhesion in Candida spp. Cell Microbiol 4(8):461–469
14. Daniel RA, Errington J (2003) Control of cell morphogenesis in bacteria: two distinct ways to make a rod-shaped cell. Cell 113(6):767–776
15. Turner RD et al (2010) Peptidoglycan architecture can specify division planes in *Staphylococcus aureus*. Nat Commun 1:26. doi:10.1038/ncomms1025
16. Hell SW (2007) Far-field optical nanoscopy. Science 316(5828):1153–1158
17. Gitai Z (2009) New fluorescence microscopy methods for microbiology: sharper, faster, and quantitative. Curr Opin Microbiol 12(3):341–346
18. Matias VRF, Beveridge TJ (2005) Cryo-electron microscopy reveals native polymeric

cell wall structure in *Bacillus subtilis* 168 and the existence of a periplasmic space. Mol Microbiol 56(1):240–251
19. Milne JLS, Subramaniam S (2009) Cryo-electron tomography of bacteria: progress, challenges and future prospects. Nat Rev Microbiol 7(9):666–675
20. Pierres A et al (2002) Dissecting streptavidin-biotin interaction with a laminar flow chamber. Biophys J 82(6):3214–3223
21. Evans EA, Calderwood DA (2007) Forces and bond dynamics in cell adhesion. Science 316(5828):1148–1153
22. Bustamante C, Macosko JC, Wuite GJL (2000) Grabbing the cat by the tail: manipulating molecules one by one. Nat Rev Mol Cell Biol 1(2):130–136
23. Sotomayor M, Schulten K (2007) Single-molecule experiments *in vitro* and *in silico*. Science 316(5828):1144–1148
24. Muller DJ, Dufrêne YF (2008) Atomic force microscopy as a multifunctional molecular toolbox in nanobiotechnology. Nat Nanotechnol 3(5):261–269
25. Hinterdorfer P, Dufrêne YF (2006) Detection and localization of single molecular recognition events using atomic force microscopy. Nat Methods 3(5):347–355
26. Dufrêne YF (2008) Towards nanomicrobiology using atomic force microscopy. Nat Rev Microbiol 6:674–680
27. Neuman KC, Nagy A (2008) Single-molecule force spectroscopy: optical tweezers, magnetic tweezers and atomic force microscopy. Nat Methods 5(6):491–505
28. Muller DJ et al (2009) Force probing surfaces of living cells to molecular resolution. Nat Chem Biol 5(6):383–390
29. Binnig G et al (1982) Tunneling through a controllable vacuum gap. Appl Phys Lett 40(2):178–180
30. Pohl DW, Denk W, Lanz M (1984) Optical stethoscopy—image recording with resolution $\lambda/20$. Appl Phys Lett 44(7):651–653
31. Lewis A et al (1984) Development of a 500 Å spatial resolution microscope: I. Light is efficiently transmitted through $\lambda/16$ diameter apertures. Ultramicroscopy 13(3):227–231
32. Binnig G, Quate CF, Gerber C (1986) Atomic force microscope. Phys Rev Lett 56(9):930–933
33. Jena BP, Hörber JKH (2002) Atomic force microscopy in cell biology. In: Wilson L, Matsudaira PT (eds) Methods in cell biology, vol 68. Academic, San Diego, CA
34. Weisenhorn AL et al (1989) Forces in atomic force microscope in air and water. Appl Phys Lett 54(26):2651–2653
35. Dufrêne YF (2004) Using nanotechniques to explore microbial surfaces. Nat Rev Microbiol 2(6):451–460
36. Magonov SN, Elings V, Whangbo MH (1997) Phase imaging and stiffness in tapping-mode atomic force microscopy. Surf Sci 375(2–3):L385–L391
37. Muller DJ et al (2009) New frontiers in atomic force microscopy: analyzing interactions from single-molecules to cells. Curr Opin Biotechnol 20(1):4–13
38. Butt H-J, Cappella B, Kappl M (2005) Force measurements with the atomic force microscope: technique, interpretation and applications. Sur Sci Rep 59(1–6):1–152
39. Gaboriaud F, Dufrêne YF (2007) Atomic force microscopy of microbial cells: application to nanomechanical properties, surface forces and molecular recognition forces. Colloids Surf B Biointerfaces 54:10–19
40. Dupres V, Verbelen C, Dufrêne YF (2007) Probing molecular recognition sites on biosurfaces using AFM. Biomaterials 28(15):2393–2402
41. Heinz WF, Hoh JH (1999) Spatially resolved force spectroscopy of biological surfaces using the atomic force microscope. Trends Biotechnol 17(4):143–150
42. Heinz WF, Hoh JH (1999) Relative surface charge density mapping with the atomic force microscope. Biophys J 76(1):528–538
43. Ludwig M, Dettmann W, Gaub HE (1997) Atomic force microscope imaging contrast based on molecular recognition. Biophys J 72(1):445–448
44. A-Hassan E et al (1998) Relative microelastic mapping of living cells by atomic force microscopy. Biophys J 74(3):1564–1578
45. Touhami A, Nysten B, Dufrêne YF (2003) Nanoscale mapping of the elasticity of microbial cells by atomic force microscopy. Langmuir 19:4539
46. Gad M, Itoh A, Ikai A (1997) Mapping cell wall polysaccharides of living microbial cells using atomic force microscopy. Cell Biol Int 21(11):697–706
47. Frisbie CD et al (1994) Functional-group imaging by chemical force microscopy. Science 265(5181):2071–2074
48. Noy A et al (1995) Chemical force microscopy exploiting chemically-modified tips to quantify adhesion, friction, and functional-group distributions in molecular assemblies. J Am Chem Soc 117(30):7943–7951
49. Lee GU, Chrisey LA, Colton RJ (1994) Direct measurement of the forces between complementary strands of DNA. Science 266(5186):771–773

50. Touhami A et al (2003) Probing specific lectin-carbohydrate interactions using atomic force microscopy imaging and force measurements. Langmuir 19(5):1745–1751
51. Berquand A et al (2005) Antigen binding forces of single antilysozyme Fv fragments explored by atomic force microscopy. Langmuir 21(12):5517–5523
52. Dupres V et al (2005) Nanoscale mapping and functional analysis of individual adhesins on living bacteria. Nat Methods 2(7):515–520
53. Hinterdorfer P et al (1996) Detection and localization of individual antibody-antigen recognition events by atomic force microscopy. Proc Natl Acad Sci U S A 93(8):3477–3481
54. Allen S et al (1997) Detection of antigen-antibody binding events with the atomic force microscope. Biochemistry 36(24):7457–7463
55. Riener CK et al (2003) Heterobifunctional crosslinkers for tethering single ligand molecules to scanning probes. Anal Chim Acta 497(1–2):101–114
56. Ebner A et al (2007) A new, simple method for linking of antibodies to atomic force microscopy tips. Bioconjug Chem 18(4):1176–1184
57. Ros R et al (1998) Antigen binding forces of individually addressed single-chain Fv antibody molecules. Proc Natl Acad Sci U S A 95(13):7402–7405
58. Schwesinger F et al (2000) Unbinding forces of single antibody-antigen complexes correlate with their thermal dissociation rates. Proc Natl Acad Sci U S A 97(18):9972–9977
59. Baumgartner W et al (2003) Ca^{2+} dependency of N-cadherin function probed by laser tweezer and atomic force microscopy. J Neurosci 23(35):11008–11014
60. Stroh C et al (2004) Single-molecule recognition imaging-microscopy. Proc Natl Acad Sci U S A 101(34):12503–12507
61. Baumgartner W et al (2000) Cadherin interaction probed by atomic force microscopy. Proc Natl Acad Sci U S A 97(8):4005–4010
62. Helenius J et al (2008) Single-cell force spectroscopy. J Cell Sci 121(11):1785–1791
63. Razatos A et al (1998) Molecular determinants of bacterial adhesion monitored by atomic force microscopy. Proc Natl Acad Sci U S A 95(19):11059–11064
64. Ong YL et al (1999) Adhesion forces between E-coli bacteria and biomaterial surfaces. Langmuir 15(8):2719–2725
65. Bowen WR et al (1998) Direct measurement of the force of adhesion of a single biological cell using an atomic force microscope. Colloids Surf Physicochem Eng Aspects 136(1–2):231–234
66. Lower SK, Hochella MF, Beveridge TJ (2001) Bacterial recognition of mineral surfaces: nanoscale interactions between *Schewanella* and α-FeOOH. Science 292:1360–1363
67. Benoit M et al (2000) Discrete interactions in cell adhesion measured by single-molecule force spectroscopy. Nat Cell Biol 2(6):313–317
68. Gad M, Ikai A (1995) Method for immobilizing microbial cells on gel surface for dynamic AFM studies. Biophys J 69(6):2226–2233
69. Radmacher M et al (1992) From molecules to cells: imaging soft samples with the atomic force microscope. Science 257(5078):1900–1905
70. Meyer RL et al (2010) Immobilisation of living bacteria for AFM imaging under physiological conditions. Ultramicroscopy 110(11):1349–1357
71. Hoh JH, Schonenberger CA (1994) Surface morphology and mechanical properties of MDCK monolayers by atomic force microscopy. J Cell Sci 107:1105–1114
72. Matzke R, Jacobson K, Radmacher M (2001) Direct, high-resolution measurement of furrow stiffening during division of adherent cells. Nat Cell Biol 3(6):607–610
73. Zhang XH et al (2004) Atomic force microscopy measurement of leukocyte-endothelial interaction. Am J Physiol Heart Circ Physiol 286(1):H359–H367
74. Parpura V, Haydon PG, Henderson E (1993) 3-Dimensional imaging of living neurons and glia with the atomic force microscope. J Cell Sci 104:427–432
75. Bolshakova AV et al (2001) Comparative studies of bacteria with an atomic force microscopy operating in different modes. Ultramicroscopy 86(1–2):121–128
76. Vadillo-Rodriguez V et al (2004) Comparison of atomic force microscopy interaction forces between bacteria and silicon nitride substrata for the three commonly used immobilization methods. Appl Environ Microbiol 70(9):5441–5446
77. Touhami A et al (2006) Nano-scale characterization and determination of adhesion forces of *Pseudomonas aeruginosa* pili using atomic force microscopy. J Bacteriol 188(2):370–377
78. Camesano TA, Logan BE (2000) Probing bacterial electrosteric interactions using atomic force microscopy. Environ Sci Technol 34:3354–3362
79. Colville K et al (2010) Effects of poly(L-lysine) substrates on attached *Escherichia coli* bacteria. Langmuir 26(4):2639–2644
80. Kasas S, Ikai A (1995) A method for anchoring round shaped cells for atomic force microscope imaging. Biophys J 68(5):1678–1680
81. Turner RD et al (2010) Improvement of the pore trapping method to immobilize vital

coccoid bacteria for high-resolution AFM: a study of *Staphylococcus aureus*. J Microsc 238(2):102–110
82. Scheuring S, Dufrêne YF (2010) Atomic force microscopy: probing the spatial organization, interactions and elasticity of microbial cell envelopes at molecular resolution. Mol Microbiol 75(6):1327–1336
83. El Kirat K, Morandat S, Dufrêne YF (2010) Nanoscale analysis of supported lipid bilayers using atomic force microscopy. Biochim Biophys Acta 1798(4):750–765
84. Liu SY, Wang YF (2010) Application of AFM in microbiology: a review. Scanning 32(2):61–73
85. Dorobantu LS, Gray MR (2010) Application of atomic force microscopy in bacterial research. Scanning 32(2):74–96
86. Butt HJ, Downing KH, Hansma PK (1990) Imaging the membrane protein bacteriorhodopsin with the atomic force microscope. Biophys J 58(6):1473–1480
87. Muller DJ et al (1995) Imaging purple membranes in aqueous solutions at subnanometer resolution by atomic force microscopy. Biophys J 68(5):1681–1686
88. Schabert FA, Henn C, Engel A (1995) Native *Escherichia coli* OMPF porin surfaces probed by atomic force microscopy. Science 268(5207):92–94
89. Scheuring S et al (1999) High resolution AFM topographs of the *Escherichia coli* water channel aquaporin Z. EMBO J 18(18):4981–4987
90. Scheuring S et al (2002) Charting and unzipping the surface layer of Corynebacterium glutamicum with the atomic force microscope. Mol Microbiol 44(3):675–684
91. Müller DJ, Baumeister W, Engel A (1996) Conformational change of the hexagonally packed intermediate layer of *Deinococcus radiodurans* monitored by atomic force microscopy. J Bacteriol 178(11):3025–3030
92. Gyorvary ES et al (2003) Self-assembly and recrystallization of bacterial S-layer proteins at silicon supports imaged in real time by atomic force microscopy. J Microsc 212:300–306
93. Müller DJ, Engel A (1999) Voltage and pH-induced channel closure of porin OmpF visualized by atomic force microscopy. J Mol Biol 285(4):1347–1351
94. Ando T et al (2001) A high-speed atomic force microscope for studying biological macromolecules. Proc Natl Acad Sci U S A 98(22):12468–12472
95. Hansma PK et al (2006) High-speed atomic force microscopy. Science 314(5799):601–602
96. Humphris ADL, Miles MJ, Hobbs JK (2005) A mechanical microscope: high-speed atomic force microscopy. Appl Phys Lett 86(3):3

97. Casuso I et al (2009) Contact-mode high-resolution high-speed atomic force microscopy movies of the purple membrane. Biophys J 97(5):1354–1361
98. Shibata M et al (2010) High-speed atomic force microscopy shows dynamic molecular processes in photoactivated bacteriorhodopsin. Nat Nanotechnol 5(3):208–212
99. Casuso I, Scheuring S (2010) Automated set-point adjustment for biological contact mode atomic force microscopy imaging. Nanotechnology 21(3):035104
100. Ahimou FO, Touhami A, Dufrêne YF (2003) Real-time imaging of the surface topography of living yeast cells by atomic force microscopy. Yeast 20(1):25–30
101. Dufrêne YF et al (1999) Direct probing of the surface ultrastructure and molecular interactions of dormant and germinating spores of *Phanerochaete chrysosporium*. J Bacteriol 181(17):5350–5354
102. Almqvist N et al (2001) Micromechanical and structural properties of a pennate diatom investigated by atomic force microscopy. J Microsc 202(3):518–532
103. Francius G et al (2008) Nanostructure and nanomechanics of live *Phaeodactylum tricornutum* morphotypes. Environ Microbiol 10(5):1344–1356
104. Boonaert CJP, Rouxhet PG (2000) Surface of lactic acid bacteria: relationships between chemical composition and physicochemical properties. Appl Environ Microbiol 66(6):2548–2554
105. Doktycz MJ et al (2003) AFM imaging of bacteria immobilized on gelatin coated mica surfaces. Ultramicroscopy 97:209–216
106. Dague E et al (2007) Chemical force microscopy of single live cells. Nano Lett 7:3026–3030
107. Dupres V et al (2009) In vivo imaging of S-layer nanoarrays on *Corynebacterium glutamicum*. Langmuir 25(17):9653–9655
108. Dague E et al (2008) High-resolution cell surface dynamics of germinating *Aspergillus fumigatus* conidia. Biophys J 94(2):656–660
109. Touhami A, Jericho MH, Beveridge TJ (2004) Atomic force microscopy of cell growth and division in *Staphylococcus aureus*. J Bacteriol 186(11):3286–3295
110. Kailas L et al (2009) Immobilizing live bacteria for AFM imaging of cellular processes. Ultramicroscopy 109(7):775–780
111. Plomp M et al (2007) In vitro high-resolution structural dynamics of single germinating bacterial spores. Proc Natl Acad Sci U S A 104(23):9644–9649
112. Andre G et al (2010) Imaging the nanoscale organization of peptidoglycan in living *Lactococcus lactis* cells. Nat Commun 1:27. doi:10.1038/ncomms1027

113. Yang L et al (2006) Atomic force microscopy study of different effects of natural and semisynthetic β-lactam on the cell envelope of *Escherichia coli*. Anal Chem 78(20): 7341–7345
114. Verbelen C et al (2006) Ethambutol-induced alterations in *Mycobacterium bovis* BCG imaged by atomic force microscopy. FEMS Microbiol Lett 264(2):192–197
115. Alsteens D et al (2008) Organization of the mycobacterial cell wall: a nanoscale view. Eur J Physiol 456:117–125
116. Francius G et al (2008) Direct observation of *Staphylococcus aureus* cell wall digestion by lysostaphin. J Bacteriol 190(24):7904–7909
117. Fantner GE et al (2010) Kinetics of antimicrobial peptide activity measured on individual bacterial cells using high-speed atomic force microscopy. Nat Nanotechnol 5(4):280–285
118. Dorobantu LS et al (2008) Atomic force microscopy measurement of heterogeneity in bacterial surface hydrophobicity. Langmuir 24(9):4944–4951
119. Ahimou F et al (2002) Probing microbial cell surface charges by atomic force microscopy. Langmuir 18(25):9937–9941
120. Kasas S, Dietler G (2008) Probing nanomechanical properties from biomolecules to living cells. Pflugers Arch 456(1):13–27
121. Pelling AE et al (2005) Nanoscale visualization and characterization of *Myxococcus xanthus* cells with atomic force microscopy. Proc Natl Acad Sci U S A 102(18):6484–6489
122. da Silva A, Teschke O (2005) Dynamics of the antimicrobial peptide PGLa action on *Escherichia coli* monitored by atomic force microscopy. World J Microbiol Biotechnol 21(6–7):1103–1110
123. Gaboriaud F et al (2005) Surface structure and nanomechanical properties of *Shewanella putrefaciens* bacteria at two pH values (4 and 10) determined by atomic force microscopy. J Bacteriol 187(11):3864–3868
124. Cerf A et al (2009) Nanomechanical properties of dead or alive single-patterned bacteria. Langmuir 25(10):5731–5736
125. Lee GU, Kidwell DA, Colton RJ (1994) Sensing discrete streptavidin biotin interactions with atomic force microscopy. Langmuir 10(2):354–357
126. Rief M et al (1997) Reversible unfolding of individual titin immunoglobulin domains by AFM. Science 276(5315):1109–1112
127. Oberhauser AF et al (1998) The molecular elasticity of the extracellular matrix protein tenascin. Nature 393(6681):181–185
128. Rief M et al (1999) Single molecule force spectroscopy of spectrin repeats: low unfolding forces in helix bundles. J Mol Biol 286(2): 553–561
129. Rief M et al (1997) Single molecule force spectroscopy on polysaccharides by atomic force microscopy. Science 275:1295–1297
130. Gilbert Y et al (2007) Single-molecule force spectroscopy and imaging of the vancomycin/D-Ala-D-Ala interaction. Nano Lett 7(3): 796–801
131. Francius G et al (2008) Detection, localization and conformational analysis of single polysaccharide molecules on live bacteria. ACS Nano 2(9):1921–1929
132. van der Aa BC et al (2001) Stretching cell surface macromolecules by atomic force microscopy. Langmuir 17(11):3116–3119
133. Abu-Lail NI, Camesano TA (2002) Elasticity of *Pseudomonas putida* KT2442 surface polymers with single-molecule force microscopy. Langmuir 18:4071–4081
134. Camesano TA, Abu-Lail NI (2002) Heterogeneity in bacterial surface polysaccharides, probed on a single-molecule basis. Biomacromolecules 3(4):661–667
135. Dupres V et al (2009) The yeast Wsc1 cell surface sensor behaves like a nanospring in vivo. Nat Chem Biol 5(11):857–862
136. Velegol SB, Logan BE (2002) Contributions of bacterial surface polymers, electrostatics, and cell elasticity to the shape of AFM force curves. Langmuir 18(13):5256–5262
137. Lee G et al (2006) Nanospring behaviour of ankyrin repeats. Nature 440(7081):246–249
138. Straede A, Heinisch JJ (2007) Functional analyses of the extra- and intracellular domains of the yeast cell wall integrity sensors Mid2 and Wsc1. FEBS Lett 581(23):4495–4500
139. Brown AEX, Discher DE (2009) Conformational changes and signaling in cell and matrix physics. Curr Biol 19(17):R781–R789
140. Vogel V, Sheetz M (2006) Local force and geometry sensing regulate cell functions. Nat Rev Mol Cell Biol 7(4):265–275
141. Friedland JC, Lee MH, Boettiger D (2009) Mechanically activated integrin switch controls alpha(5)beta(1) function. Science 323(5914):642–644
142. Geiger B, Spatz JP, Bershadsky AD (2009) Environmental sensing through focal adhesions. Nat Rev Mol Cell Biol 10(1):21–33
143. Bershadsky AD, Kozlov M, Geiger B (2006) Adhesion-mediated mechanosensitivity: a time to experiment, and a time to theorize. Curr Opin Cell Biol 18(5):472–481
144. Smith AS et al (2008) Force-induced growth of adhesion domains is controlled by receptor mobility. Proc Natl Acad Sci U S A 105(19): 6906–6911

145. Alsteens D et al (2009) Unfolding individual Als5p adhesion proteins on live cells. ACS Nano 3:1677–1682
146. Alsteens D et al (2010) Force-induced formation and propagation of adhesion nanodomains in living fungal cells. Proc Natl Acad Sci U S A 107(48):20744–20749
147. Rauceo JM et al (2006) Threonine-rich repeats increase fibronectin binding in the *Candida albicans* adhesin Als5p. Eukaryot Cell 5(10):1664–1673
148. Otoo HN et al (2008) *Candida albicans* Als adhesins have conserved amyloid-forming sequences. Eukaryot Cell 7(5):768–782

Chapter 11

X-Ray Microscopy for Neuroscience: Novel Opportunities by Coherent Optics

Tim Salditt and Tanja Dučić

Abstract

X-ray microscopy and tomography can provide the three-dimensional density distribution within cells and tissues without staining and slicing. In addition, chemical information—i.e. the elemental distribution—can be retrieved by X-ray spectro-microscopy based on contrast variation around photon absorption edges and X-ray fluorescence. For a long time, X-ray microscopy has been limited in resolution by the fabrication of zone plate lenses, in particular for the hard X-ray range, which is needed to penetrate multicellular samples. Recent progress in X-ray optics and lensless coherent imaging now pave the way for enhanced imaging tools in neuroscience.

Key words X-ray microscopy, Coherent imaging, Tomography, Cellular diffraction, Nano X-ray optics

1 Introduction and Scope

In solving the structure of its macromolecular constituents by X-ray crystallography, neuroscience takes ample advantage of the atomic resolution resulting from the small wavelength of X-rays. On the other end of length scales in the nervous system, neurology and clinical research routinely use X-rays to image the brain by virtue of the high penetration depth of X-rays in the organ. In between these two extreme cases, *X-ray microscopy* (XM) constitutes a rather new and yet uncommon tool in neuroscience. In this chapter we present an introduction to XM and review selected applications to illustrate different approaches. We discuss persisting challenges and the potential for future progress. XM is defined here in the wider sense as X-ray imaging techniques which offer a resolution better than the values currently achieved by standard micro-tomography instruments in the hospital or laboratory. This leaves aside the emerging phase contrast techniques which are capable to visualize soft tissues on the μm scale and above, based on either grating interferometry [1] or free space propagation.

The resolution in such instruments is limited by macroscopic detector pixels, optics, and/or source size, but interesting neuroscience application on the tissue and organ scale has become apparent [2]. Aside from resolution, different contrast mechanisms, including the quantification of small density variations in native unstained cells or tissues, are major asset of XM.

To date molecular resolution is only achieved by X-ray diffraction, but not X-ray microscopy. Hence, high resolution is achieved in reciprocal (Fourier) space by diffracting from an ensemble of constituents, not in real space as in microscopic imaging. In view of its importance as a workhorse for high resolution analysis, and since it can be combined with real space resolution, we will include X-ray diffraction of neural organelles, cells, or tissues in this review of X-ray microscopy. The review, however, does not cover crystallographic structure analysis of crystallized biological macromolecules commonly referred to as macromolecular diffraction (MD). Instead, diffraction from multi-component non-crystalline biomolecular assemblies will be partly included, in particular the so-called small-angle X-ray scattering (SAXS) regime and its generalization to a scanning nanobeam technique. In the latter case, diffraction, i.e. the recording of far-field diffraction patterns is combined with real space imaging, i.e. by scanning a micro- or nano-focused X-ray beam over the sample. Diffraction can even be combined with full field imaging methods, as a special form of contrast method revealing structural information on length scales much smaller than the pixel size. Note that the word *scattering* is often used synonymously with *diffraction from non-crystalline samples*, even if this use is unfortunate, since X-ray diffraction denotes the coherent and elastic interaction of X-ray photons *via* the Thompson process, leading to the interference effects known from classical wave fields. Diffraction is thus a subset of the more general class of scattering processes, which can also be based on other interaction mechanisms, such as resonant or non-resonant absorption, inelastic (Compton) scattering, or more complex interactions with electrons involving special atomic or molecular transitions, see [3, 4] for an advanced textbook level treatment.

Below, we first discuss the advantages and disadvantages of X-ray microscopy in the context of more standard microscopy techniques. We then give an overview over the technique in Sect. 3. Selected references to the development of the optical methods are intended to provide a useful bibliography for the reader interested in a deeper understanding of the method. This description of the methods as such is independent of the application field, but discussed with an emphasis on the topics relevant to biological imaging. Section 3 finishes with the presentation of recent cellular X-ray microscopy results based on lensless coherent imaging with iterative phase reconstruction. Since no examples of neuroscience applications for these emerging techniques are available yet, we will resort to general applications of biological imaging. In Chaps. 4 and 5 we then focus on few exemplary X-ray studies relevant to neuroscience, illustrating the different variants and approaches, and what can be learned from each of them on different length scales. Starting from biomolecular assemblies and model systems, this will include studies at the level of the organelle, neuron, and nervous system tissue, respectively. Note that instead of providing a systematic compilation, we focus on our own recent work, in particular in the choice of the visualized examples, simply for reasons of access and being most familiar. We will start in Sect. 4 on the organelle level reviewing recent diffraction analysis of synaptic vesicles and their interactions with lipid membranes as studied by SAXS. In the same section, we then discuss scanning SAXS of cells and tissues, i.e. the simplest combination of diffraction and scanning microscopy. Section 5 reviews a multi-modal X-ray microscopy study of single isolated myelinated axons. Section 6 presents a brief synopsis and outlook.

2 X-Ray Microscopy in Neuroscience: Why Bother?

Let us briefly address X-ray microscopy (XM) in the context of other microscopy techniques. Electron microscopy (EM) and tomography [5] certainly offers the highest resolution in fixed specimens, however, at the price of sample size limitations and invasive and lengthy sample preparation. Chemical fixation, drying, cutting and staining is prone to compromise the native hydrated structure. Moreover, the optical properties and contrast values in EM are very hard to quantify due to problems of multiple scattering. Contrarily, optical fluorescence microscopy, in particular confocal microscopy, can detect the 3D distribution of specific markers in living cells, and single molecule optical spectroscopy yields very quantitative information on many observables such as molecular distance, orientation, and diffusion constants.

With the enabling development of super-resolution optical fluorescence microscopy or *optical nanoscopy* by Stefan Hell and coworkers [6], the situation has changed dramatically. The distribution of fluorescently labeled biomolecules in cells can now be imaged at isotropic resolution of 20 nm in fixed cells, and ongoing developments (instrumentation, algorithms, marker molecules) are headed towards live cell imaging at similar resolution. The basic principle is based on saturation of a transition between two marker states. A light distribution exhibiting an intensity zero is used to switch off fluorescence except of the markers around the zeros of the intensity distribution, leading to a significantly reduced focal volume. The use of such reversible saturable optical fluorescence transitions, first developed in form of stimulated emission depletion (STED), overcomes the Abbé diffraction limit in a fundamental way.

Many new variants of nanoscopy based on different switching strategies have emerged, which can be adapted and optimized for neuroscience applications, such as the observation of synaptic vesicles in live neurons [7]. However, not all structural problems can be solved based on the distribution of fluorescence markers. Some samples such as thick tissues may not be sufficiently transparent for visible light, suffer from auto-fluorescence, or cannot be transfected or stained without artifacts. More fundamentally still, entirely different types of contrast are needed depending on the biological question. Therefore, it is well understood that most biological problems necessitate the combined information of several microscopy techniques.

2.1 It's Contrast, Not Just Resolution!

But then, what type of images can X-rays deliver? Just as electron density of biomolecules as retrieved in MD provides an important backbone for structural biology, the native three-dimensional density distribution on the organelle level is needed for a complete understanding of "spatial cellular biology." XM in combination with tomography offers a unique potential for quantitative three-dimensional (3D) determination of density in unstained and unsliced biological cells and tissues [8]. The distribution of density contrast in the sample along with the reconstructed volumes, shapes, and topologies provides a nondestructive "ultra-centrifuge" at the organelle and supra-molecular level. As pointed out in [8], the need to probe the native three-dimensional density of biological cells and tissues at high resolution and under hydrated conditions has motivated a continuous and long-lasting effort to develop X-ray microscopy techniques, first started with a lens-based approach [4, 9, 10]. Specific advantages of X-rays concern (1) the resolution [11], (2) the kinematic nature of the scattering process enabling quantitative image analysis [12], (3) element-specific contrast variation [13], and (4) compatibility with unsliced (three-dimensionally extended), unstained, and hydrated specimens [14, 15]. Not withstanding significant achievements, X-ray microscopy based on Fresnel zone plate (FZP) lenses is severely limited by the comparatively small numerical apertures, leading to a resolution which is one or two orders of magnitude away from the classical diffraction barrier associated with the wavelength λ. The fabrication of high resolution lenses poses severe technological problems, in particular for higher photon energies where the tomographic reconstruction of extended specimens can best be achieved. To circumvent the practical resolution barrier associated with X-ray lens fabrication, different forms of lensless coherent X-ray diffractive imaging (CXDI) methods, based on iterative object reconstruction from the measured coherent diffraction patterns, have been developed [16–18]. CXDI has been demonstrated on biological unstained specimens using soft [19, 20] and hard X-rays [8, 21, 22].

Despite its potential, X-ray microscopy plays only a minor role in neuroscience to date. Since these techniques are largely restricted to experiments at large-scale research facilities such as synchrotron radiation sources or even the novel free electron lasers (FEL), impact is restricted to the solution of very specific problems. At the same time, a dramatic increase in X-ray brilliance achieved by these sources may now enable unprecedented experimental capabilities. Single-pulse exposures may possibly enable time-resolved imaging of rapidly fluctuating samples and dynamic processes. The vision of structure analysis of non-crystallized biomolecules and biomolecular assemblies down to the near-molecular resolution with short femtosecond X-ray pulses may soon turn to reality. In parallel to the realization of large FEL facilities, compact laser-driven X-ray sources have made short X-ray pulses available on a small laboratory-scale, opening up a window to study at least some biomolecular processes with high spatial and temporal resolution using laboratory scale equipment. But does this progress automatically imply a new tool at reach to solve any neuroscience problems? If a strong scientific case exists, the example of protein crystallography shows that large-scale instruments can indeed be used very efficiently at high throughput. Resolving protein folding, misfolding, and pathological aggregation in vitro may be worth the effort. On the cellular and tissue level, the improved instrumentation and focusing optics of third generation synchrotron beamlines enables 3D structure analysis at increasing resolution. Presently, the most relevant neuroscience application of XM is scanning X-ray fluorescence microscopy (XRF-microscopy), yielding the quantitative elemental distribution of a biological cell or tissue by scanning a highly focused X-ray beam and collecting an X-ray fluorescence spectrum in each pixel [23].

2.2 X-Ray Optical Properties of Biological Samples and Spectral Range: Go Soft or Go Hard?

The soft X-ray range in particular the so-called water window with photon energies $284\,\text{eV} < E < 534\,\text{eV}$ in between the C and the O absorption edges is well suited for microscopy and tomography of single cells in absorption contrast. Contrarily, the medium and hard X-ray spectral range of 4 keV and above is needed to penetrate larger samples beyond a single biological cell, from tissues up to the entire organ. In this case, smaller features in soft biological tissues can best be detected in phase contrast. Describing the interaction of X-ray waves with the sample in terms of the local index of refraction $n(\vec{r}) = 1 - \delta(\vec{r}) + i\beta(\vec{r})$, the small dispersion $\delta \ll 1$ and absorption corrections $\beta \ll 1$ determine phase shifts and absorption in the sample, respectively. For a given element and atom density ρ_a the refractive index is given in terms of the atomic form factors

$$n = 1 - \delta + i\beta = -\frac{r_e \lambda^2}{2\pi}\rho_a(\vec{r})[Z + f'(\omega)] + i\frac{r_e \lambda^2}{2\pi}\rho_a(\vec{r})f''(\omega), \quad (1)$$

where Z is the atomic number, $r_e = 2.82 \cdot 10^{-15}$ m the Thompson scattering length, and $f'(\omega) \ll Z$ and $f''(\omega) \ll Z$ are the dispersion and absorption corrections, respectively. For mixed elemental composition, the indices of refraction are weighted averages according to the local stoichiometry of elements. Given n, the wave propagation in the sample can be described in simple terms. For example, assuming a plane wave e^{ikz} incident at $z=0$ onto a homogeneous sample of thickness Δz, the wavefield behind the object (outgoing wave) can thus be written as

$$\tau = e^{ik\Delta z(1-\delta+i\beta)} = e^{ik\Delta z} \cdot e^{ik\delta\Delta z} e^{-k\beta\Delta z}, \quad (2)$$

where $e^{ik\delta\Delta z}$ describes the phase shift and $e^{-k\beta\Delta z} = e^{-\mu\Delta z/2}$ the absorption of the wave in the medium (absorption coefficient μ). For an inhomogeneous medium and if diffraction in the specimen is neglected, the complex-valued transmission function of the sample is expressed by the integral along the optical path

$$\tau(x,y) = \exp\left[-ik\int_0^{\Delta z}\left(\delta(\vec{r}) - i\beta(\vec{r})\right)dz\right]. \quad (3)$$

The underlying approximation is known as the projection approximation which consists of assuming that the value of the wavefield is entirely determined by the phase and amplitude shifts that are accumulated along streamlines of the unscattered beam, neglecting the spread of the wave by diffraction within the sample. This is correct only down to a critical resolution (pixel size), depending on sample thickness and photon energy E. The higher the resolution, and the larger the sample, the higher E must be chosen. With the Rayleigh resolution given in terms of the numerical aperture NA as $r = 0.61\lambda/\text{NA}$, and the depth of focus as $\text{DOF} = \lambda/\text{NA}^2$ [4], the maximum allowable thickness of the sample from $\Delta z \leq \text{DOF}$ in terms of the number of resolution elements is $N_z = \text{DOF}/r = 1.64(r/\lambda)$. In other words, for fixed resolution r, the wavelength has to be decreased as $\lambda \propto 1/\Delta z$ in order to warrant the projection approximation needed in particular for tomographic reconstruction.

2.3 X-Ray Optics and Focusing

Progress in X-ray optics and focusing has been significant in recent years, comprising all the three different classes of focusing optics: diffractive [11], refractive [24], and reflective optics [25, 26]. Experimental needs and parameters such as beam size, intensity, coherence, bandpath, and eventually pulse length must be considered for the best choice in a given experiment. Focusing optics has provided enabling technology for X-ray microscopy in all its variants. Even if still far from the diffraction limit, the recently measured record value of 4.3 × 4.7 nm (FWHM of Gaussian fit) achieved by a sliced multilayer zone plate (MZP) lens indicates that sub-5 nm spot sizes can be reached. Up to recently, this had been conjectured to be impossible based on fundamental limits in X-ray optical material properties [27]. The record value was obtained based on the concept of a combined optical system with efficient pre-focusing onto the lens [28]. It extends the current race for ultra-small X-ray beams, opened a few years ago by one-dimensional focusing to 7 nm spot size using atomically flat, elliptical multilayer mirrors [26], as well as by crossed multilayer Laue lenses (MLL) [29]. MLL and MZP are rather novel forms of diffractive optics based on the FZP scheme, which enable high aspect ratios needed for hard X-ray focusing. FZPs fabricated by e-beam lithography have in turn also reached the resolution range of 10 nm lines and spaces in the water window spectral range [30], while 15 nm lines and spaces have been resolved in hard X-ray scanning transmission X-ray microscopy (STXM) [31]. Compound refractive lenses (CRL) are well suited for scanning X-ray microscopy at high photon energies above 15 keV, also referred to as nanoprobe [24]. Finally, beam confinement down to 10 nm in two dimensions has been realized by X-ray waveguides [32], which also provide coherent filtering of X-rays [33]. However, outside specialized record experiments, the resolution reliably provided at FZP-based microscopes is more typically in the range of 20–40 nm for soft X-rays, and 60–100 nm for hard X-rays. The actual resolution achieved on biological samples (as opposed to test structures) may be still lower. As a rather unusual optical element, let us close with a comment on X-ray waveguides. X-ray waveguides offer a quasi-point source for X-ray propagation imaging, with very clean wavefronts [34, 35]. Using waveguides in combination with focusing devices, it is possible to decouple the coherence and spot size of the exit beam from the primary source [36]. Similar to other reflective optical components, waveguides are essentially non-dispersive and can be adapted to a wide range of photon energies and bandpass. Two planar one-dimensionally confining waveguides (1DWG) can be combined to confine the beam in two orthogonal directions. Specific values achieved based on this crossed waveguide approach were a beam confinement to 10.0 nm and 9.8 nm beam confinement (full width at half maximum, FWHM) in the respective horizontal and vertical focal planes, with an integrated photon flux of $2.0 \cdot 10^7$ ph/s, measured at 15 keV photon energy [32]. More recent improvements resulted in an even higher flux of $1.0 \cdot 10^8$ ph/s with a cross section of 10.7 nm and 11.4 nm FWHM in horizontal and vertical direction, respectively, at 13.8 keV photon energy [37]. The high spatial coherence and small beam cross section down to the 10 nm range of X-ray yields very clean in-line holograms of biological specimen [35, 38], which can be used for tomography at low dose [34].

2.4 Sample Environment and Radiation Damage

Radiation damage is a major concern in X-ray microscopy, precluding live cell imaging at high resolution, since detrimental chemical reactions are triggered by free radicals and photo-electrons in aqueous environment. At the same time, reducing the dose $D \propto r^\varepsilon$

results in loss of resolution, according to a dose–resolution relationship with an exponent $\varepsilon \simeq 3 - 4$, as derived from basic analytical theory and experimental data [39]. Most current synchrotron X-ray studies of biological specimen work with chemically fixed or freeze-dried samples. Best isomorphous structure preservation under hydrated conditions is achieved by rapid cooling of the sample below the glass transition, so-called vitrification. The new cryo-stages available on several synchrotron beamlines enable the investigation of cells and tissues after cryo-preservation (plunging in liquid ethane or high pressure freezing), in this so-called frozen-hydrated state. For these conditions, no modification of the oxidation state and a minimization of beam damage have been reported [14, 40–42].

3 X-Ray Microscopy with and Without Lenses

We will distinguish three approaches to X-ray microscopy, which we denote by (Type I) the FZP or optical microscopy type, (Type II) the diffractive imaging type, and (Type III) the projection imaging type, as sketched in Fig. 1. The three different approaches can be best understood by considering from where they depart. Type I can be understood as the X-ray version of the classical optical microscope consisting of condenser, sample, and objective lens. It uses the same scheme but replaces the refractive optics of a standard microscope by FZP lenses, which constitute reasonable substitutes in particular for soft X-rays. This was the first practical form of X-ray microscopy and undergoes continuous progress, notably in the fabrication of FZP lenses paving the way for higher resolution. Type II is closer to a diffraction experiment than a classical microscope, with a plane wave impinging on the sample, and the diffraction pattern recorded in the far field. However, the ensemble average of a standard diffraction experiment with beam sizes larger than the coherence length is dispensed with, since the beam is fully coherent. It turns out that full coherence in conjunction with oversampling or support constraints in real space is sufficient to solve the phase problem, so that the unmeasurable phases of the diffracted wave can be retrieved, enabling image reconstruction. Type III can be considered as an extension of full field radiographic imaging with extra magnification achieved by projection. Historically projection imaging with X-rays is probably the oldest form of X-ray microscopy, but has gained relevance only after conceptual improvements replacing simplified geometric description by a full optical treatment of wave propagation for absorption and phase contrast samples. Moreover, practical applications necessitate quasi-spherical illumination by point-like sources which have been realized only more recently by the progress in X-ray nanofocusing. Type II and Type III are both based on wave optical

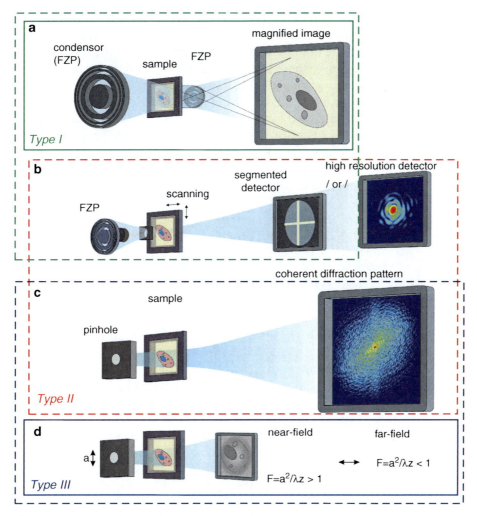

Fig. 1 Schematic overview of different types of X-ray microscopy: (**a**) The full field X-ray microscope with condenser, sample, and objective resembles the schemes of a visible light compound microscope (type I, *green rectangle*). It is termed transmission X-ray microscope (TXM) and is distinguished from the scanning X-ray microscopy (STXM), shown in (**b**). STXM is characterized by a high resolution focusing optics, in which the sample is scanned, to record fluorescence, absorption, and phase shift (e.g., by a segmented detector). The resolution of a classical STXM is given by the spot size. Together TXM and STXM are the two classical well-developed forms of XM (*green dashed rectangle*). If the STXM operates under coherent focusing conditions, and is upgraded with a suitable detector to record the far-field pattern at high enough sampling, super-resolution below the focus size can be achieved. This is enabled by inversion of the diffraction pattern via ptychography algorithms. Inversion of the far-field diffraction pattern as recorded by a high resolution detector is at the heart of a coherent diffraction imaging (type II, *red dashed rectangle*). (**c**) The scheme of a coherent diffraction imaging experiment with a full field illumination, e.g. by a pinhole select a coherent field of view, and no optics in between sample and detector. In most CDI experiments, the detector is placed in the (Fraunhofer) far-field, but near-field coherent imaging as sketched in (**d**) based on high resolution detectors and/or magnification is an attractive alternative. The two cases are distinguished by the Fresnel number F. Near-field (propagation) imaging can be considered to be a wave-optical generalization of radiography (type III, *blue rectangle*). Together, (**c**), (**d**), and the ptychographic extension of (**b**) form the set of coherent imaging techniques (*dashed blue rectangle*) (**c**) and the ptychographic extension of STXM shown in (**b**) shares the detection of a far-field diffraction pattern (type II, *red dashed rectangle*). Similarly, TXM as shown in (**a**) and STXM without super-resolution can be categorized as classical XM schemes (type I)

propagation in free space, transforming the exit wave behind the sample to a diffraction pattern, the difference being that Type II can be described by far-field diffraction in the so-called Fraunhofer regime, where the diffraction pattern is much larger than the primary beam size and has become stationary with respect to the optical axis z except for a simple dilation. Contrarily, Type III can be described by near-field diffraction in the Fresnel regime, where the features of the diffraction pattern are smaller than the beam size, leading to interference between the scattered and primary wave amplitude. Next, we will briefly address each of the approaches one by one.

3.1 FZP-Based X-Ray Microscopy

FZP-based X-ray microscopy has been developed since the 1970 and 1980, in form offull field transmission X-ray microscopy (TXM) mainly at the University of Göttingen by Günter Schmahl and coworkers [9], and in form of STXM mainly by Janos Kirz and coworkers at Brookhaven National Laboratory [10]. The names TXM and STXM mimic the EM nomenclature of transmission electron microcopy (TEM) and scanning electron microscopy (SEM). Classically, the term *X-ray microscopy* was almost exclusively reserved to TXM and STXM, before the more recent developments of many different X-ray microscopy approaches. TXM is based on FZP optics working as an objective lenses placed in between the sample and the detector, as well as a condenser to focus the light onto the sample. Typically, FZP with larger diameters, mirrors, or capillaries are used as condensers. STXMs work with only a condenser system, typically an FZP in front of the sample, equipped with an order sorting aperture. Since FZPs are essentially circular gratings with a particular radial gradient in the grating periodicity, the basic properties are similar to a diffraction grating: (1) multiple diffraction orders occur corresponding to the Fourier expansion of the height profile and (2) interference leads to sharp diffraction orders and thus dispersion. In other words, the focus changes with wavelength. In contrast to reflective/refractive optics which involve only two well-separated beams, diffractive optics is thus always somewhat limited by the presence of other orders/foci, reducing the efficiency and/or increasing stray light. Alike refractive optics and crystal optics, diffractive X-ray optics is dispersive, while reflective optics are essentially non-dispersive, i.e., compatible with broader bandpass. Since the early phase, FZP-based microscopy is now a mature technique after undergoing slow but steady progress, in particular concerning the fabrication of the FZP lenses as the crucial optical element determining the resolution. FZP are concentric rings of alternating absorbing/non-absorbing or phase shifting zones. Working as lenses for a given diffraction order m, the Rayleigh (full period) resolution δ of an FZP depends on the numerical aperture $NA \simeq \lambda/(2\ dr_N)$, which in turn depends on the outermost zone width dr_N [4], so that

$$\delta = \frac{c\,\lambda}{m\,\mathrm{NA}} \simeq \frac{2c\,dr_N}{m} \qquad (4)$$

where c is a prefactor depending on illumination (coherence and inclination), and m is the diffraction order, in which the FZP operate. In standard cases we have $m=1$ for reasons of efficiency, and $k=0.61$ for normal incidence incoherent illumination. Thus the smaller zones one can fabricate, the better the resolution. Unfortunately, there is a limit not so much in terms of absolute $d\,r_N$, but in the aspect ratio dr_N/t, where t as the FZP thickness needed to induce sufficient absorption or phase shift strongly increases with photon energy E. That's why FZP microscopes usually work well in the soft but not so well in the hard X-ray spectral range. Most TXMs operate in the so-called water window, i.e. at photon energies in between the K-edge of C and O, where water layers in the range of several μm are still transparent, and organic (hydrocarbon) molecules exhibit a high absorption contrast [10]. TXMs mostly work in absorption contrast, which is particularly strong in the soft X-ray regime. Extension to phase contrast has been achieved in the same way as in light microscopy, namely by placing Zernicke-like phase rings in the back focal plane of the objective [43]. This again shows the close analogy with light microscopy, in contrast to lensless phase reconstruction schemes which achieve quantitative sensitivity to the complex-valued transmission function of the sample by completely different means, such as free space propagation, interferometry, or diffraction followed by phase reconstruction from different constraints. One of the first working XRMs was installed by Schmahl and coworkers at the BESSY storage ring, initially at a bending magnet beamline at BESSY I, then at an undulator source at BESSY II in Berlin. Note that a TXM does not need coherent illumination, just as full field light microscopy is not carried out with laser sources. In fact the BESSY II TXM requires a condenser system which reduces the coherence, based on movement of optical parts (dynamical aperture synthesis). Without movement, partially coherent undulator radiation would lead to an interference pattern spoiling the image of the sample. It thus lends itself to operation at rather incoherent sources, including laser plasma sources. Significant progress has been achieved in the instrumentation of laser-driven compact TXMs, with benchmark values in resolution and data acquisition time achieved by Hans Hertz and coworkers at the KTH in Stockholm [44]. Let us briefly consider the new BESSY full field transmission X-ray microscope at the undulator U41 [30, 45], as a leading TXM instrument at synchrotron radiation sources. The microscope enables high spatial and spectral resolution $\lambda/\Delta\lambda \simeq 9{,}000$ and tomography of cryogenic samples, using a tilt holder adapted from EM. A single bounce ellipsoidal capillary condenser is used for hollow cone illumination, typically of a 20 × 20 μm field of view (FOV) on the specimen. A high resolution micro-zone plate with an outermost zone width of 40 nm typical value is used to magnify the sample onto the cooled back-illuminated soft X-ray CCD camera. Important in the optical design of TXMs is the matching of condenser and objective, i.e. the condenser must provide a hollow cone illumination of the sample, which is matched to the aperture of the micro-zone plate. The zero order beam encircles the CCD, so that the image is formed just with the first order in the center of the detection plane.

Typical Rayleigh resolution in absorption contrast soft X-ray microscopy (in first order imaging) is in the range of 20–40 nm. Higher resolution can be achieved by third order imaging, at the price of reduced efficiency [30]. Record values of sub-15 nm (full period or Rayleigh) resolution for metal test structures have been obtained by Chao and coworkers at Berkeley [11] by a multilayer lithography approach. For higher photon energies, the fabrication of zone-doubled FZP by atomic layer deposition has opened up a way to resolve 15 nm lines and spaces, (in metal test structures), corresponding to 30 nm Rayleigh resolution [31]. 3D images of vitrified 5 μm thick biological cells have been demonstrated at a (full period or Rayleigh) resolution of 36 nm [45]. Much progress in X-ray microscopy and tomography of unsliced biological cells immobilized by structure preserving rapid cryogenic plunging has been driven by Carolyn Larabell and coworkers using a TXM installed at the Advanced Light Source in Berkeley [46, 47].

3.2 X-Ray Fluorescence Microscopy and X-Ray Spectro-Microscopy

Current resolutions are in the range of a few 100 nm with standard instrumentation, and down to below 50 nm for the most advanced settings. Sensitivity down to the range of 10^2 atoms of an element in a pixel can be reached. In contrast to electron microscopy-based X-ray microanalysis (EDX) of elemental distributions in neurons, XRF can provide spectra without risk of changing the cell organization by cutting or chemical fixation. Moreover, XRF collects signal from the interior of the cell, while EDX is sensitive only to the outer molecular layers. This makes XRF-microscopy a unique technique for trace element mapping of different neural cells and tissues [42, 48, 49], in view of their suspected role in several neurodegenerative diseases [50]. For example, the concentration, distribution, and elemental speciation of Fe, Cu, and Mn is believed to be important in redox regulation processes of cellular stress and even signaling. By X-ray fluorescence microscopy it was discovered that neural cells redistribute significant pools of copper from their cell bodies to the periphery upon calcium activation [51]. Further state-of-the-art examples of cellular X-ray fluorescence microscopy studies can be found in [52], while [53] gives an example of the use of X-ray fluorescence microscopy on brain tissue slices, and the associated methodologies of sample preparation. Finally, it should be mentioned that if augmented by X-ray spectro-microscopy, the energy-dependent fluorescence yield (near edge X-ray absorption spectrum) can be analyzed to retrieve the oxidation state of the element [54]. Advanced analysis tools for quantitative spectro-microscopic identification of components are available [55].

3.3 Coherent X-Ray Diffraction Imaging

In FZP microscopy, the resolution is limited by lens fabrication. To overcome these limits, coherent X-ray imaging (CXI) techniques have been developed. The seminal work of Miao and coworkers [16], who have first shown how the unknown phase of a coherent X-ray diffraction experiment can be reconstructed by inversion schemes based on iterative algorithms, has started the development of coherent X-ray diffractive imaging (CXDI) based on the so-called plane wave illumination, which was then quickly taken up by number of groups utilizing the unique opportunities of the high brilliance (and hence coherence) of the novel third generation synchrotron radiation sources. At about the same time, coherent X-ray propagation imaging (CXPI) emerged, based on the recording of the Fresnel diffraction behind the sample, in other words of the near-field intensity distribution. This approach was pioneered by Peter Cloetens and coworkers at the European Synchrotron Radiation Facility [56, 57], and first carried out without magnification in an enlarged partially coherent beam, with phase reconstruction based on the recording of multiple intensity distributions at various detector-to-sample positions. In combination with geometric magnification, phase contrast by illumination with spherical waves had been achieved even earlier by Steve Wilkins and coworkers [58]. Image formation and phase reconstruction both for parallel and divergent beams treated in a series of influential papers by the Australian researchers Timur Gureyev, David Paganin, Steve Wilkins, and coworkers [59, 60]. Coherent X-ray imaging (CXI) now thus comprises two different approaches, namely CXDI and CXPI (notation introduced in this work), which can be distinguished by the image recording regimes, as well as by the phase reconstruction algorithms. The quantitative reconstruction of the electron density [63] can be achieved in both cases by a numerical inversion of a measured intensity pattern diffraction pattern, without any lenses between sample and detector. In other words, the beam fluence and the sample scattering power alone determine the achievable resolution given by the highest numerical aperture at which signal is recorded above noise (and can be phased) on the detector. A comprehensive overview of CXI is found in [64], a tutorial to CDXI in [65], finally a textbook treatment of coherent X-ray optics in [66].

Let us briefly address the enabling phase reconstruction strategies in this field, in other words the solution of the phase problem. In CXDI, uniqueness is achieved by imposing additional ("harmless") constraint sets such as sample support, positive definiteness of electron density, contrast constraints, or overlap constraints between several exposures with the sample shifted in a beam which is not translationally invariant. Such constraints are not very restrictive with respect to the imaging task, but can be powerful enough to solve the phase problem under condition that the coherence of the incoming beam is high enough to illuminate the entire sample coherently. Reconstruction algorithms based on the support constraint constitute the most important class of CDXI algorithms, implemented in form of the simple error reduction scheme first proposed in electron microscopy [67] or on more elaborate schemes [68]. The use of the overlap constraint is at the heart of the so-called ptychographic algorithms [69], which in its most advanced form are capable to reconstruct the unknown object by scanning the sample through the beam with partial overlap as well as the unknown complex-valued illumination function (so-called probe retrieval) [70–73]. The ptychographic algorithms have thus made two things possible: Firstly, an unlimited FOV by scanning and replacing the support constraint by the overlap constraint [69]. Secondly, by extensions of the original algorithm [70, 71], a simultaneous reconstruction of object and the probing beam (illumination). This solves a second significant problem: the illumination wave front does not need to be assumed as an ideal plane wave, lifting the experimental constraints and preconditions significantly. In the meantime, ptychographic reconstruction of biological cells [8] has been demonstrated, with applicability in the hard as well as the soft spectral range [74], and in combination with anomalous diffraction to gain chemical sensitivity [75, 76].

Concerning CXPI, phase information is more directly contained in the intensities by interference of the primary beam with the diffracted amplitudes. By this holographic interference, a weak scattering amplitude can be amplified high above background signals of residual scatter, detector dark current or readout noise. CXPI is therefore, in particular at larger sample-to-detector distances, correctly denoted as in-line X-ray (Gabor) holography. Other forms of X-ray holography such as Fourier holography [77] and off-axis X-ray holography [78] have also been devised, but are more restrictive in terms of sample and experimental geometry, and thus less relevant to biological imaging. Compared to plane wave CXDI, CXPI is not a priori limited to samples of finite support. An important difference of CXDI with respect to CXDI is in the detection scheme. In CXDI, the pixels can be filled much more evenly than for far field diffraction, avoiding complications associated with a high dynamic range of the signal, in particular pixel saturation and loss of information due to beamstops. However, a clean and quantitative image reconstruction has been a major challenge. In-line (holographic) image recording can always be followed by holographic reconstruction in form of a direct (single-step) backpropagation. But in this case the image quality is severely limited by the well-known twin image problem of holography. Experimental remedies used to overcome this problem include the recording of multiple holograms at various detector-to-sample positions [56, 57], or iterative phase retrieval with knowledge of the illumination function [18], as well as for short detector distances, where the holographic regime has not fully developed, by the transport-of-intensity (TIE) equation [66]. An alternative approach valid for all defocus distances and based on a single sample exposure plus a corresponding empty beam has been proposed by [38], based on an adaptation of CXDI algorithms (HIO and Gerchberg–Saxton). Overcoming the twin image problem completely, this approach also removed the remaining artifacts due to well-known zeros in the contrast transfer function (CTF) offree-space propagation in the Fresnel regime [59], if some empty beam regions are present around the sample. In the experiment, the quasi-spherical wave front exiting from a two-dimensional X-ray waveguide (WG) [32, 63, 79] leading to sub-20 nm beam confinement was used for holographic image recording from an unstained biological cell, followed by a robust and quickly converging iterative reconstruction scheme applied to a single recording. Stable reconstructions of weakly scattering biological specimen were thus obtained without exact knowledge of the illumination function, even in the presence of intensity fluctuations. The method proved to be dose efficient, and is well adapted for the experimental situation of an essentially pure phase contrast specimens, and takes photon noise effects into account quantitatively [38], providing quantitative 3D density reconstructions of biological cells in combination with tomography [34]. To probe the three-dimensional structure of a larger specimen, CXPI as a full field technique may prove to be of significant advantage over ptychographic CDXI, since overhead in detector read out becomes almost untolerably high, when scanning three degrees offreedom (two translations, one tomographic rotation). In contrast to scanning techniques such as STXM or ptychographic CXDI, a multicellular organism, a tissue up to an organ can be covered in one or a few exposures, eventually enlarged by stitching. Most importantly, using either zoom magnification in the detection scheme and/or cone beam projection, the magnification and the FOV can easily be adapted and combined. The experimental requirement for all CXI techniques are sufficiently high coherence properties of the beam, which, however, are less stringent for CXPI than for CXDI.

3.4 Cellular Imaging by Coherent X-Ray Microscopy and Tomography

As far as biological imaging is concerned, the following studies should be pointed out: Cellular images of a yeast cell at 30 nm resolution by plane wave CDI at 750 eV [20], images of unstained single viruses recorded at 5 keV [21], the 3D density distribution of a human chromosome by plane wave tomographic CDI

[22], quantitative 3D imaging of yeast spore cells at 50–60 nm resolution [80]. Furthermore, the highest resolution estimate of sub-20 nm was reported for specifically labeled yeast cells [81]. However, the resolution was claimed for the gold beads and was obtained after convolution. It certainly does not apply for the unlabeled parts of the cell. The first CDI imaging of vitrified cells in frozen-hydrated as opposed to freeze-dried state was reported in [82, 83].

Next, we turn to a specific example, the gram-positive bacterium *Deinococcus radiodurans* (*D. radiodurans*), which can serve as a benchmark for cellular imaging, and compare the two alternative approaches, ptychographic CXDI and CXPI. Both are capable to circumvent the restrictions of a compact support, and thus represent particularly well suited for biological imaging. The ptychographic CXDI study offreeze-dried *D. radiodurans* cells was the first ptychography study beyond simple test specimen [8]. Projection images of the electron density with quantified errors were obtained by scanning a simple pinhole followed by ptychographic phasing, as shown in Fig. 2a, b. The pronounced dense regions in the images were interpreted in view of the highly disputed nucleoid structure of bacterial cells. The estimated resolution for the cells was 85 nm (half-period length), corresponding to a Rayleigh resolution around 160 nm, but the invested dose of $1.3 \cdot 10^5$ Gy was orders of magnitude lower than in the above CDI studies. With respect to the diameter of the pinhole used to illuminate the samples, a super-resolution of about 15 was achieved. Importantly, the reconstruction did not rely on any support information, and quantitative electron density was facilitated since no beamstop was employed. Within just 2 years, improvements in the experiment and data analysis resulted in a full 3D structure of *D. radiodurans* [84]. The projection image shown in Fig. 2c and tomogram (d) exhibit globular dense regions of ca. 1.6 g/cm^3 in the bacterial cell with approx. isotropic shape, attributed to DNA-rich regions. Reconstructions of a single projection image reached a (half-pitch) resolution of about 50 nm for *D. radiodurans*, while with the same setup 10 nm was obtained for the tantalum test pattern used as a benchmark. This illustrates nicely the resolution gap in XM between weakly and strongly scattering samples. This gap also exists for FZP-type X-ray microscopes, even if in most cases authors tend to cite just the maximum (theoretical) resolution of the FZP. In contrast to the 2D case, the resolution of the tomogram, shown in Fig. 2d, was found to be limited by radiation-induced damage, which is not surprising in view of the dose in the range of $5 \cdot 10^8$ Gy, as opposed, for example, to the dose of $4.9 \cdot 10^6$ Gy deposited when recording the single projection shown in Fig. 2c. To overcome the loss in structural integrity, frozen-hydrated samples have in the meantime successfully been imaged by ptychography. In addition to the ptychographic reconstruction,

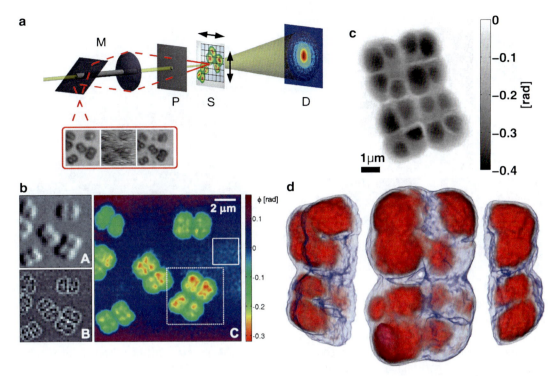

Fig. 2 Ptychographic imaging of freeze-dried bacteria (*D. radiodurans*), realized by (**a**) scanning a micron-sized pinhole over the sample. Conventional differential phase contrast and dark field contrast images are shown in the *upper left* and *lower left sub-graphic* of (**b**), respectively, at the ordinary resolution corresponding to the beam size (pinhole diameter). The enhanced super-resolution image shown in the *right sub-graphic* of (**b**) is calculated based on algorithmic inversion of the coherent diffraction patterns based on the overlap constraints. The retrieved X-ray phase shift can be converted to projected mass density. The dense regions are associated with DNA-rich regions of the bacterial nucleoid, from [8]. This first ptychographic study on cells was soon extended to higher resolution and a full 3D analysis, see the projection image in (**c**) and the tomographic rendering of the bacteria depicted in (**d**), from [84]

the local SAXS diffraction patterns have been recorded up to a momentum transfer of $q \leq 2$ nm^{-1} which is to date unachievable by CDI. Averaging over the speckle pattern, the diffraction decays with an exponent q^γ in the range $-4 \leq \gamma \leq -3$.

The same bacterial cells were then studied by propagation imaging in cone beam projection geometry [34], i.e., cone beam projection tomography with phase contrast formation by free space propagation, see Fig. 3. The sample was illuminated by the highly curved wave fronts emitted from a virtual quasi-point source with 10 nm cross section, realized by two crossed X-ray waveguides. The sample was placed at a (defocus) distance $z_1 = 8$ mm from the WG exit plane, where the divergent WG beam has broadened to a FOV of $40 \times 40\,\mu m^2$, see Fig. 3a. The image formation leading to the measured hologram in the detector plane at distance $z_2 \gg z_1$ behind the sample can then be described in an equivalent parallel-beam geometry with a demagnified (effective) detector pixel size

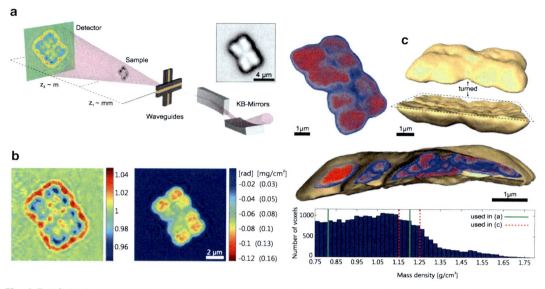

Fig. 3 Full field X-ray propagation imaging and tomography of freeze-dried bacteria (*D. radiodurans*). (**a**) The sample is placed at a distance z_1 behind the waveguide exit. The area pixel detector is placed at a distance $z_2 \gg z_1$ away from the sample to collect the diffraction pattern. (**b**) (*Left*) X-ray in-line hologram of a cell, recorded after geometric magnification by cone beam illumination using a quasi-point source realized by a X-ray waveguide. (*Right*) image reconstruction by the iterative algorithm (modified hybrid-input-output algorithm), based on the constraint that the object is completely transparent (phase contrast constraint) for hard X-rays, as well as (automated) support determination. (**c**) Direct volume rendering of the 3D effective mass density. The coloring indicates densities from 0.8 g/cm^3 (*blue*) to 1.2 g/cm^3 (*red*). (*Right*) surface rendering obtained by choosing 0.75 g/cm^3 as a threshold value. (*Bottom*) combined direct volume rendering (1.15–1.25 g/cm^3) and surface rendering along with the histogram of the effective mass density inside the cell. From [34]

of Δ_D/M and a (de)magnification factor of $M = (z_1 + z_2)/z_1$ as well as an effective sample-detector distance $z_{\text{eff}} = z_1 z_2/(z_1 + z_2) = z_2/M$, resulting in an effective detector pixel size of $\simeq 83$ nm, see Fig. 3b. The tomographic structure displayed in (c) was reconstructed from 83 projection images I_ϕ collected over 162° with a total exposure time of 10 min for each angle ϕ. The total fluence for each projection was $\simeq 4.38 \cdot 10^6$ photons/µm^2, corresponding to a dose of $\simeq 1.9 \cdot 10^3$ Gy, based on calculations presented in [39]. To this end, the normalized intensity distribution of each projection $\overline{I}_\phi(x,y) = I_\phi / I_0$ was obtained by division with the empty beam intensity distribution $I_0(x,y)$, resulting in a corrected hologram as illustrated in the left sub-figure of (b). Using a modified HIP reconstruction adapted to Fresnel propagation and phase pure contrast samples [38] in combination with automated support constraint, the right image of (b) was obtained. Clearly, the experimental scheme thus allows for a particularly dose efficient determination of the 3D cellular structure. FOV and resolution are easily varied by z_1. Current efforts are directed at a scaled resolution.

CXI has not really been applied to the field of neuroscience, despite its significant potential. It should be considered in all cases where the 3D density of unstained and unsliced neural cells and small scale tissues (up to a few cellular layers) needs to be imaged at sub-cellular resolution in the range of 20–50 nm. The ultrastructure of the synapse is one example where the method could be useful. Compared to FZP-type X-ray microscopy, nearly the same resolution values can be achieved at comparatively low dose. In future, even higher resolution than in FZP-type X-ray microscopy are likely to be achieved. Thicker specimen are possible without compromise in the efficiency, which is typical for FZP optics at multi-keV photon energy, and quantitative density contrast can be achieved. On the downside, coherence requirements limit CXI to a much more restrictive set of X-ray sources (notably undulators and FEL than FZP X-ray microscopy). Furthermore, images cannot yet be reconstructed in real time, during acquisition.

4 From Diffraction to Imaging: Synaptic Vesicles, Model Membranes, and Membrane Fusion

Synaptic vesicles (SV) are small membranous organelles at the nerve terminal with a radius of about 20 nm, encapsulating neurotransmitters and enabling a highly controlled fusion event with the synaptic membrane, an essential process in nerve conduction. To enable this function, the SV membrane is covered with a dense layer of proteins, in particular fusiogenic SNARE proteins [85]. Structural characterization of SVs, e.g., by electron microscopy (EM) is at the limit of the spatial resolution required to identify constituents of the functional structure. Furthermore, an independent assessment of SV structure compatible with more physiological conditions than in EM, and possibly with higher (near molecular) resolution was needed. To fill this gap, synchrotron-based SAXS was used to determine the average radial density profile $\rho(r)$ and the size polydispersity of SV, proving that X-ray diffraction can be used to study the supra-molecular architecture of an entire functional organelle under physiological conditions [86], see Fig. 4. The resulting profile $\rho(r)$ of SV yielded structural parameters of the protein layers, as well as the polydispersity function $p(R)$, derived with no free prefactors on absolute scale, and confirmed the main aspects of recent numerical modeling [85], based on the crystal structures of the constituent proteins and stoichiometric analysis. By using mixed suspensions of SV and proteoliposomes, fusion and interactions of SV can be evidenced on a structural level by SAXS [86, 87]. Vesicle–membrane interaction was studied using X-ray diffraction and a Langmuir film balance with lipid monolayers of controlled composition, adjustable pH and salt concentration, and the SV solution injected in the sub-phase

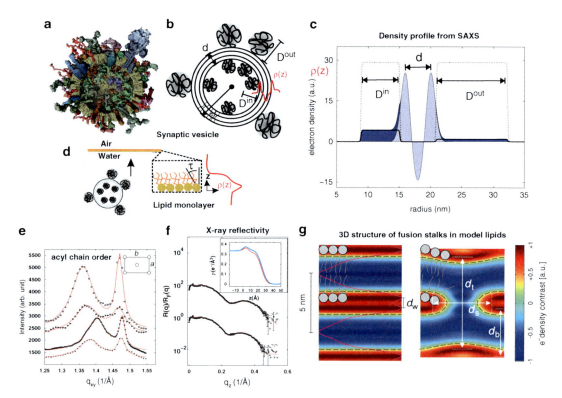

Fig. 4 (**a**) Visualization of a synaptic vesicle (SV) containing neurotransmitters with the densely packed layer of proteins, which control the fusion reaction, from [85]. The visualization is based on the lipid and protein stoichiometry from biochemical analysis. (**b**) Structural model used to parameterize the protein layers with vesicle radius, and the protein radii of gyration and densities as free fitting parameters in the SAXS analysis. (**c**) The density profile of the SV membrane with the adjacent protein layers as obtained from a least square fit of the SAXS curve, (**b**, **c**) from [86]. SV suspensions in buffer solution (as purified from mouse) are kept in a macroscopic glass capillary for the measurement. (**d**) Schematic of the monolayer experiments probing the interaction of synaptic vesicles with the lipid film, as in [87, 88]. (**e**) Grazing incidence diffraction results indicating the acyl chain ordering in the film, for different lipid mixtures (shifted for clarity, *from top to bottom*): pure DPPC monolayer, DPPC after injection of SVs, DPPC/PIP2, and DPPC/PIP2 with SVs. The Bragg peaks are fitted by two Lorentzians (*solid lines*). The centered rectangular unit cell is illustrated in the inset. From [88]. (**f**) X-ray reflectivity measurements (fits indicated by *solid red line*) indicate structural changes of the DPPC/PIP2 monolayer before (*top*) and after (*bottom*) injection of SVs into the subphase, see the corresponding electron density profiles shown in the inset. From [88]. (**g**) 3D structure of fusion stalks in model lipids

[88], a model system contributing to a molecular understanding of the interaction of synaptic vesicles with the inner leaflet of the synaptic membrane. Finally, membrane fusion intermediate structures have been studied at near molecular resolution in a protein-free multi-lamellar model lipid bilayer system [89], elucidating how lipid composition controls the formation of stalk structures. Some of these examples of non-crystallographic X-ray diffraction are illustrated in Fig. 4. Such studies are to be distinguished from X-ray microscopy. They probe Fourier components of the average structure in hydrated physiological states and are compatible with a multitude of sample environments. They yield only the ensemble

averaged structure at high resolution. In contrast to microscopy, local deviations from the mean and complex nonhomogeneous samples with a hierarchy of length scales are generally not accessible by diffraction. Let us consider the example of a lipid membrane: Small-angle X-ray diffraction from membrane suspensions, for example, requires several microliters of relatively homogeneous samples. Contrarily, optical microscopy enables a direct visualization of membranes, however, not at the resolution needed to probe the molecular structure of the bilayer. Finally, electron microscopy offers the desired resolution, but is limited by high vacuum conditions and invasive sample preparation, including cryogenic or chemical fixation, staining, and/or sectioning. What has been missing up to now is an experiment capable of deducing quantitative structural information down to the molecular scale, while at the same time preserving local real space representation without the usual ensemble averaging. The recent developments in X-ray microscopy may provide a solution. However, due to radiation damage, image quality, to date, has been satisfactory only for freeze-dried or frozen-hydrated samples, and moreover lipid bilayer structure has almost completely eluded X-ray microscopy. We have recently introduced a hybrid approach, combining dose efficient free space X-ray propagation imaging and near-field diffraction to locally resolve thickness, density, and more generally the density profile of membranes along the membrane normal [90, 91]. The dose in the phase contrast projection imaging scheme can be significantly reduced. By locally averaging the structure in the plane of the membrane over a length scale on the order of 5–20 μm, the molecular structure along the bilayer normal can be detected over extended illumination periods, and with relatively high resolution even at a photon fluence and dose fully compatible with room temperature, hydrated sample environment [90]. Importantly, local deviations and profiles become accessible. In this sense it differs from a simple diffraction experiment since it gives a real space visualization of the membrane contour and is thus compatible with external control parameters such as exerted forces, out-of-equilibrium transport, or local fields. The use of this method in combination with microfluidics can yield structural information on biomolecular interfaces during (hydro)dynamic processes, such as the formation of bilayers, thinning or bulging, as well as membrane fusion. The examples of X-ray diffraction experiments on synaptic vesicles and model membranes are illustrated in Fig. 4, while Fig. 5 gives a graphical example of the bilayer imaging technique.

Another conceptually straightforward combination of diffraction and imaging is based on scanning a micro- or nano-focused X-ray beam across the sample, and recording of a SAXS pattern in each pixel. Maps of the samples with contrast generated from the diffraction patterns can then be computed. Pioneered in the field of biomaterials [92], this scheme has now evolved and been demonstrated on the tissue up to the organ level [93], as well as for

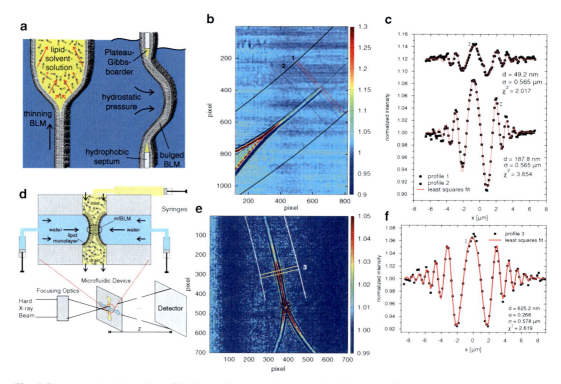

Fig. 5 Phase contrast imaging of lipid membranes using an adapted setup of black lipid membranes (BLMs). The goal of the experiment is to combine such a setup equipped for electrophysiology with a in-situ structural X-ray probe. To this end, a highly focused X-ray beam (17.5 keV) coherently illuminates a spherically bulged BLM, spanned over a micro-machined hole and located downstream the focal plane of a KB-mirror system. The image is formed by free propagation of the wave field between the sample and the detector, where the intensity profile is recorded. (**a**) Schematic of the formation and bulging process of a BLM. Organic solvent, used to dissolve the lipid molecules, diffuses towards the outer rim of the aperture supporting the membrane. The BLM starts to thin until the two monolayers at the oil–water interfaces approach to finally form a bilayer lipid membrane. Application of hydrostatic pressure to one side of the BLM leads to bulging of the interface. (**b**) Typical dataset showing the BLM contour and Fresnel oscillations, shown in (**c**) for the indicated cuts, which are then fitted to a model of the bilayer profile, yielding the local bilayer structure. Analysis of the Fresnel fringes of a completely thinned native bilayer indicating a thickness of around 3 nm (not shown). (**d**) Microfluidic chamber for monolayer fusion experiments, and positioning in the X-ray beam. (**e**) Image of two adhering monolayers, along with the associated (**f**) analysis, before the onset of bilayer formation, which proceeds via "zippering effect." Adapted from [91]

single cells [84, 94]. Structural analysis of pathological protein aggregates in neural cells is an example which could take advantage of this approach.

5 Myelinated Nerve Fibers Studied by Different X-Ray Microscopy Modalities

In this section, we review our recent X-ray microscopy study of myelinated sciatic neurons isolated from wild-type mice [42]. Myelin is well known as the spiral structure formed by extensions of the plasma membrane of the myelinating glial cells, the oligodendrocytes in the central nerve system (CNS), and the Schwann cells in the peripheral nervous system (PNS) [95, 96]. Several structural features characterize myelin [97]. Its periodic structure, with alternating concentric layers is easily detected after staining and slicing by electron microscopy [98], or by X-ray diffraction from entire nerves yielding the averaged electron density profile of myelin. The major dense line (dark layer in electron microscopy) forms as the cytoplasmic surfaces of the expanding extensions of the myelinating glial cell are brought into close apposition. The fused two outer leaflets (extracellular apposition) form the intraperiodic lines (or minor dense lines). The periodicity of the lamellae depends on the species and neuron types, and it is around 12 nm. The typical length of each myelin sheath segment or internode is in the range of 150–200 μm in the CNS and up to 1 mm in the PNS [97, 99]. Internodes are separated by spaces where myelin is lacking, the nodes of Ranvier. The nodes of Ranvier play a major role in nerve impulse conduction. The myelin sheath around most axons constitutes the most abundant membrane structure in the vertebrate nervous system. Myelin ensures the electrical insulation of axons, and its unique segmental structure enables the saltatory conduction of nerve impulses necessary for fast nerve conduction in the thin fibers [96], in contrast to the slow propagation of nerve signals along the axon in unmyelinated or demyelinated fibers [96]. The importance of myelin for signal conduction is highlighted by its role in a number of different neurological diseases such as leukodystrophies, multiple sclerosis (MS) in the CNS, and peripheral neuropathies in the PNS or hypermyelination (tomaculous neuropathy—HNPP) [100]. Important questions relate to the myelin assembly and its spatial organization, such as the exact three-dimensional sequence of the topological fold of this organelle, as well as to the chemical composition and distribution of metabolites.

Single nerve fibers, including both freeze-dried and cryopreserved myelinated axons, have been studied by four different imaging modalities using synchrotron radiation: (1) Scanning X-ray fluorescence imaging to map out the elemental distribution in the myelin sheath. Notably, the elemental distribution of P, S, Cl, Na, K, Fe, Mn, and Cu was studied in view of characteristic differences in the spatial distribution of ions and metabolites at internodes and the RN. (2) Single nerve diffraction was demonstrated using hard X-ray beams focused by elliptical mirrors to about 150 nm. The lamellar periodicity averaged locally over the area defined by the nano-focused synchrotron beam can be determined from the diffraction signal. Studying the local multilamellar structure in a single nerve fiber by diffraction translates the classical work of myelin diffraction [98, 99] from the level of an entire (averaged) nerve to an individual fiber, with the possibility to distinguish characteristic regions like the node of Ranvier from the internode. Straightforward extensions of the method, using X-ray nanobeams to scan single myelinated axons of nerve cells, yield scanning diffraction data, from which the local periodicity, lamellar ordering, number of bilayers can be inferred. The common analysis of (1) and (2) is sometimes denoted as the so-called hard X-ray nanoprobe. (3) Soft X-ray microscopy and tomography based on absorption contrast was carried out using a X-ray microscope based on FZP lenses, in order to elucidate the three-dimensional distribution of membranes near the RN. (4) Finally, synchrotron radiation-based Fourier transform infrared (FTIR) spectromicroscopy was applied. The study of [42] was meant as a proof of concept and will need follow-up experiments to fully reach its goals, but the feasibility of chemical and diffraction mappings down to the scale of 150 nm has clearly been demonstrated. Note that in the case of nanobeam diffraction, the resolution in reciprocal space averaged over the size of the beam was one to two orders of magnitude higher, reaching the 5 nm scale. Below, we will briefly illustrate imaging modalities (1), (2), and (3), as applied to single nerve fibers with exemplary images, and summarize the results.

Figure 6 shows the absorption and fluorescence map of a freeze-dried isolated myelinated axon of a wild-type mouse, close to a node of Ranvier, obtained at the FZP focusing at energy 2.9 keV settings at beamline ID21/ESRF [42]. The fluorescence maps of Na, P, Cl, and S are shown in sub-figures (a, b, c, d) respectively, as collected by scanning a FOV of 50 µm × 25 µm with 246 × 125 pixels, and dwell time 180 ms per pixel. The corresponding absorption image and summed fluorescence spectrum are shown in (e) and (f), respectively. The Na and P concentrations show a clear enhancement around the node of Ranvier, by almost a factor of two compared to the internode. Contrarily, Cl and S are distributed quite homogeneously over the whole axon. For further verifications, untreated cryo-fixated (vitrified frozen-hydrated) axons have also been imaged, using the cryo-stage and transfer system at ID21/ESRF at cryogenic conditions, at a photon energy of 7.2 keV and KB-mirror focusing, to cover all elemental lines up to Fe. The higher energy and photon flux enabled us to map out trace metals such as Fe, Cu, and Mn, see Fig. 6g.

Next, we have investigated whether the multilamellar ordering of the myelin sheath can be mapped out by diffraction with nanometer-sized hard X-ray beams, extending the resolution of scanning SAXS from the micron range [101, 102] to the range of 100 nm and below. In principle, simultaneous mappings of the lamellar periodicity d (from radial peak position), membrane orientation (from the angular peak position), and possibly the number of adjacent membranes N (from the peak width) are

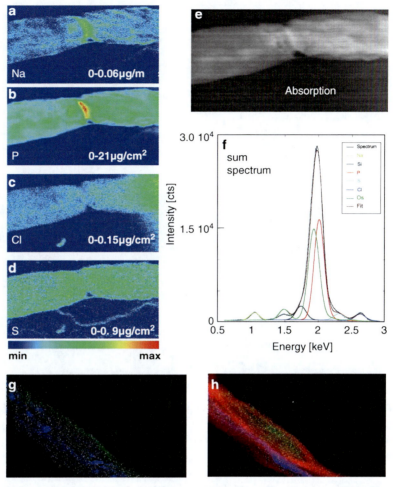

Fig. 6 X-ray fluorescence images of a single isolated myelinated axon, with region of interest around a node of Ranvier. Distribution of (**a**) sodium, (**b**) phosphorus, (**c**) chloride, (**d**) sulfur, and (**e**) the corresponding absorption X-ray image, as recorded by scanning X-ray microscopy at 2.9 keV photon energy. Scales indicate elements concentration in μg/cm². Image size 50 × 25 μm, 246 × 125 pixels with dwell time 180 ms per pixel. (**f**) The summed fluorescence spectrum of the scanned area along with a fit (PyMCA software). (**g, h**) X-ray fluorescence of cryo-preserved myelinated neurons. (**g**) Overlay distribution of microelements (Mn (*red*), Fe (*blue*), and Cu (*green*), and (**h**) the distribution of macroelements: P (*red*), Cl (*green*), and K (*blue*). Image size 70 × 40 μm, 280 × 160 pixels with dwell time 150 ms per pixel, recorded at E = 7.2 keV. From [42]

conceivable, i.e. similar structural information as from classical myelin diffraction experiments which average over an entire nerve. Figure 7a, b gives an example of such a diffraction measurement using both (a) synchrotron and (b) an in-house rotating anode

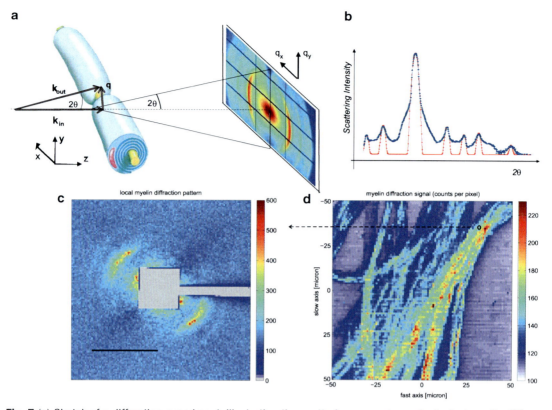

Fig. 7 (**a**) Sketch of a diffraction experiment, illustrating the scattering geometry and a typical myelin diffraction intensity distribution, as recorded for an oriented optical nerve of wild-type mouse with 6.2 keV synchrotron radiation. The multilamellar ordering of the myelin membranes is reflected by one or several diffraction maxima. The peak position reflects the membrane spacing, while the intensities carry the information on the electron density profile. Finally the peak width is governed by the lamellar ordering, and azimuthal distribution corresponds to the orientational distribution. Importantly, the diffraction signal is an average over the entire nerve. (**b**) Scattering intensity curve for a similar sample (optical nerve) as recorded at a sealed tube X-ray generator by scanning a point detector with slit collimation. (**c**, **d**) In contrast to studies of the entire nerve as in (**a**, **b**), the local myelin structure can be probed by scanning a nano-focused synchrotron beam over a single axon (teased fibre) preparation, see [42] for details on preparation. (**c**) Lamellar diffraction pattern of myelin corresponding to the pixel marked by the *black circle* in (**d**). The *gray structure* is a shadow of the beamstop. Scale bar is 1 nm^{-1}. (**d**) Myelin scattering intensity map in units of average photon number per 50 ms acquisition time and detector pixel (region of interest around the lamellar diffraction). The real space pixel size in (**d**) is 1 μm. The image shows several parallel fibers as deposited on a polyimide foil. From the data, a multitude of structural variables can be calculated for each real space pixel, such as fiber orientation, lamellar spacing, as well as an order parameter

source, taken from a series of nerves, investigated in view of correlation of sample treatment and structural integrity (T. Salditt, M. Bartels, A. Saab, W. Moebius, unpublished). The data in (b) was recorded immediately after nerve preparation, keeping the sample in buffer solution in between thin polypropylene windows, without any fixation. Contrarily to these conventional experiments, diffraction using nano-focused beams can be carried out on single axons, offering—in addition to the reciprocal space resolution—high resolution in real space. This is useful to distinguish interesting regions, e.g., around the node of Ranvier and could help to unravel the structural organization of myelin and its evolution as well as myelin-associated pathologies. However, to this end significant experimental challenges have to be met regarding the preparation of clean tails in focusing detector read-out rate and dynamic range, as well as protection from radiation damage (e.g., by cryogenic conditions), as detailed in [42]. Figure 7c, d shows the typical signal recorded in a preliminary single fiber diffraction experiment, carried out at the GINIX nano-focus instrument at beamline P10 of the PETRAIII storage ring in Hamburg, scanning a 7.9 keV photon energy with a cross section of about 280 nm (vertical) and 360 nm (horizontal). The beam was scanned over a freeze-dried single axon (teased fibre preparation), see [42] for details on preparation. To minimize damage, the sample was kept in a cryogenic nitrogen gas jet (Oxford CryoSystems). At each scan point, a diffraction image is recorded by a single photon sensitive pixel detector (Pilatus), positioned 329 mm behind the sample, see the myelin diffraction pattern shown in Fig. 7c, corresponding to the scan point indicated by a black circle in Fig. 7d. The two opposing diffraction orders result from an oriented myelin distribution and peaks at a distance of $0.7\,\text{nm}^{-1}$ reflecting the lamellar spacing of myelin. The entire (dark field) intensity map of the $5 \times 50\,\mu$ scan is shown in (d).

Finally, a lead-off experiment on soft X-ray tomography has been carried out demonstrating that single fiber tomography under cryogenic conditions has now become possible with the new BESSY X-ray microscope. In contrast to electron tomography, which still is superior in resolution for thin samples, sectioning and staining is not required. Figure 8 shows a cryo-transmission X-ray imaging and tomography of a myelinated fiber. The interior structure becomes visible in absorption contrast, showing an intricate structure at node of Ranvier imaged, which can be well visualized in absorption (density) contrast without staining or slicing. Reconstruction of the 108 images collected at 0.5° intervals through 54° of rotation in the tomographic data set yields the depth resolved structure. A representative slice out of the stack of images is shown in (c). The preliminary reconstruction outlines the great potential of soft X-ray tomography to unravel the native 3D myelin architecture at the node of Ranvier [42].

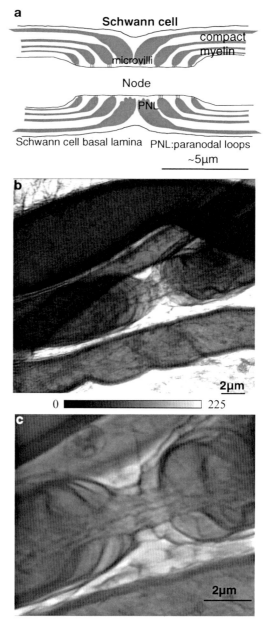

Fig. 8 (**a**) Sketch of the myelin structure at a node of Ranvier. (**b**) Transmission X-ray micrograph of a fiber prepared and imaged with soft X-rays under cryogenic conditions. The interior membrane organization near a node of Ranvier is visualized by a projection image in absorption contrast (transmitted intensity in linear grayscale). (**c**) Representative reconstruction slice from a tilt series showing the absorption coefficient. From [42]

6 Challenges Ahead

With significant progress in the last decade and a multitude of technical variants which have evolved, X-ray microscopy is now ready for challenging applications in neuroscience. Future experiments may address the ultrastructure of neurons and the three-dimensional structure of neural tissues, with contrast mechanisms not available previously: elemental mapping and elemental speciation, absorption and phase contrast which can be translated in quantitative terms to electron density, and X-ray dark field contrast which reveals the presence of structural variations on length scales smaller than the resolution.

For cellular structure analysis, coherent X-ray microscopy and X-ray diffraction with nanometer beams will undergo a synergistic relationship, since both can be carried out on the same sample with the same setup, yielding complementary information in real and reciprocal space, respectively. However, even if further progress, e.g., in resolution is very likely, compared to light and electron microscopy, cellular imaging with X-rays will probably remain a rather small activity used for selected problems. Significant challenges will concern sample preparation issues, and the issue of radiation damage. Rapid vitrification and structural preservation under cryogenic conditions required dedicated sample environments. Dose efficient phase retrieval algorithms which deliver as much information as possible from as few photons as possible will need to take the Poissonian distribution of the shot noise into account. These efforts are worthwhile. The ultrastructure of the synapse, the structural assembly of the myelinated axon, amyloid protein deposits in cellular and extracellular spaces are examples for the many formidable problems in cellular structure, which could take significant advantage of XM. Finally, correlative microscopy combining the unique information collected by complementary microscopy techniques is a promising strategy.

An arguably larger field of application which starts to unfold is X-ray imaging of neural tissues up to the entire organ, since penetration power and contrast make X-rays a unique probe to map out the three-dimensional structure of specimen by tomography. Phase contrast is currently introduced in biomedical imaging, complementing conventional radiography by an unprecedented sensitivity to weakly absorbing tissues, and increasing its resolution. On scales of the $10\,\mu m$ range and above, this field has been revolutionized by phase sensitive imaging based on the Talbot effect. For higher nanoscale resolution, i.e. the veritable domain of microscopy, propagation imaging and ptychography provide the most promising approach. To further develop this field, important challenges are region-of-interest tomography, as well as simultaneous two-dimensional phase retrieval and tomographic reconstruction. The future will show to which extent larger tissue sections can be investigated at cellular and sub-cellular resolution, enabling a zoom into a region of interest. If it was possible to resolve the dendrites in the unsliced central nervous system of small animals, the problem of neural connectivity could be solved, i.e. a connectivity map between all neurons could be established.

Acknowledgments

We thank all of our collaborators on the original work reviewed here, in particular Simon Castorph, Sajal Ghosh, André Beelink, Klaus Giewekemeyer, Robin Wilke, and Matthias Bartels. We are thankful to the European Synchrotron Radiation Facility (ESRF), the Swiss Light Source (SLS) and Desy Photon Science, Hamburg, for generous beam time allocation and the financial support by DFG Research Center 103 Molecular Physiology of the Brain (CMPB) within the Cluster of Excellence 171 "Microscopy at the Nanometer Range," Germany.

References

1. Pfeiffer F, Kottler C, Bunk O, David C (2007) Hard x-ray phase tomography with low-brilliance sources. Phys Rev Lett 98:108105
2. Bartels M, Hernandez VH, Krenkel M, Moser T, Salditt T (2013) Phase contrast tomography of the mouse cochlea at microfocus x-ray sources. Appl Phys Lett 103:083703
3. Als-Nielsen J, McMorrow D (2001) Elements of modern X-ray physics. Wiley, London
4. Attwood DT (2000) Soft X-rays and extreme ultraviolet radiation: principles and applications. Cambridge University Press, Cambridge
5. Medalia O, Weber I, Frangakis AS, Nicastro D, Gerisch G, Baumeister W (2002) Macromolecular architecture in eukaryotic cells visualized by cryoelectron tomography. Science 298:1209–1213
6. Hell SW (2007) Far-field optical nanoscopy. Science 316:1153–1158
7. Willig KI, Rizzoli SO, Westphal V, Jahn R, Hell SW (2006) Sted microscopy reveals that synaptotagmin remains clustered after synaptic vesicle exocytosis. Nature 440:935–939
8. Giewekemeyer K, Thibault P, Kalbfleisch S, Beerlink A, Kewish CM, Dierolf M, Pfeiffer F, Salditt T (2010) Quantitative biological imaging by ptychographic x-ray diffraction microscopy. Proc Natl Acad Sci 107:529–534
9. Schmahl G, Rudolph D, Niemann B, Christ O (1980) Zone-plate x-ray microscopy. Q Rev Biophys 13:297–315
10. Kirz J, Jacobsen C, Howells M (1995) Soft x-ray microscopes and their biological applications. Q Rev Biophys 28:33–130
11. Chao W, Harteneck BD, Liddle JA, Anderson EH, Attwood DT (2005) Soft x-ray microscopy at a spatial resolution better than 15 nm. Nature 435:1210–1213
12. Hornberger B, Feser M, Jacobsen C (2007) Quantitative amplitude and phase contrast imaging in a scanning transmission x-ray microscope. Ultramicroscopy 107:644–655
13. Yang L, McRae R, Henary M, Patel R, Lai B, Vogt S, Fahrni C (2005) Imaging of the intracellular topography of copper with a fluorescent sensor and by synchrotron x-ray fluorescence microscopy. Proc Natl Acad Sci USA 102:11179
14. Larabell C, Le Gros M (2004) X-ray tomography generates 3-d reconstructions of the yeast, saccharomyces cerevisiae, at 60-nm resolution. Mol Biol Cell 15:957
15. Weiss D, Schneider G, Niemann B, Guttmann P, Rudolph D, Schmahl G et al (2000) Computed tomography of cryogenic biological specimens based on x-ray microscopic images. Ultramicroscopy 84:185–197
16. Miao J, Charalambous P, Kirz J, Sayre D (1999) Extending the methodology of x-ray crystallography to allow imaging of micrometre-sized non-crystalline specimens. Nature 400:342–344

17. Williams G, Pfeifer M, Vartanyants I, Robinson I (2003) Three-dimensional imaging of microstructure in au nanocrystals. Phys Rev Lett 90:175501
18. Williams GJ, Quiney HM, Dhal BB, Tran CQ, Nugent KA, Peele AG, Paterson D, de Jonge MD (2006) Fresnel coherent diffractive imaging. Phys Rev Lett 97:025506
19. Miao J, Hodgson KO, Ishikawa T, Larabell CA, LeGros MA, Nishino Y (2003) Imaging whole *Escherichia coli* bacteria by using single-particle x-ray diffraction. Proc Natl Acad Sci 100:110–112
20. Shapiro D, Thibault P, Beetz T, Elser V, Howells M, Jacobsen C, Kirz J, Lima E, Miao H, Neiman AM, Sayre D (2005) Biological imaging by soft x-ray diffraction microscopy. Proc Natl Acad Sci 102:15343–15346
21. Song C, Jiang H, Mancuso A, Amirbekian B, Peng L, Sun R, Shah SS, Zhou ZH, Ishikawa T, Miao J (2008) Quantitative imaging of single, unstained viruses with coherent x rays. Phys Rev Lett 101:158101
22. Nishino Y, Takahashi Y, Imamoto N, Ishikawa T, Maeshima K (2009) Three-dimensional visualization of a human chromosome using coherent x-ray diffraction. Phys Rev Lett 102:018101
23. Fahrni C (2007) Biological applications of x-ray fluorescence microscopy: exploring the subcellular topography and speciation of transition metals. Curr Opin Chem Biol 11:121–127
24. Schroer CG, Kurapova O, Patommel J, Boye P, Feldkamp J, Lengeler B, Burghammer M, Riekel C, Vincze L, van der Hart A, Kuchler M (2005) Hard x-ray nanoprobe based on refractive x-ray lenses. Appl Phys Lett 87:3 (124103)
25. Hignette O, Cloetens P, Rostaing G, Bernard P, Morawe C (2005) Efficient sub 100 nm focusing of hard x rays. Rev Sci Instrum 76:063709
26. Mimura H, Handa S, Kimura T, Yumoto H, Yamakawa D, Yokoyama H, Matsuyama S, Inagaki K, Yamamura K, Sano Y, Tamasaku K, Nishino Y, Yabashi M, Ishikawa T, Yamauchi K (2010) Breaking the 10 nm barrier in hard-x-ray focusing. Nat Phys 6:122–125
27. Bergemann C, Keymeulen H, van der Veen JF (2003) Focusing X-ray beams to nanometer dimensions. Phys Rev Lett 91:204801
28. Ruhlandt A, Liese T, Radisch V, Krüger SP, Osterhoff M, Giewekemeyer K, Krebs HU, Salditt T (2012) A combined Kirkpatrick-Baez mirror and multilayer lens for sub-10 nm x-ray focusing. AIP Adv 2:7 (012175)
29. Yan H, Rose V, Shu D, Lima E, Kang H, Conley R, Liu C, Jahedi N, Macrander A, Stephenson G et al (2011) Two dimensional hard x-ray nanofocusing with crossed multilayer laue lenses. Opt Express 19:15069–15076
30. Rehbein S, Heim S, Guttmann P, Werner S, Schneider G (2009) Ultrahigh-resolution soft-x-ray microscopy with zone plates in high orders of diffraction. Phys Rev Lett 103:110801
31. Vila-Comamala J, Gorelick S, Färm E, Kewish CM, Diaz A, Barrett R, Guzenko VA, Ritala M, David C (2011) Ultra-high resolution zone-doubled diffractive x-ray optics for the multi-kev regime. Opt Express 19:175–184
32. Krüger SP, Giewekemeyer K, Kalbfleisch S, Bartels M, Neubauer H, Salditt T (2010) Sub-15 nm beam confinement by twocrossed x-ray waveguides. Opt. Express 18:13492–13501
33. Osterhoff M, Salditt T (2011) Coherence filtering of x-ray waveguides: analytical and numerical approach. New J Phys 13:103026
34. Bartels M, Priebe M, Wilke R, Kruger S, Giewekemeyer K, Kalbfleisch S, Olendrowits C, Sprung M, Salditt T (2012) Low-dose three-dimensional hard x-ray imaging of bacterial cells. Opt Nanoscopy 1:10
35. Krenkel M, Bartels M, Salditt T (2013) Transport of intensity phase reconstruction to solve the twin image problem in holographic x-ray imaging. Opt Express 21:2220–2235
36. Jarre A, Fuhse C, Ollinger C, Seeger J, Tucoulou R, Salditt T (2005) Two-dimensional hard x-ray beam compression by combined focusing and waveguide optics. Phys Rev Lett 94:074801
37. Kruger S, Neubauer H, Bartels M, Kalbfleisch S, Giewekemeyer K, Wilbrandt P, Sprung M, Salditt T (2012) Sub-10 nm beam confinement by x-ray waveguides: design, fabrication and characterization of optical properties. J Synchrotron Radiat 19:227–236
38. Giewekemeyer K, Krüger SP, Kalbfleisch S, Bartels M, Beta C, Salditt T (2011) X-ray propagation microscopy of biological cells using waveguides as a quasipoint source. Phys Rev A 83:023804
39. Howells M, Beetz T, Chapman H, Cui C, Holton J, Jacobsen C, Kirz J, Lima E, Marchesini S, Miao H, Sayre D, Shapiro D, Spence J, Starodub D (2009) An assessment of the resolution limitation due to radiation-damage in x-ray diffraction microscopy. J Electron Spectrosc Related Phenomena 170:4–12
40. Schneider G (1998) Cryo x-ray microscopy with high spatial resolution in amplitude and phase contrast. Ultramicroscopy 75:85
41. Carmona A, Cloetens P, Devès G, Bohic S, Ortega R (2008) Nano-imaging of trace metals

by synchrotron x-ray fluorescence into dopaminergic single cells and neurite-like processes. J Anal At Spectrom 23: 1083–1088
42. Dučić T, Quintes S, Nave K, Susini J, Rak M, Tucoulou R, Alevra M, Guttmann P, Salditt T (2011) Structure and composition of myelinated axons: a multimodal synchrotron spectromicroscopy study. J Struct Biol 173:202–212
43. Schmahl G, Rudolph D, Guttmann P, Schneider G, Thieme J, Niemann B (1995) Phase contrast studies of biological specimens with the x-ray microscope at bessy (invited). Rev Sci Instrum 66:1282–1286
44. Bertilson M, von Hofsten O, Vogt U, Holmberg A, Christakou AE, Hertz HM (2011) Laboratory soft-x-ray microscope for cryotomography of biological specimens. Opt Lett 36:2728–2730
45. Schneider G, Guttmann P, Heim S, Rehbein S, Mueller F, Nagashima K, Heymann JB, Müller WG, McNally JG (2010) Three-dimensional cellular ultrastructure resolved by x-ray microscopy. Nat Methods 7:985–987
46. Larabell CA, Nugent KA (2010) Imaging cellular architecture with x-rays. Curr Opin Struct Biol 20:623–631
47. Uchida M, McDermott G, Wetzler M, Le Gros MA, Myllys M, Knoechel C, Barron AE, Larabell CA (2009) Soft x-ray tomography of phenotypic switching and the cellular response to antifungal peptides in candida albicans. Proc Natl Acad Sci 106:19375–19380
48. Bohic S, Murphy K, Paulus W, Cloetens P, Salomé M, Susini J, Double K (2008) Intracellular chemical imaging of the developmental phases of human neuromelanin using synchrotron x-ray microspectroscopy. Anal Chem 80:9557–9566
49. Szczerbowska-Boruchowska M (2008) X-ray fluorescence spectrometry, an analytical tool in neurochemical research. X-Ray Spectrom 37:21–31
50. Crichton R, Ward R (2005) Metal-based neurodegeneration: from molecular mechanisms to therapeutic strategies. Wiley, London
51. Dodani S, Domaille D, Nam C, Miller E, Finney L, Vogt S, Chang C (2011) Calcium-dependent copper redistributions in neuronal cells revealed by a fluorescent copper sensor and x-ray fluorescence microscopy. Proc Natl Acad Sci 108:5980
52. Bohic S, Cotte M, Salomé M, Fayard B, Kuehbacher M, Cloetens P, Martinez-Criado G, Tucoulou R, Susini J (2011) Biomedical applications of the esrf synchrotron-based microspectroscopy platform. J Struct Biol 177:248–258
53. James S, Myers D, de Jonge M, Vogt S, Ryan C, Sexton B, Hoobin P, Paterson D, Howard D, Mayo S et al (2011) Quantitative comparison of preparation methodologies for x-ray fluorescence microscopy of brain tissue. Anal Bioanal Chem 401:1–12
54. Bacquart T, Deves G, Carmona A, Tucoulou R, Bohic S, Ortega R (2007) Subcellular speciation analysis of trace element oxidation states using synchrotron radiation micro-x-ray absorption near-edge structure. Anal Chem 79:7353–7359
55. Vogt S, Maser J, Jacobsen C (2003) Data analysis for x-ray fluorescence imaging. J Phys IV France 104:617–622
56. Cloetens P, Ludwig W, Baruchel J, Dyck DV, Landuyt JV, Guigay JP, Schlenker M (1999) Holotomography: quantitative phase tomography with micrometer resolution using hard synchrotron radiation x rays. Appl Phys Lett 75:2912–2914
57. Cloetens P, Mache R, Schlenker M, Lerbs-Mache S (2006) Quantitative phase tomography of arabidopsis seeds reveals intercellular void network. Proc Natl Acad Sci 103: 14626–14630
58. Wilkins SW, Gureyev TE, Gao D, Pogany A, Stevenson AW (1996) Phase-contrast imaging using polychromatic hard x-rays. Nature 384:335–338
59. Mayo S, Miller P, Wilkins S, Davis T, Gao D, Gureyev T, Paganin D, Parry D, Pogany A, Stevenson A (2002) Quantitative x-ray projection microscopy: phase-contrast and multi-spectral imaging. J Microsc 207:79–96
60. Gureyev TE, Mayo SC, Myers DE, Nesterets Y, Paganin DM, Pogany A, Stevenson AW, Wilkins SW (2009) Refracting röntgen's rays: propagation-based x-ray phase contrast for biomedical imaging. J Appl Phys 105:102005
61. Abbey B, Nugent KA, Williams GJ, Clark JN, Peele AG, Pfeifer MA, de Jonge M, McNulty I (2008) Keyhole coherent diffractive imaging. Nat Phys 4:394–398
62. Williams GJ, Quiney HM, Peele AG, Nugent KA (2010) Fresnel coherent diffractive imaging: treatment and analysis of data. New J Phys 12:035020
63. Giewekemeyer K, Neubauer H, Kalbfleisch S, Krüger SP, Salditt T (2010) Holographic and diffractive x-ray imaging using waveguides as quasi-point sources. New J Phys 12:035008
64. Nugent KA (2010) Coherent methods in the x-ray sciences. Adv Phys 59:1–99

65. Quiney H (2010) Coherent diffractive imaging using short wavelength light sources. J Mod Optic 57:1109–1149
66. Paganin DM (2006) Coherent X-ray optics. Oxford University Press, Oxford
67. Fienup JR (1982) Phase retrieval algorithms: a comparison. Appl Opt 21:2758–2769
68. Marchesini S (2007) Invited article: a unified evaluation of iterative projection algorithms for phase retrieval. Rev Sci Instrum 78:011301
69. Rodenburg JM, Hurst AC, Cullis AG, Dobson BR, Pfeiffer F, Bunk O, David C, Jefimovs K, Johnson I (2007) Hard-x-ray lensless imaging of extended objects. Phys Rev Lett 98:034801
70. Thibault P, Dierolf M, Menzel A, Bunk O, David C, Pfeiffer F (2008) High-resolution scanning x-ray diffraction microscopy. Science 321:379–382
71. Guizar-Sicairos M, Fienup JR (2008) Phase retrieval with transverse translation diversity: a nonlinearoptimization approach. Opt Express 16:7264–7278
72. Schropp A, Boye P, Feldkamp JM, Hoppe R, Patommel J, Samberg D, Stephan S, Giewekemeyer K, Wilke RN, Salditt T, Gulden J, Mancuso AP, Vartanyants IA, Weckert E, Schoder S, Burghammer M, Schroer CG (2010) Hard x-ray nanobeam characterization by coherent diffraction microscopy. Appl Phys Lett 96:3 (091102)
73. Kewish CM, Thibault P, Dierolf M, Bunk O, Menzel A, Vila-Comamala J, Jefimovs K, Pfeiffer F (2010) Ptychographic characterization of the wavefield in the focus of reflective hard x-ray optics. Ultramicroscopy 110:325–329
74. Giewekemeyer K, Beckers M, Gorniak T, Grunze M, Salditt T, Rosenhahn A (2011) Ptychographic coherent x-ray diffractive imaging in the water window. Opt Express 19:1037–1050
75. Beckers M, Senkbeil T, Gorniak T, Reese M, Giewekemeyer K, Gleber S, Salditt T, Rosenhahn A (2011) Chemical contrast in soft x-ray ptychography. Phys Rev Lett 107:208101
76. Takahashi Y (2011) Multiscale element mapping of buried structures by ptychographic x-ray diffraction microscopy using anomalous scattering. Appl Phys Lett 99:131905
77. Eisebitt S, Luning J, Schlotter WF, Lorgen M, Hellwig O, Eberhardt W, Stohr J (2004) Lensless imaging of magnetic nanostructures by x-ray spectro-holography. Nature 432: 885–888
78. Fuhse C, Ollinger C, Salditt T (2006) Waveguide-based off-axis holography with hard x rays. Phys Rev Lett 97:254801
79. Salditt T, Krüger S, Fuhse C, Bähtz C (2008) High-transmission planar x-ray waveguides. Phys Rev Lett 100:184801
80. Jiang H, Song C, Chen C-C, Xu R, Raines KS, Fahimian BP, Lu C-H, Lee T-K, Nakashima A, Urano J, Ishikawa T, Tamanoi F, Miao J (2010) Quantitative 3d imaging of whole, unstained cells by using x-ray diffraction microscopy. Proc Natl Acad Sci 107:11234–11239
81. Nelson J, Huang X, Steinbrener J, Shapiro D, Kirz J, Marchesini S, Neiman AM, Turner JJ, Jacobsen C (2010) High-resolution x-ray diffraction microscopy of specifically labeled yeast cells. Proc Natl Acad Sci 107:7235–7239
82. Lima E, Wiegart L, Pernot P, Howells M, Timmins J, Zontone F, Madsen A (2009) Cryogenic x-ray diffraction microscopy for biological samples. Phys Rev Lett 103:198102
83. Huang X, Nelson J, Kirz J, Lima E, Marchesini S, Miao H, Neiman AM, Shapiro D, Steinbrener J, Stewart A, Turner JJ, Jacobsen C (2009) Soft x-ray diffraction microscopy of a frozen hydrated yeast cell. Phys Rev Lett 103:198101
84. Wilke RN, Priebe M, Bartels M, Giewekemeyer K, Diaz A, Karvinen P, Salditt T (2012) Hard X-ray imaging of bacterial cells: nano-diffraction and ptychographic reconstruction. Opt Express 20:19232–19254
85. Takamori S, Holt M, Stenius K, Lemke E, Grønborg M, Riedel D, Urlaub H, Schenck S, Brügger B, Ringler P et al (2006) Molecular anatomy of a trafficking organelle. Cell 127:831–846
86. Castorph S, Riedel D, Arleth L, Sztucki M, Jahn R, Holt M, Salditt T (2010) Structure parameters of synaptic vesicles quantified by small-angle x-ray scattering. Biophys J 98:1200–1208
87. Ghosh S, Castorph S, Konovalov O, Jahn R, Holt M, Salditt T (2010) In vitro study of interaction of synaptic vesicles with lipid membranes. New J Phys 12:105004
88. Pfeiffer F, Kottler C, Bunk O, David C (2012) Measuring ca^{2+}-induced structural changes in lipid monolayers: implications for synaptic vesicle exocytosis. Biophys J 102:1394–1402
89. Aeffner S, Reusch T, Weinhausen B, Salditt T (2012) Energetics of stalk intermediates in membrane fusion are controlled by lipid composition. Proc Natl Acad Sci 109:E1609–E1618
90. Beerlink A, Mell M, Tolkiehn M, Salditt T (2009) Hard x-ray phase contrast imaging of

black lipid membranes. Appl Phys Lett 95:203703
91. Beerlink A, Shashi Thutupalli S, Mell M, Bartels M, Cloetens P, Herminghaus S, Salditt T (2012) X-ray propagation imaging of a lipid bilayer in solution. Soft Matter 8:4595–4601
92. Fratzl P, Jakob HF, Rinnerthaler S, Roschger P, Klaushofer K (1997) Position-resolved small-angle x-ray scattering of complex biological materials. J Appl Cryst 30:765–769
93. Jensen T, Bech M, Bunk O, Menzel A, Bouchet A, Le Duc G, Feidenhans'l R, Pfeiffer F (2011) Molecular x-ray computed tomography of myelin in a rat brain. NeuroImage 57:124–129
94. Weinhausen B, Nolting J-F, Olendrowitz C, Langfahl-Klabes J, Reynolds M, Salditt T, Köster S (2012) X-ray nano-diffraction on cytoskeletal networks. New J Phys 14:085013
95. Quarles R, Macklin W, Morell P (2006) Myelin formation, structure, and biochemistry. Academic, London
96. Baumann N, Pham-Dinh D (2001) Biology of oligodendrocyte and myelin in the mammalian central nervous system. Physiol Rev 81:871–927
97. Siegel G (2006) Basic neurochemistry: molecular, cellular and medical aspects. Elsevier Academic
98. Blaurock AE (1976) Myelin x-ray patterns reconciled. Biophys J 16:491–501
99. Kirschner D, Blaurock A (1992) Organization, phylogenetic variations, and dynamic transitions of myelin. In: Martinson RE (ed.), Myelin: biology and chemistry. CRC Press, Boca Raton, pp. 3–80
100. Chance PF (2001) Molecular basis of hereditary neuropathies. Phys Med Rehabil Clin N Am 12:277–291
101. Fratzl P (2003) Small-angle scattering in materials science - a short review of applications in alloys, ceramics and composite materials. J Appl Crystallogr 36:397–404
102. Aichmayer B, Margolis H, Sigel R, Yamakoshi Y, Simmer J, Fratzl P (2005) The onset of amelogenin nanosphere aggregation studied by small-angle x-ray scattering and dynamic light scattering. J Struct Biol 151:239–249

Chapter 12

Nonlinear Optics Approaches Towards Subdiffraction Resolution in CARS Imaging

Klaus-Jochen Boller, Willem P. Beeker, Carsten Cleff, Kai Kruse, Chris J. Lee, Petra Groß, Herman L. Offerhaus, Carsten Fallnich, and Jennifer L. Herek

Abstract

In theoretical investigations, we review several nonlinear optical approaches towards subdiffraction-limited resolution in label-free imaging via coherent anti-Stokes Raman scattering (CARS). Using a density matrix model and numerical integration, we investigate various level schemes and combinations of the light fields that induce CARS along with additional control laser fields. As the key to techniques that gain far-field resolution below the diffraction limit, we identify the inhibition of the buildup of vibrational molecular coherence via saturation or depletion of population (Beeker et al. Opt Express 17:22632–22638, 2009) or the generation of Stark broadening and spatially dependent Rabi sideband generation (Beeker et al. Phys Rev A 81(1), 2010). Depending on the coherence and population decay rates offered by a particular molecular energy level scheme, we identify various different regimes. In the first case, where an additional state (called the control state) and a vibrational state are able to rapidly exchange population via incoherent processes, a prepopulation of the upper vibrational state inhibits the buildup of vibrational coherence. With increasing control laser intensity, this suppresses CARS emission via an incoherent, saturation type of nonlinear process. Using an intense, donut-shaped control laser beam, similar to stimulated emission depletion (STED) microscopy, this can suppress CARS emission from all sample locations except within a subdiffraction-sized range around the central node. Scanning the control beams across the sample provides subdiffraction-limited resolution imaging. An alternative, which does not require a rapid exchange of population with the control state, applies a control beam that only partially depletes the vibrational ground state. Thereby, a CARS point spread function containing a subdiffraction-limited component is generated. Subdiffraction images can then be retrieved through deconvolution. Further approaches are based on the coherent, nonlinear, resonant response of the sample. In this case, CARS signal depletion by Stark splitting of the weakly populated upper vibrational state or the observation of spatially dependent Rabi oscillation may increase the resolution beyond the diffraction limit.

Key words Nonlinear polarizability, Stimulated Raman, Multiphoton numerical modeling, Subdiffraction imaging, Coherent anti-Stokes Raman (CARS)

Eugenio F. Fornasiero and Silvio O. Rizzoli (eds.), *Super-Resolution Microscopy Techniques in the Neurosciences*, Neuromethods, vol. 86, DOI 10.1007/978-1-62703-983-3_12, © Springer Science+Business Media, LLC 2014

1 Introduction

Optical microscopy is at the heart of analysis in biology, where it is used at scales ranging from systemic to subcellular. Other fields are also heavily reliant on optical microscopy, such as material science and microfabrication techniques. Important recent developments in resolution enhancement are far-field subdiffraction-limited resolution imaging techniques that have opened up new vistas for quantifying subcellular features.

Notable examples of far-field subdiffraction-limited resolution imaging are stimulated emission depletion (STED) microscopy [1, 2], photoactivated localization microscopy (PALM) [3], and stochastic optical reconstruction microscopy (STORM) [4–6]. STORM and PALM require that the fluorescent emitters be sufficiently spaced so that by careful activation and deactivation, single-photon detection and counting statistics can be used to fit the center locations of individual emitters far more accurately than the diffraction-limited spot. STORM and PALM have long image acquisition times, so that cells must be fixed before imaging. Finally, the careful activation and deactivation process requires rather specialized fluorescent labels [7].

STED microscopy [1] exploits the incoherent optical nonlinear process of saturation. First, a population of electronic states of fluorescent markers (fluorophores) is excited by a light field. Then, before spontaneous fluorescence occurs, the excited population is selectively depleted by another light field via stimulated emission. The stimulating field, which may also be termed the control field, suppresses the spontaneous fluorescence from the fluorophore population, effectively limiting it to locations where the control light intensity is less than the saturation intensity of the label molecule. Using a control field with a node at the center of the beam limits the spontaneous fluorescence to emission from a subdiffraction-limited volume at the beam center. Furthermore, the resolution increases with the square root of the intensity of the additional laser beam, allowing one to tune the resolution. Recent results show that STED can provide images with a lateral resolution of 6 nm in solids [8], enabling, for example, the all-optical recording of electron spin resonances [9]. In liquids, which are of relevance for biological samples, the resolution has reached a value around 30 nm [10].

Label-free imaging techniques, such as coherent anti-Stokes Raman scattering (CARS) microscopy, stimulated Raman scattering (SRS), or coherent Stokes Raman scattering (CSRS [11]) have the benefit of not requiring fluorescent labels for imaging. These techniques offer chemical specificity via recording resonances of the exciting light fields with characteristic vibrational molecular transitions. These so-called fingerprint modes can be used to identify characteristic chemical bonds for what is called chemical imaging. State-of-the-art CARS microscopy can provide high-contrast video rate imaging while effectively limiting photo damage as the drive lasers are far off-resonant from electronic transitions [12]. Notably, CARS microscopy is now also beginning to be applied for the chemical imaging of artificial nanostructures and for characterizing the influence of nanoparticles on biological systems [13]. The investigation of inorganic samples seems particularly promising due to a much lower susceptibility to photo damage. This should allow the incident intensities and average powers to be increased. Furthermore, it would enable shorter drive wavelengths, possibly in the ultraviolet (UV) range, to be used to increase the spatial resolution. Indeed, it is of considerable interest to achieve subdiffraction-limited CARS imaging, and several research groups are actively investigating linear and nonlinear techniques [14, 15]. However, linear techniques cannot be used to increase the spatial frequency cutoff of an imaging system to obtain subdiffraction-limited resolution, leaving only nonlinear optical processes.

In this chapter, using a theoretical analysis, we propose several nonlinear optical approaches that might enable subdiffraction-limited resolution in CARS imaging. It turns out that, depending on the specific properties of the molecules or species to be imaged, different schemes appear to be suitable. By presenting these schemes and the corresponding molecular properties required for a proper working, we intend to stimulate experimental work, since, so far, experimental results are still lacking.

The modeling results show that the presence of an appropriate molecular level scheme is central for each of the approaches, such as suitably high (or low) decay or exchange rates of the population of excited molecular energy levels. A common feature of the presented scheme is that, along with the CARS drive lasers, an additional, strong laser field is injected, which we call the control field. In the simplest case, the control field consists of quasi-monochromatic light pulses, but we also discuss schemes where the control field comprises two or more pulses at different frequencies in order to increase the resolution. The purpose of the control field is to induce population changes of the molecular states that are probed by the CARS process to increase the resolution of CARS imaging, similar to the use of excited state depletion in a STED microscope.

We base our modeling on the density matrix description [16] of various prototype molecular systems [17]. The motivation for this somewhat more elaborate approach is that it allows both types of nonlinear processes—incoherent as well as coherent—and light intensities to be modeled in a unified framework that goes beyond perturbative approaches. Incoherent processes, such as the transition of population densities between states via nonradiative decay, collisional excitation, or spontaneous emission, are described via the diagonal elements of the density matrix, similar to the well-known rate equations. At the same time, the off-diagonal matrix elements (also called coherences) describe the coherent response of the molecules to the driving optical fields. These are based on an optically induced, linear or nonlinear, polarization of the medium, leading to optically induced transitions, such as absorption, stimulated emission, Rabi oscillation, or parametric nonlinear optical processes. The coherences are proportional to the optically induced polarization of the molecules and, thus, generally act as source terms for all radiation emitted or scattered by the molecules.

Besides the choice of the driving light frequencies and intensities, equally important parameters that determine the molecules' optical response are the damping constants for the diagonal and off-diagonal elements. The diagonal damping rates describe the strength of the incoherent processes named above. As in rate equations, they determine at which drive intensities (in optical pumping, for example) the population of the molecular states become significantly modified. The off-diagonal damping constants, also called dephasing rates, describe how rapidly a given transition loses its memory for the phase of optical excitation, which is determined by the optical phases of the drive laser and control fields. The relative values of the population damping rates, dephasing rates, and the applied light fields (the latter expressed as Rabi frequencies) essentially determine the type of the molecular response—they determine to what extent the irradiated molecules show a coherent response, such as in CARS emission or dynamic Stark splitting, and to what extent the response is incoherent, such as in power broadening or saturation (bleaching) phenomena.

To enable subdiffraction resolution in CARS imaging, the energy level scheme and decay rates of molecules to be imaged cannot be selected because they are inherent to the type of molecule and its structure and to the physical and chemical environment, such as the presence of a solvent. Therefore, each type of molecule to be imaged will have an individual set of parameters. This excludes the possibility that the same approach to subdiffraction imaging can be applied to all types of molecules. What can, however, be selected is the incident light frequencies and intensities. Specifically, the frequencies and intensities of a control field can be selected to match them to the properties of a given type of molecule, i.e., for maximizing the resolution for a given set of molecular parameters. Here, we present a set of prototype molecular systems: each characterized by its own level scheme, optical and incoherent transitions and associated damping parameters, to demonstrate various options for gaining subdiffraction resolution in CARS imaging.

As a generic situation, we consider the molecular level scheme in Fig. 1, and various modifications to this scheme will be discussed

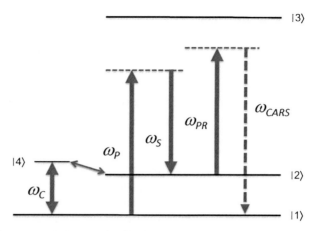

Fig. 1 Energy level diagram for CARS. Level $|1\rangle$ is the ground level and initially fully occupied, $|2\rangle$ is a vibrational level of the medium, and $|3\rangle$ is an electronically excited level. The *arrows* indicate the frequencies of the incident laser fields. The frequencies, ω_P, ω_S, and ω_{PR}, induce a nonlinear polarization (*dashed arrow*) at a frequency $\omega_{CARS} = \omega_P - \omega_S + \omega_{PR}$. This polarization radiates an electromagnetic field at ω_{CARS}, a process called coherent anti-Stokes Raman scattering (CARS). To influence the CARS process, a control field at ω_C is applied. In the shown case, the control field excites an additional level (control level $|4\rangle$) which shows a strong, incoherent mutual coupling to $|2\rangle$. This leads to a control intensity-dependent suppression of the CARS emission, which can be exploited for gaining a spatial resolution below the diffraction limit

in the following sections. The *rhs.* of the figure shows the standard description of CARS. A light field at frequency ω_{CARS} is generated by driving the molecules with three other light fields of high intensity, called the pump (ω_P), Stokes (ω_S), and probe fields (ω_{PR}). In addition, as shown on the *lhs.*, we consider a control field, which couples another level, referred to as the control level, $|4\rangle$ to the ground state $|1\rangle$. We note that the simplification to just a few states is a strong approximation, as real molecular systems comprise a large multitude of ro-vibrational and electronic resonances. However, for sake of clarity regarding basic dynamical aspects of such systems, we use the simplified system and have verified the basic applicability of our schemes to situations with an increased number of states.

Before discussing the various approaches in more detail, we note that we have identified two fundamental mechanisms that may lead to subdiffraction-limited CARS microscopy. These mechanisms may be termed incoherent and coherent approaches. A case of an incoherent approach is to select the control laser frequency so that it excites a control level that rapidly exchanges its population with the CARS vibrational level, $|2\rangle$, via an incoherent process [18]. An incoherent excitation of $|2\rangle$ may also be achieved using more than one control frequency [19]. In both cases, the buildup of the vibrational coherence (on the $|1\rangle$–$|2\rangle$ transition) is suppressed, which reduces the CARS emission with increasing control laser intensity. Similarly, transferring a part of the population to a control state that does not exchange population with the CARS vibrational level, using single or multiple pulses, reduces the vibrational coherence as well, giving rise to a resolution improvement. For the coherent approach, we consider the case where molecules can be found with a control state that has a dephasing rate of the $|1\rangle$–$|4\rangle$ control transition sufficiently small compared to the transition rate (or Rabi frequency) imposed by the control laser. In this case, there will be a coherent response in the form of Rabi oscillations between $|1\rangle$ and $|4\rangle$, inducing a Rabi splitting of the CARS emission [17]. The Rabi splitting is intensity dependent and, thus, varies locally within the diffraction-limited control beam, which can be used to identify features within a diffraction-limited volume (reminiscent of the work performed by Gardner et al. [20]). We also investigate the effect of Rabi splitting (dynamic Stark splitting) without an additional control level, using, instead, a $|2\rangle$–$|3\rangle$ resonant control laser. In this case, the splitting decreases the CARS signal with increasing control laser intensity.

2 Theoretical Model

The level scheme of the medium in Fig. 1 depicts the standard CARS scheme in which the pump, Stokes, and probe lasers (ω_P, ω_S, and ω_{PR}) generate a CARS field (ω_{CARS}) provided that ω_P and ω_S are tuned to Raman (two-photon) resonance with the vibrational transition from the ground state ($|1\rangle$) to state ($|2\rangle$). The excitation of CARS emission occurs in the far detuned wing of an electronically excited state, i.e., in the wings of the transitions $|1\rangle$–$|3\rangle$ and $|3\rangle$–$|2\rangle$, which we consider as dipole allowed. For the discussion of our first scheme, the additional control state, $|4\rangle$, allows an infrared-active (dipole allowed) transition, such that $|4\rangle$ can be populated with a control laser pulse comprising a single frequency (ω_C). All other transitions are considered dipole forbidden.

In the standard perturbative description of CARS, the generation of light at ω_{CARS} is seen as a partially resonant four-wave mixing process, enabled by the third-order nonlinear susceptibility of the sample medium. At sufficient intensity, the driving fields at frequencies ω_P, ω_S, and ω_{PR} induce a third-order nonlinear polarization at frequency ω_{CARS}. This nonresonant molecular polarization acts as a source term in the Maxwell equations to generate an electromagnetic wave of the same frequency, called the CARS signal. The chemical specificity of the CARS imaging arises from the circumstance that the nonlinear susceptibility and thus the CARS signal are resonantly enhanced at the characteristic molecular vibration frequencies, i.e., when tuning the pump–Stokes frequency difference across vibrational resonances.

The perturbative description is widely used because it provides an excellent and relatively straightforward interpretation of the signals commonly observed in CARS microscopy. There is, however, an inherent limitation to this model in that certain additional processes are difficult to include, which are necessary to determine whether these processes have the potential to provide a subwavelength-resolved CARS response. For instance, incoherent processes involving significant repopulation of the levels (saturation or bleaching processes), such as that which STED microscopy is based upon, is beyond the perturbative description. Other examples that are difficult to include are coherent nonlinear optical processes with intense, resonant light fields. A situation of this kind is present if an intense control laser is tuned to a molecular resonance. Similarly, CARS inherently involves a two-photon resonant excitation of the vibrational state $|2\rangle$ from the ground state. This results in population transfer and in a sensitivity of CARS to the $|1\rangle$–$|2\rangle$ population difference which is not contained in the perturbative description. In particular, the perturbative approach cannot properly take into account that, for each pair of molecular states, there are two distinctively different types of damping mechanisms. As was indicated in the introduction, this is, on the one hand, damping by transition of population to other states, which usually occurs along with a change of energy in the system and may, thus, also be addressed as inelastic damping. The other mechanism is the dephasing of the nonradiating, radiating, and laser-driven macroscopic molecular polarization, i.e., the dephasing of the oscillatory motion of the molecular ensemble through collisions or inhomogeneous broadening, for example. The dephasing rates in molecules are usually rather big, in the order of 0.1 THz, which is why they often determine the spectral bandwidth of optical transitions. As dephasing usually does not change the energy of the system, it may be addressed as a damping by elastic (phase changing rather than energy changing) processes.

A model that includes the described phenomena, providing a more complete description of CARS, is based on the Maxwell–Bloch equations. These comprise the quantum mechanical density matrix equations to evaluate the response of the sample to the incident fields and damping phenomena, and Maxwell's equations to calculate the generated electromagnetic waves. In this framework, the generation of CARS signals in an optically thin sample may be viewed using a two-step approximation. In the first step, the pump and Stokes fields, when tuned to the $|1\rangle$–$|2\rangle$ resonance, excite a resonant vibrational coherence at that transition. When looking more closely, this occurs only if the driving fields encounter a population difference at this transition. The far detuning of the pump and Stokes fields from their electronic resonances warrants that the electronic excitation of the molecules (i.e., population and electronic coherence involving state $|3\rangle$) remains small. In biological (organic) samples, this is important because it limits photo damage of the sample. For inorganic samples, photo damage should be much less of an issue and may allow resonant excitation of $|3\rangle$. In the second step, the probe laser, acting on the vibrational coherence, induces nonresonant sidebands on the dipole allowed electronic coherences $|1\rangle$–$|3\rangle$ and $|2\rangle$–$|3\rangle$. The radiation generated thereby appears as optical sidebands around the probe frequency, and the upper sideband is called the CARS emission frequency ($\omega_{CARS} = \omega_P - \omega_S + \omega_{PR}$). In an experimental setting, the laser fields ω_P and ω_{PR} are often chosen to be degenerate in frequency, by generating them with a single laser. In our model, however, for more clarity, we use different light frequencies for all fields.

To explore how an additional control laser influences CARS emission, we calculate the temporal development of the envelopes of the density matrix elements with the Liouville equation, $d\rho/dt = -i/\hbar [H, \rho] - (\Gamma_{ij}\rho_{ij})$ [21]. In the case of the four-level system of Fig. 1, with the control field chosen as resonant with the $|1\rangle$–$|4\rangle$ transition, the temporal development of the matrix elements reads

$$\dot{\rho}_{11} = -\frac{i}{2}\left(\chi_{13}\rho_{13} - \chi_{13}^*\rho_{13}^* + \chi_{14}\rho_{14} - \chi_{14}^*\rho_{14}^*\right) + \rho_{22}R_{21} + \rho_{33}R_{31} + \rho_{44}R_{41} \quad (1)$$

$$\dot{\rho}_{22} = -\frac{i}{2}\left(\chi_{23}\rho_{23} - \chi_{23}^*\rho_{23}^* + \chi_{24}\rho_{24} - \chi_{24}^*\rho_{24}^*\right) + \rho_{33}R_{32} + \rho_{44}R_{41} - \rho_{22}R_{24} - \rho_{22}R_{21} \quad (2)$$

$$\dot{\rho}_{33} = \frac{i}{2}\left(\chi_{13}\rho_{13} - \chi_{13}^*\rho_{13}^* + \chi_{23}\rho_{23} - \chi_{23}^*\rho_{23}^*\right) - \left(R_{31} + R_{32} + R_{34}\right)\rho_{33} \quad (3)$$

$$\dot{\rho}_{44} = \frac{i}{2}\left(\chi_{14}\rho_{14} - \chi_{14}^*\rho_{14}^* + \chi_{24}\rho_{24} - \chi_{24}^*\rho_{24}^*\right) + \rho_{33}R_{34} - \rho_{44}R_{41} - \rho_{44}R_{42} + \rho_{22}R_{24} \quad (4)$$

$$\dot{\rho}_{12} = \frac{i}{2}\left(\chi_{13}^*\rho_{23}^* - \chi_{23}\rho_{13} + \chi_{14}^*\rho_{24}^* - \chi_{24}\rho_{14}\right) - \Gamma_{12}\rho_{12} \quad (5)$$

$$\dot{\rho}_{13} = \frac{i}{2}\left(\chi_{13}^*(\rho_{33} - \rho_{11}) + \chi_{14}^*\rho_{34}^* - \chi_{23}^*\rho_{12}\right) - \Gamma_{13}\rho_{13} \quad (6)$$

$$\dot{\rho}_{14} = \frac{i}{2}\left(\chi_{14}^*(\rho_{44} - \rho_{11}) + \chi_{13}^*\rho_{34} - \chi_{24}^*\rho_{12}\right) - \Gamma_{14}\rho_{14} \quad (7)$$

$$\dot{\rho}_{23} = \frac{i}{2}\left(\chi_{23}^*(\rho_{33} - \rho_{22}) + \chi_{24}^*\rho_{34}^* - \chi_{13}^*\rho_{12}^*\right) - \Gamma_{23}\rho_{23} \quad (8)$$

$$\dot{\rho}_{24} = \frac{i}{2}\left(\chi_{24}^*(\rho_{44} - \rho_{22}) + \chi_{23}^*\rho_{34} - \chi_{14}^*\rho_{12}^*\right) - \Gamma_{24}\rho_{24} \quad (9)$$

$$\dot{\rho}_{34} = \frac{i}{2}\left(\chi_{13}\rho_{14} + \chi_{23}\rho_{24} - \chi_{14}^*\rho_{13}^* - \chi_{24}^*\rho_{23}^*\right) - \Gamma_{34}\rho_{34} \quad (10)$$

In these equations, $\chi_{ij} = e \cdot E_{ij} \cdot r_{ij}/\hbar$ depicts the electric field, E_{ij}, of the driving light waves at the dipole allowed transitions, expressed as Rabi frequencies, where r_{ij} is the appropriate dipole moment. Dipole forbidden transitions have $r_{ij}=0$ (i.e., $\chi_{ij}=0$) and therefore do not occur in the equations. The damping of the system that describe, e.g., the decay of population (diagonal elements, ρ_{jj}) of the excited states, is described by rate coefficients, R_{ij}, that transfer population from states $|i\rangle$ to $|j\rangle$. The damping of the molecular polarization (coherences, ρ_{ij}) is given by Γ_{ij}. To ensure that all Fourier components present in the driving field can act on any of the allowed transitions, the incident light fields at each transition, E_{ij}, are taken as the sum of the envelopes from the pump, Stokes, probe, and control light fields: $E_{ij} = \Sigma_n A_n(t) \exp(i\Delta_{ij} t)$, where $n = \{P, S, PR, C\}$. As is common in experiments, pulsed light fields are used, with $A_n(t)$ the Gaussian-shaped pulse envelopes. The carrier waves of the fields are retained via their detunings, $\Delta_{ij} = \omega_n - \omega_{ij}$, from the molecular transition frequencies, ω_{ij}. Unless stated otherwise, we have assumed that there is no temporal delay between the pump, probe, and Stokes pulses, and that these have a pulse duration in the order of a few ps. In certain cases, however, we prepare the state of the medium by letting the control pulses arrive and terminate before the onset of the other pulses.

The diagonal elements display the population distribution across the available states created by the drive pulses. The normalization is such that full occupation of a state $|j\rangle$ corresponds to $\rho_{jj}=1$. The off-diagonal density matrix elements (coherences), $\rho_{14}(t)$, $\rho_{24}(t)$, $\rho_{13}(t)$, and $\rho_{23}(t)$, are normalized to attaining a maximum magnitude of 0.5 and act as the source terms of radiation in the Maxwell equations [16]. We obtain the full temporal development of these elements by multiplying the envelopes $\rho_{ij}(t)$ with their corresponding carrier wave which is the (Bohr) frequency of the considered transition, ω_{ij}. The Fourier transform of the sum of the coherences provides the amplitude spectrum of the emitted radiation, while the square of the spectrum, evaluated at the CARS frequency, is proportional to the radiated CARS intensity.

Considering the large complexity of molecules and also their physical–chemical environment, the dynamical properties of molecules vary extremely strongly. Specifically, the resonance frequencies present, the strengths of these resonances, which molecular states exchange population at which rates, and the type of spectral broadening mechanisms all vary over a very wide range. As mentioned above, it cannot be expected that a single specific method for increasing the resolution for a certain type of molecule would also work for molecules with different properties. To extend subdiffraction CARS imaging to a larger variety of molecules, we have investigated various molecular prototype systems with different dynamical properties and address them with one or more approaches, each consisting of a matched choice of the frequencies and intensities of the control pulses. In the following, we present several schemes that may result in CARS imaging with a spatial resolution below the diffraction limit.

In Sect. 3, we describe a scheme in which the incoherent population exchange rate between the control state, $|4\rangle$, and the vibrational state, $|2\rangle$, is high, as is often the case for states with close energetic proximity. Then, a control field, resonant with the $|1\rangle$–$|4\rangle$ transition may be used to rapidly equalize the population between $|4\rangle$ and $|2\rangle$, to suppress the CARS signal. In Sect. 4, we consider molecules in which a control state, $|4\rangle$, can be found, which is only weakly coupled to the vibrational state, $|2\rangle$, (Sect. 4) meaning that there is no significant population exchange with $|2\rangle$. This situation is of broad interest because it can be fulfilled by a large variety of molecular systems. As was mentioned above, when the control laser is resonant with the $|1\rangle$–$|4\rangle$ transition, this yields a subdiffraction resolution, but on a diffraction-limited background. We show that the background can be removed by deconvolution, provided that the signal-to-noise level is sufficient. Further improvements are achieved by exciting additional (weakly coupled) control states with simultaneous or sequential control pulses.

In Sect. 5, we discuss a similar case, which results in a strongly reduced background. This case would be realized by applying a pair of infrared control pulses such that they are doubly resonant with the $|1\rangle$–$|2\rangle$ transition via an intermediate state.

Section 6 considers a case of CARS suppression with a single control laser, without an additional control state, by inducing a Stark splitting of the upper state of the CARS vibrational transition ($|2\rangle$). We found that in this case almost all molecules remain in the ground state, which might be an advantage in terms of resistance to photo damage.

Finally, in Sect. 7, we consider a coherent scheme using an infrared control laser, tuned to ground state transition ($|1\rangle$–$|4\rangle$) that possesses a relatively small dephasing rate. In this case, the control field can induce Rabi oscillations with spectrally distinguishable sidebands, which are used as a spectral label for the spatial position of the molecule.

To investigate these options to suppress and manipulate CARS signals, we have chosen parameters that are typical of, rather than specific to, a particular molecule. The $|1\rangle$–$|3\rangle$, $|1\rangle$–$|2\rangle$, and $|1\rangle$–$|4\rangle$ transition frequencies are taken to be 1,000 THz (~300 nm, ~33,000 cm^{-1}), 47 THz (~6.4 μm or ~1,550 cm^{-1}), and 97 THz (~3.1 μm or ~3,200 cm^{-1}), respectively. The pump and Stokes light fields are set to be two-photon resonant with the $|1\rangle$–$|2\rangle$ transition, with the pump and Stokes detunings, $\Delta_P = \Delta_S = 353$ THz, measured with respect to the $|1\rangle$–$|3\rangle$ and $|2\rangle$–$|3\rangle$ transitions, respectively. The detuning of the probe, Δ_{PR}, is set to 200 THz with respect to the $|1\rangle$–$|3\rangle$ transition. This corresponds to pump, Stokes, and probe wavelengths of 463, 500, and 375 nm.

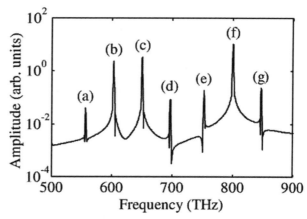

Fig. 2 Emission spectrum calculated with the control field turned off. The features are (**a**) the Stokes shifted field of ω_S, (**b, c, f**) Rayleigh scattering of ω_S, ω_P, and ω_{PR}, respectively, (**d**) the anti-Stokes shifted field of ω_P, (**e**) coherent Stokes Raman scattering (CSRS) and (**g**) coherent anti-Stokes Raman scattering (CARS) of ω_{PR} occurring at $\omega_{CARS} = \omega_P - \omega_S + \omega_{PR}$

The remaining parameters are chosen as specified in the following sections. The simulations were performed numerically, using a fourth-order Runge–Kutta solver, with steps size chosen between a few femtoseconds and a few attoseconds, over time intervals of up to a few 100 ps. Initially, all population is assumed to be in the ground state ($\rho_{11}=1$ and all other $\rho_{ij}=0$).

First, the stability and accuracy of the numerical solutions were confirmed by reproducing well-known results (see Fig. 2), such as the various Raman shifted and Rayleigh scattered fields that occur upon excitation with multiple light frequencies. Note that the calculation for Fig. 2 is performed in absence of a control field. Instead, we used this calculation to verify the generation of a CARS signal with the expected properties. These are, e.g., a signal power which increases with the product intensities of the pump, Stokes, and probe fields, and that the signal is maximized with an appropriate temporal overlap of the drive pulses. The population dynamics found in these initial calculations confirm that the two-photon resonance of the pump and Stokes light fields generates vibrational coherence, which radiates CARS emission as a sideband of the probe field.

3 CARS Suppression with Incoherently Coupled Control State

To achieve suppression of the CARS signal with increasing control beam intensity, we consider strong, incoherent coupling between $|4\rangle$ and $|2\rangle$, such as indicated in Fig. 1. With such coupling we designate a fast nonradiative, incoherent exchange of population between $|4\rangle$ and $|2\rangle$. Additionally, for operation of this approach, a long lifetime is required for population to decay from the states $|4\rangle$ and $|2\rangle$ back to the ground state $|1\rangle$. Although such requirements might seem extraordinary at first glance, they have been found fulfilled in several rather generic organic molecules in the liquid state. Potentially suitable examples are given in Table 1, and additional examples can be found, e.g., in the comparative studies of Deak et al. [22]. Organometallic complexes, especially, provide long vibrational lifetimes (with regard to relaxation of the population to the ground state) and, simultaneously, a strong coupling between the ligand vibrations (common examples for ligands are carbonyl or pyrryl groups), which seems to be a consequence of a widespread, shared electron orbital with the metal center ($d\pi$-π* bonding) [23]. Such organometallic complexes play an essential role in biological processes; among these are hemoglobin (iron ion center, four pyrryl ligands) and chlorophyll (magnesium ion center, four pyrryl ligands).

With regard to the feasibility of an experimental demonstration, we note that our scheme to suppress CARS with excitation of closely related states is quite similar to an established technique called 2D IR-Raman (or IR echo spectroscopy) [24]. Also there, a strong mid-IR pump pulse is used to populate vibrational states in order to influence a Raman signal, which is, however, incoherent Raman scattering.

In our calculations, to represent rates as found in the last column of Table 1, we used a population exchange rate of 0.1 THz. We also included the effect of a larger number of vibrational states involved in this cross-relaxation. We found that this reduces the amount of population contributing to the CARS signal, i.e., it leaves the basic approach unaffected, which is a suppression of the CARS signal with increasing control laser intensity.

The total lifetime of the upper electronic state, $|3\rangle$, is taken to be short, in the order of picoseconds. The actual value was not found to matter much, because no significant population enters that state, due to the large detuning of all light frequencies from the $|1\rangle$–$|3\rangle$ and $|2\rangle$–$|3\rangle$ transitions. The population lifetimes of $|2\rangle$ and $|4\rangle$, with regard to decay back to the ground state, are taken as long, in the order of a nanosecond, while the coherence lifetime between each other and $|1\rangle$ is of the order of picoseconds [25–27]. The laser pulse durations, τ, are set to 2 ps ($1/e$ full width of the field envelopes), except for the control pulse, which is 35 ps in duration and arrives 60 ps in advance of the pump, Stokes, and probe pulses.

Figure 3a shows how the control pulse prepares the medium by populating state $|4\rangle$ to a density that depends on the pulse area (maximum Rabi frequency multiplied by the pulse duration) of the control pulse. The high population exchange rate between $|2\rangle$ and $|4\rangle$ ensures that the population density of $|2\rangle$ closely follows that of $|4\rangle$,

Table 1
Molecular examples showing long vibrational lifetimes (with regard to the ground state) and a strong coupling (population exchange rate) with other vibrational states

Chemical group	Molecule	IR-excited vibration (cm^{-1})	Vibr. life time (ps)	Population exchange rate (THz)
(Poly-) hydrocarbons	Polyethylene	CH_2 stretch 2,924	260 ± 100 [36]	~0.200 [36]
(Organo-)halides	Methylene dichloride CH_2Cl_2	C–H stretch 2,985, 3,050 [37]; 2,985, 3,048 [38]	40 ± 10[a] (CCl_4 sol.) [37]; 50 [38]; 12 [39]	~0.125 [37]
(Organo-)halides	Trichloroethylene CH_3CCl_3	C–H stretch 2,939 (degenerate) [37]	100[a] (CCl_4 sol.) [37]	~0.067 [37]
Alcohols	Ethanol, CH_3CH_2OH	C–H stretch 2,928 (degenerate) [37]	40[a] [37]; 18 ± 9 [38]	–
Ketones	Acetone, CH_3COCH_3	C–H stretch 2,843 ± 2 [38]	52 ± 30 [38]	–
Carotenes	Beta-carotene $C_{40}H_{56}$	C=C bend 1,150, 1,520 [40]	–	11.000 [40]
(Organic) nitriles	Acetonitrile, C_2H_3N	C≡N 2,173 [22]	80	
(Organic) nitriles	Pyrrole, C_4H_5N	N–H stretch 3,400 [41]	42 ± 3 (neat), 220 ± 20[a] (diethyl ether sol.) [41]	–
(Organo-)-metal carbonyl complexes	Tungsten hexacarbonyl $W(CO)_6$	Carbonyl (C–O) stretch 1,920–1,985 [42]	800 ± 200[a] (CCl_4 sol.)	–
		~1,980 [43]	480 ± 50[a] ($CHCl_3$ sol.) 140 ± 15 (n-hexane sol.) 60 ± 6 (benzene) [42] 700[a] (CCl_4 sol.) 370[a] ($CHCl_3$ sol.) [43]	
(Organo-)-metal carbonyl complexes	Chromium hexacarbonyl, $Cr(CO)_6$	Carbonyl (C–O) stretch 1,920–1,985 [42]	440 ± 70[a] (CCl_4 sol.)	–
		~1,980 [43]	295 ± 30[a] ($CHCl_3$ sol.) 145 ± 25 (n-hexane sol.) 59 ± 6 (benzene sol.) [42] 260[a] (CCl_4 sol.) 340[a] ($CHCl_3$ sol.) [43]	
(Organo-)-metal carbonyl complexes	$Rh(CO)_2(C_5H_7O_2)$	Carbonyl (C≡O) stretch	–	2,083 [23]
(Organo-)-metal carbonyl complexes	Carboxy-hemoglobin	Carbonyl (C–O) stretch 1,920 [30]	25 [30]	–

All processes are measured at room temperature, unless otherwise stated
[a]Measured in solution

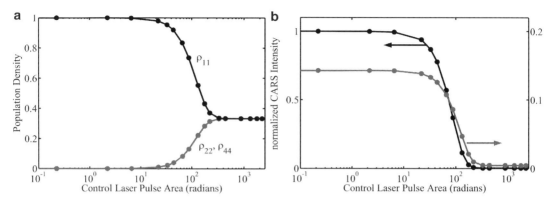

Fig. 3 (**a**) Typical population densities as a function of control laser pulse area. A significant fraction of the ground state population (*black trace*) is transferred to |4⟩ (*grey*) and, via nonradiative transitions, to |2⟩ (*grey*) once the control pulse area is large. (**b**) Vibrational coherence ρ_{12} (*grey*) and corresponding CARS emission intensity (*black*) as a function of control laser pulse area. Beyond a control pulse area of ~100 rad, the vibrational coherence is suppressed and with it the CARS emission, in a manner that resembles saturation, such as is present also in STED microscopy

which lets the traces for ρ_{44} and ρ_{22} almost overlap. This process starts to saturate for a pulse area of about 100 rad for this particular case, i.e., beyond this value the population of the states |1⟩, |4⟩, and |2⟩ equalize, rendering an almost vanishing population difference on the vibrational transition |1⟩ and |2⟩.

The consequence of this saturation process is shown in Fig. 3b, where the control pulse has the effect of suppressing CARS emission. The nature of the suppression is as follows: the equalized populations of the states |1⟩ and |2⟩ ($\rho_{22}-\rho_{11}=0$) make sure that, although the pump and Stokes fields are present, no two-photon coherence, ρ_{12}, can build up, i.e., due to $d/dt(\rho_{12})=0$, ρ_{12} remains at 0. This can be shown via an adiabatic elimination of |3⟩ [28], rendering $d/dt(\rho_{12})$ proportional to $\rho_{22}-\rho_{11}$ and the two-photon Rabi frequency (product of the pump and Stokes Rabi frequencies, χ_{12} and χ_{32}). With ρ_{12} remaining 0, sideband generation on ρ_{12} via ω_{PR} is prevented and, thus, CARS emission is suppressed. For our parameters, we observe that CARS emission can be suppressed to a maximum of 99.8 % compared to the nondepleted intensity of CARS emission.

From the depletion process, a saturation intensity can be defined as the control pulse intensity that frustrates CARS emission to one-half of its maximum value, in analogy with the definition for fluorescence in a STED scheme [29]. In our case, this saturation intensity corresponds to a 25 % reduction of the ground state population density, as can also be seen in Fig. 3. We note that vibrational population inversion and 50 % ground state depletion of biologically relevant samples by direct excitation using a mid-IR laser has been observed by Ventalon et al. [30]. We used the reported pulse duration, pulse energy, and the bandwidths of the vibrational states to estimate that the populations of their ground state and first vibrational state became equalized at intensities in the order of 70 GW cm^{-2} with the 100 fs mid-IR pulses they had applied. Although their intensities are high, we note that for our scheme, due to the long population lifetime of the control state, it is the pulse *area* and pulse *energy* deposited into the $|4\rangle-|1\rangle$ transition within that lifetime that determines the degree of saturation. Specifically, if a control pulse with a longer pulse duration is used, such as the 35 ps duration used in our example, the required intensity is much less, in the order of 200 MW cm^{-2}. This is well below the threshold for multiphoton ionization or electronic excitation in the mid-infrared and, thereby, indicates the feasibility of the suggested scheme.

To estimate the resolution enhancement that corresponds to the calculations in Fig. 3, we consider an experimental setup wherein the control laser illuminates the sample with a donut mode, focused to a diffraction-limited spot. We define the size of the node of the control laser beam to be the area where the control laser's intensity is lower than the saturation intensity. CARS emission from within the node is unsuppressed, and, thus, largely unaltered, while outside the node it is suppressed by up to 99.8 %. As the control laser intensity is further increased, the node becomes smaller and the total CARS emission from within the node decreases. However, since the intensity outside the node saturates at a residual intensity background of 0.2 %, this places an upper limit on the resolution. For a measurement with a signal-to-noise ratio of one, the best resolution can be defined by the radius of the node for which the signal contribution from within the node is greater than the signal contribution from outside the node. This allows us to derive an upper limit to the improved resolution of approximately $\lambda/(22\,\mathrm{NA})$, where NA is the numerical aperture of the imaging system and λ is the wavelength of the *probe* beam.

The calculations described so far show that subdiffraction-limited resolution images can be obtained by extending the standard CARS microscopy setup with a mid-IR control laser beam having a donut-shaped transverse intensity profile. This laser should be resonant with an appropriate vibrational transition and have an intensity above the saturation intensity of the vibrational transition. The node leaves the CARS process unaffected in a subdiffraction-limited area around the center of the common focus, whereas CARS emission from that area outside the node is suppressed. The measured CARS signal can then be attributed to this subdiffraction-limited region, where the intensity is less than the saturation intensity. CARS images with a resolution below the diffraction limit can then be recorded by scanning the control beam across the sample.

An advantage of the described mid-IR excitation scheme is that the molecules remain in their electronic ground states. Further, we note that the intensities also of all other nonlinear emission processes (Fig. 2, peaks (a), (d) and (e)) were found to be suppressed in a similar manner. This might open the way to increasing the resolution of other types of nonlinear microscopy as well.

4 Ground State Depleted CARS

The working principle of the previous case was to equalize the initial population difference at the vibrational transition so that $\rho_{22} - \rho_{11} = 0$, with the goal to maintain ρ_{12} close to 0. In this section, we discuss a more general approach for CARS suppression [31] in which the ground state population is reduced by moving part of it to another level. More specifically, we consider cases where control lasers are tuned to a vibrational transition (mid-IR control laser) or to electronic transitions (visible or UV control laser), as is depicted in Fig. 4, or that two or more control laser pulses are injected simultaneously or sequentially.

The main difference with the previous approach (Sect. 3) is that the excited control level $|4\rangle$ does not need to fulfill specific requirements (which was a rapid population exchange with state $|2\rangle$), or that there is no additional control level required at all (such as when using a control pulse resonant with the $|1\rangle$–$|3\rangle$ transition). This relaxing of requirements means a significantly increased generality of the scheme, i.e., it may be applicable to more types of molecules.

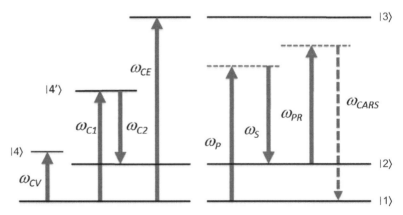

Fig. 4 Reducing the CARS signal by partially depleting the ground state with one or more preceding control laser pulses. The control pulses can excite a vibrational transition (such as ω_{CV} resonant with $|1\rangle$–$|4\rangle$), an electronic transition (such as ω_{CE} resonant with $|1\rangle$–$|3\rangle$), or two or more such control pulses are injected simultaneously or sequentially

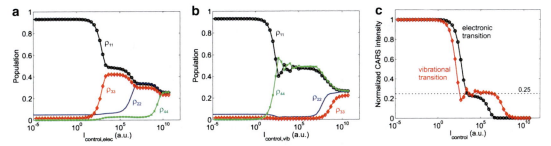

Fig. 5 Population densities of the molecular states vs. control laser intensity (**a**) when the control laser is resonant with the |1⟩–|3⟩ transition, and (**b**) resonant with the |1⟩–|4⟩ transition. In both cases, the ground state population is reduced to about 50 % once the control laser intensity exceeds a certain saturation intensity. (**c**) The corresponding intensity of the CARS signal as a function of control laser intensity. The CARS signal is reduced to about 25 % of its initial value. Additional saturation occurs with further increases in intensity

In our calculations, we assume that the CARS driving light pulses (frequencies $\omega_P = 650$ THz, $\omega_S = 603$ THz, and $\omega_{PR} = 800$ THz) are incident without mutual delay and each have a 2 ps (half-width at $1/e^2$ amplitude) duration. To prepare the sample, we take control pulses (either at ω_{CV} or ω_{CE}) arriving 30 ps prior to the CARS drive pulses with duration of 10 ps. The dephasing times of the |1⟩–|3⟩, |2⟩–|3⟩, |1⟩–|4⟩, and |1⟩–|2⟩ transitions are chosen as 1, 1, 5, and 5 ps, respectively, with population lifetimes of 1 ns. The Runge–Kutta algorithm of fourth order used a fixed step size of 0.5 fs, carried out over a time interval of 100 ps.

The main results of the calculations are summarized in Fig. 5 for the case of a single control laser pulse exciting either |4⟩ (with control frequency ω_{CV}) or |3⟩ (with ω_{CE}). It can be seen that in both cases, once the control laser is beyond a certain saturation intensity, a reduction of the ground state population to 50 %, by exciting half of the population into the targeted upper state (here either |4⟩ or |3⟩), is observed. In Fig. 5c, it can be seen that the corresponding CARS signal intensity drops to a value of about 25 %. This can be understood as a result of the coherent nature of the CARS process, making the CARS intensity proportional to the square of the number of emitters in the probed volume. It might be suggested, by comparing the traces in Fig. 5c, that using a control laser at the vibrational transition (|1⟩–|4⟩) yields an easier depletion of the CARS signal, because depletion is observed at lower intensities. However, this effect results from our particular choice of dephasing and dipole moments only. Choosing the same dipole moments at both transitions provides equal Rabi frequencies at the same intensities, but a shorter dephasing time spectrally broadens the |1⟩–|3⟩ transition. This reduces the spectral match between the control laser and the transition, such that higher intensities are required for the same population excitation rate. We cannot give absolute values for the saturation intensity because this would strongly depend on the particular molecule and transition chosen. For example, for a mid-infrared control pulse, as described in Sect. 3, the value should lie in the order of hundreds of MW cm^{-2}.

For further increased intensity, Fig. 5 shows additional saturation phenomena that eventually lead to the population being distributed equally over all the available states. In the calculations, this occurs when the control laser Rabi frequency is chosen to assume very high values that are comparable with the detunings from the other transitions. We believe that this situation is of low relevance for an experimental realization, because it only occurs at extreme intensities that are several orders of magnitude above the first saturation.

In the following, we investigate schemes that use more than a single control pulse. Although more complicated, it is possible to achieve more than 50 % ground state depletion and, thereby, more than 75 % suppression of CARS, without having to rely on extreme intensities. Generally, one can consider the injection of several (N) control laser pulses each exciting a different upper state from the ground state. In this case, the relative delay between the control pulses becomes important, though we consider all of them to precede the CARS drive pulses. Using rate equations, one can easily show that N simultaneous control pulses, each saturating its corresponding transition to one of the N excited states, distribute the population equally over $N+1$ states, thereby reducing the ground state to a fraction $1/(N+1)$. For instance, simultaneously applying ω_{CV} and ω_{CE} ($N=2$) should equalize the populations of the states $|1\rangle$, $|4\rangle$, and $|3\rangle$, and, thereby, deplete the ground state population to 1/3. Due to the square dependence on the number of emitters, this yields a suppression of the CARS signal by a factor of 9. In the sequential case, the ground state depletion would become even stronger, reaching a fraction of $1/N^2$, which would reduce the CARS intensity to $1/N^4$ (i.e., to 1/16 for $N=2$).

We have investigated this approach for $N=2$ with numerical calculations (injection of two pulses, at ω_{CV} and ω_{CE}) and the results are compared in Fig. 6 with the previous case of $N=1$ (only ω_{CE} injected). It can be seen that, with simultaneous excitation, the CARS signal is depleted by about 90 % (trace a). The sequential excitation (trace b, pulse with ω_{CV} delayed by 10 ps with regard to ω_{CE}) provides an even stronger suppression (about 95 %), compared to single pulse suppression reaching about 75 %.

The above discussion shows that a limited amount of CARS suppression should be achievable in many molecular systems. It is, thus, important to investigate what imaging resolution can be obtained if the suppression factor is limited. For instance, a ground state depletion by 50 % might seem unattractive because, naively, one might expect only a small factor in resolution improvement. In the following, we show that with a sufficiently good signal-to-noise ratio, such as by increasing the recording times, a spatial resolution well below the diffraction limit can nevertheless be obtained.

Nonlinear Optics Approaches Towards Subdiffraction Resolution in CARS Imaging

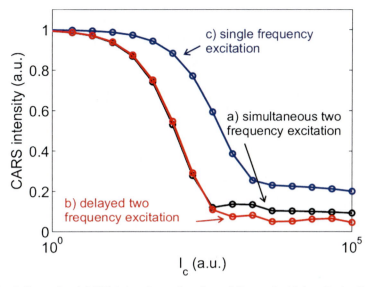

Fig. 6 Normalized CARS intensity as function of the control intensity for three cases. With increasing intensity, beyond the first saturation process, the CARS intensity becomes suppressed by a single control pulse (here at frequency ω_{CE}) to a value of 25 %. (*a*) Two simultaneously control pulses (at frequencies ω_{CE} and ω_{CV}) lead to a suppression of about 90 %. (*b*) Applying the control pulses sequentially (the ω_{CV} pulse delayed by 10 ps with respect to the ω_{CE} pulse) suppresses the CARS intensity to about 95 %

To illustrate this, we calculate the CARS emission profiles for a maximum of 50 % ground state depletion. To demonstrate the resolution enhancement in a CARS image, the spatial beam profiles of the exciting lasers and the resulting spatial profiles of CARS emission were calculated in two dimensions, in the focus plane (x, y) perpendicular to the beam propagation. In the following, we assume a control laser with frequency ω_{CE} (resonant with $|1\rangle$–$|3\rangle$). The case with a ω_{CV} as control laser frequency (resonant with $|1\rangle$–$|4\rangle$) gives very similar results, though the absolute resolution is lower by the ratio of frequencies. For generating CARS emission profiles narrower than the diffraction, we consider the control laser beam having a donut-shaped field distribution. Such a beam can be generated, e.g., from a Gaussian beam via a π-phase plate (depicted in Fig. 7) [2]. The CARS drive beams (pump, Stokes, and probe) are taken with Gaussian shape. Figure 7 displays the transverse intensity profile of the beams in the focus of a microscope objective with large numerical aperture (NA = 0.65).

Examples of the resulting CARS emission profiles are displayed in Fig. 8, with the control laser intensity stepwise increased from zero to beyond the first saturation in Fig. 5. It can be seen that the initially diffraction-limited profile (widest curve, control intensity 0) reshapes into a profile that consists of a rather narrow peak, with

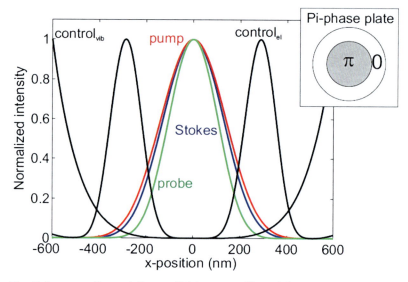

Fig. 7 Cross sections of the spatial beam profiles of the pump (*red*), Stokes (*blue*), probe (*green*), and control (*black*) light fields. The *inset* schematically shows a phase plate that might be used to generate the donut-shaped beam profile of the control light field

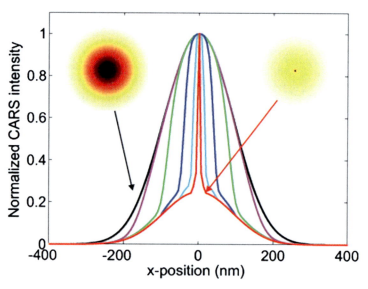

Fig. 8 CARS emission profile when a donut-shaped control beam is applied with stepwise increased peak intensity. From the initially diffraction-limited profile (*widest curve*, control intensity $I=0$), a subdiffraction-limited peak forms on top of a diffraction-limited background of 25 % (*narrowest curve*, with a control intensity $I=10^5$ a.u.). The *FWHM* of the narrow central peak is below 10 nm. The *insets* show 2D views of the broadest and the narrowest CARS emission profile

Nonlinear Optics Approaches Towards Subdiffraction Resolution in CARS Imaging 311

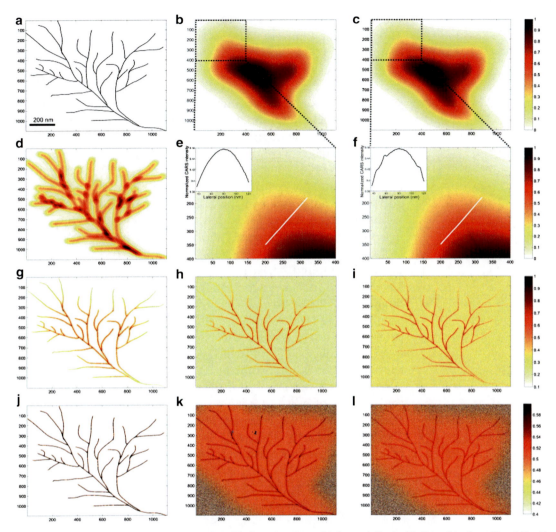

Fig. 9 (**a**) Computer generated test sample with fine structure. (**b**) Convolution with the widest profile of Fig. 7 (i.e., control field turned off) and (**c**) convolution with the narrowest emission profile of Fig. 7 (i.e., with the donut-shaped control light field applied). Cross section through zoomed-in parts of the images [*white lines* and *insets* in (**e**) and (**f**)] show a subdiffraction-limited fine structure on a diffraction-limited background. High spatial resolution images are retrieved by subtracting the square root of image (**e**) from the square root of (**f**), assuming different levels of intensity noise in (**e**) and (**f**): 0 % (**j**), 0.5 % (**k**), and 1 % (**l**)

a width well below the diffraction limit (FWHM below 10 nm), superimposed on a diffraction-limited background of 25 %.

To illustrate that such profile is suitable to provide CARS images with a resolution well below the diffraction limit, we simulate CARS images of a test image (Fig. 9a) with a dendritic line structure. Such a test image was chosen (numerically generated) because it provides 2D lines with varying spacing and density, such as found, e.g., in brain tissue. The size of the test image is chosen as roughly 1 × 1 μm in which the drawn structure possesses a line thickness of 4 nm, with spatial distances between 4 and 400 nm. The emitters were assumed to be homogeneously and continuously distributed along the lines. For generating a CARS image, we perform a convolution of the test image and one of the excitation profiles of Fig. 7. As the coherent nature of CARS emission is to be taken into account, the convolution is performed with the electric field amplitudes and subsequently the square value is taken to obtain the intensity image.

CARS images were calculated for the control beam turned off (Fig. 9b) and the control beam turned on (Fig. 9c). Although the figures (b) and (c) appear equivalent, there is an important difference. This can be seen in the enlarged images in (e) and (f) by taking intensity cross sections (white lines and corresponding insets). Comparing the insets shows that the image with the control laser turned on contains subdiffraction features, though on a diffraction-limited background. To retrieve the fine features, we have plotted in Fig. 9j the squared field-difference between the image with the control beam on (Fig. 9c) and the diffraction-limited image (control beam off, Fig. 9b). It can be seen that the original test image is well retrieved. The spatial resolution obtained is on the order of 15 nm, which is the width of the sharp peak in the smallest emission profile in Fig. 8.

Of central importance for the quality of the retrieved images is the amount of noise that is present in recording the two images that are to be subtracted. To obtain an impression of the noise present in an experiment, we first estimate the maximum number of photons that can be expected from CARS emission in a given measurement interval. We then compare this with the corresponding number of noise photons (shot noise, quantum noise) and with noise from the detector (dark count noise).

Using an 80-MHz laser system, a typical CARS signal of 10^6 photons/s can be generated [32]. With an optimized laser system having the same average power at reduced repetition rate of 1 MHz, a maximum CARS signal of about 6×10^9 photons/s should be possible without introducing multiphoton damage. Assuming the recording of images as in Fig. 9 with a resolution of $10^3 \times 10^3$ pixels in a recording time of about 60 s then provides a maximum of 10^5 photons per pixel. Applying Poisson's square root law gives a quantum limit of 3×10^2 photons, which corresponds to a relative intensity noise of about 0.3 %. A second source of noise to be considered is dark counts from the detector. State-of-the-art photon counting modules are available with less than 10 dark counts/s when operating at 10^6 counts/s (MP993 Excelitas technologies) which gives a relative intensity noise of 0.001 %.

While often detectors with a lower signal-to-noise ratio might be used it can be seen that the noise levels can be low (0.3 % for quantum noise and less than 1 % for dark count noise) and can further be reduced by longer integration times. In an experiment, an optimum recording time might be found where quantum and dark count noise are balanced, reaching noise levels around 0.5 %. The noise could be reduced further by reducing the number of pixels in an image with a smaller field of view.

The effect of different noise levels is displayed in Fig. 9, where (j) is calculated with quantum noise only, (k) with additional 0.5 % rms detector noise, and (l) with additional 2 % rms detector noise. It can be seen that increasing noise adds a diffraction-limited background; however, the images still clearly display the structure of the test image with a resolution well below the diffraction limit, in this case with a resolution of 10–20 nm.

5 Doubly Resonant Suppression of CARS

It is also possible to improve the resolution in CARS microscopy by preparing the sample with two simultaneously applied control beams (both in the form donut beams) [19] in a manner that depletes the ground state and increases the population of the vibrational state, $|2\rangle$. To simplify the implementation of the scheme in (1–10), we have assumed a scheme as in Fig. 10, i.e., where no control state is used. Instead, the control frequencies, ω_{C1} and ω_{C2}, are taken as stepwise resonant with the $|1\rangle$–$|3\rangle$ and the $|2\rangle$–$|3\rangle$ transitions and are, thereby, also two-photon resonant with the $|1\rangle$–$|2\rangle$ transition. We note that such a doubly resonant excitation of $|2\rangle$ can also be achieved via an additional control state as intermediate state.

The calculations are carried out with a pulse duration of the CARS drive lasers of 2 ps. The dephasing time of the vibrational state $|2\rangle$ is 5 ps, and all other dephasing times are taken as 1 ps, and

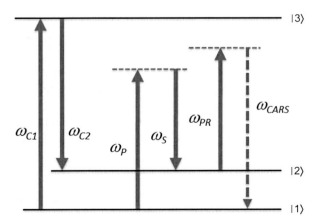

Fig. 10 Level scheme for CARS depletion with doubly resonant control lasers

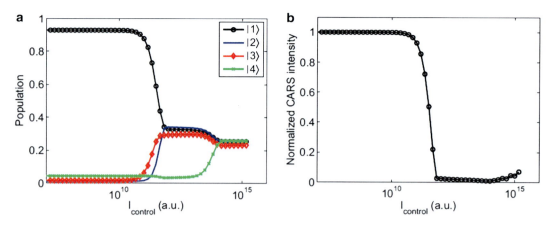

Fig. 11 (**a**) Calculated population distribution and (**b**) suppression of CARS signal when preparing the sample with a preceding pair of control pulses of increasing intensity (ω_{C1} and ω_{C2} are resonant with the transitions $|1\rangle$–$|3\rangle$ and $|2\rangle$–$|3\rangle$, respectively). The control pulses equalize the populations on the CARS transition $|1\rangle$–$|2\rangle$. (**b**) Equalizing the populations in $|1\rangle$ and $|2\rangle$ leads to a suppression of the CARS signal; in this case, by about 99 %

a population lifetime of $|2\rangle$ and $|3\rangle$ much longer (1 ns) than the dephasing time. The duration of the two control laser pulses is set to 10 ps and the drive pulses are delayed by 30 ps with respect to the control pulses. It is important that the control pulse duration and the drive pulse delay are both much longer than the dephasing times, which leads to an incoherent population of state $|2\rangle$, i.e., without Rabi oscillations. Typical results of the calculations are summarized in Fig. 11 and show that, at sufficient control intensity, this leads to an equalized population in $|1\rangle$ and $|2\rangle$.

The potential of this technique for gaining subdiffraction-limited resolution was again inspected with calculations of CARS emission profiles assuming donut-shaped beams as in Fig. 7. Figure 12 shows the calculated cross section of emission profiles. It can be seen that CARS emission is suppressed, in this case by about 99 %, except for a narrow range around the node of the control beams.

Using again a computer-generated test sample and convolution with the widest and narrowest emission profiles of Fig. 12, CARS images were generated as would be obtained without and with the control lasers applied. Figure 13a shows the test image together with a diffraction-limited image, (b), and the subdiffraction image, (c). Clearly, the image displays a resolution below the diffraction limit. In the case shown here, corresponding to more than 99.9 % suppression, the resolution improves roughly by a factor of 30 with respect to the diffraction limit.

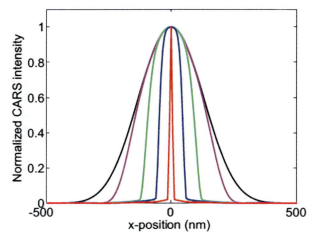

Fig. 12 CARS emission intensity profile calculated for two (doubly resonant) control pulses having a donut-shaped beam profile. With increasing control laser intensity, the initially Gaussian-shaped and diffraction-limited profile (*widest curve*) narrows to sharp peak (*narrowest curve*)

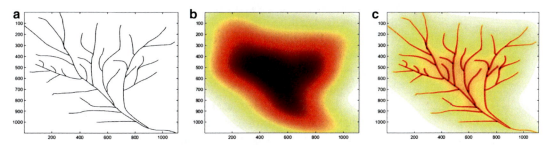

Fig. 13 (**a**) Computer generated test sample extending over ~ 1 × 1 µm size, (**b**) its diffraction-limited CARS image, and (**c**) the subdiffraction image obtained with a doubly resonant pair of control lasers yielding a maximum CARS suppression of 99 %

6 Stark-Splitting Suppressed CARS

The approaches discussed so far are based on addressing a substantial part of the ground state population with a control field. The motivation was to either shift a larger amount of population to state $|2\rangle$ to reduce the population difference and coherence buildup on the $|1\rangle$–$|2\rangle$ CARS transition or to shift a larger part of population to other states that are not involved in CARS signal generation. In all cases, the amount of ground state population that needs to be removed is quite significant, in the order of 50 % and more. As was discussed in Sect. 3, such excitation seems feasible in organic systems with mid-infrared control lasers without photo damage. Using electronic transitions (with visible and UV control lasers), a similar degree of excitation might still be applicable in inorganic, solid-state systems, which are much less susceptible to damage.

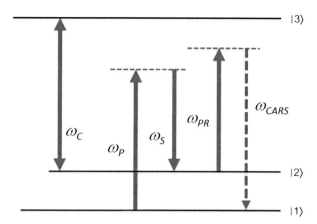

Fig. 14 Scheme for CARS suppression by upper state depletion. In addition to the CARS driving lasers, a control laser is tuned to resonance with the $|2\rangle$–$|3\rangle$ transition

Nevertheless, it is of great interest to identify approaches for suppressing CARS emission that leave the ground state population almost unchanged. In the following, we describe such an approach and, for simplicity, we illustrate its essential properties in a three-level system as depicted in Fig. 14.

In the calculations, the lifetime of state $|3\rangle$ is set to 1 ns, the dephasing times are set to 1 ps, and the laser pulse durations are set to 10 ps. The CARS drive laser pulses and the control laser pulse are chosen to arrive simultaneously. Similar to that seen in Fig. 3b, we observe a strong suppression of the CARS signal if the Rabi frequency χ_{23} is increased above values in the order of 10^{13} rad/s, i.e., when the Rabi frequency exceeds the dephasing rate. However, there is an important difference with the previous schemes. To emphasize the difference, we have calculated the maximum energy per molecule, E_M, where $E_M = \rho^M_{33}(h\nu_{13}) + \rho^M_{22}(h\nu_{12})$, where ρ^M_{ii} is the maximum probability of population of the excited states (either $|3\rangle$ or $|2\rangle$) and where $h\nu_{ij}$ is the energy of the excited states with regard to the ground state. In this notation, the absolute maximum energy to which a molecule can be excited, if brought to the highest energetic state, $|3\rangle$, is 6.5×10^{-19} J. In Fig. 15, we show both the CARS emission amplitude and average energy per molecule vs. the light frequency of the control laser.

As can be observed, there are two control laser frequencies at which the CARS signal is significantly suppressed. These are the $|2\rangle$–$|3\rangle$ and the $|1\rangle$–$|3\rangle$ transition frequencies at control laser (angular) frequencies $\omega_C = 4.4$ and 4.72×10^{15} rad/s. Compared to the unperturbed CARS amplitude, the suppression amounts to factors of about 10^2 and 10^3, which corresponds to suppressing the intensity by a factor of 10^4 and 10^6, respectively. A spatial resolution well below the diffraction limit should thus be obtainable at both control laser frequencies.

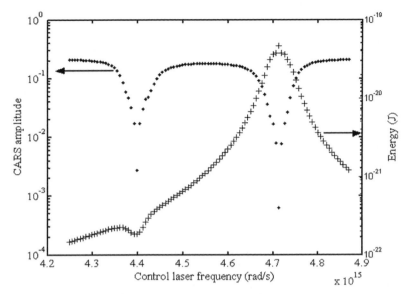

Fig. 15 Amplitude of the CARS signal (*left axis*) and maximum energy per molecule as a function of the control laser frequency, ρ_C, while maintaining the Rabi frequency of the control laser constant ($\chi_C = 10^{13}$ rad/s). The CARS amplitude becomes suppressed by more than two orders of magnitude at $\omega_C = 4.4$ and 4.72×10^{15} rad/s, which corresponds to resonance with the upper state transition $|2\rangle$–$|3\rangle$ and the ground state transition $|1\rangle$–$|3\rangle$. The average energy per molecule varies over wide ranges, from weak excitation (at $|2\rangle$–$|3\rangle$ resonance) to strong excitation (at $|1\rangle$–$|3\rangle$ resonance)

In order to identify the working principle of this suppression we have, at a fixed control laser Rabi frequency, tuned the pump–Stokes difference frequency through the vibrational resonance frequency, $|1\rangle$–$|2\rangle$. We found that the control laser introduces a dynamic (AC) Stark splitting, also called Autler–Townes splitting or Rabi splitting of this resonance. Essentially, the control laser, via Stark splitting, shifts the vibrational transition frequency out of resonance with the CARS drive frequencies and, thereby, suppresses the CARS signal. This working principle is common to both cases, i.e., when the control laser is tuned to the $|2\rangle$–$|3\rangle$ or the $|1\rangle$–$|3\rangle$ transition frequencies.

The main difference of the two cases can be seen in the average energy that a molecule attains in these excitations, which is displayed at the right axis of Fig. 15. If the control laser is resonant with the transition from the ground state, $|1\rangle$–$|3\rangle$, about half of the population is pumped into the excited state. In a real experiment, this might easily damage the sample if it consists of organic molecules. However, if the control laser is tuned to the transition between the excited states, $|2\rangle$–$|3\rangle$, the total absorbed energy is much smaller. With the parameters considered here, the average energy per molecule is found to be 3×10^3-times smaller. This value depends mostly on the CARS drive laser intensities—the CARS drive laser intensities set the population ratio between $|1\rangle$ and $|2\rangle$, and, therefore, the probability of a molecule being available for excitation from $|2\rangle$ to $|3\rangle$—rather than on the control laser intensity. Such strongly reduced excitation is certainly of great interest for avoiding photo damage in subdiffraction CARS microscopy.

It can further be seen that the control laser even somewhat reduces the degree of molecular excitation when tuned to the $|2\rangle$–$|3\rangle$ resonance, compared to tuning near this resonance. This observation resembles what is found in electromagnetically induced transparency (EIT), where a laser that couples excited states can suppress the absorption of light at other frequencies [33]. In essence, investigating CARS with a resolution better than the diffraction limit looks quite promising when applying a control beam resonant with the transition between the electronically excited state and the upper vibrational CARS level.

7 Rabi Modulation of CARS

Finally, we consider a technique, which might be called Rabi labeling to resolve features and locations of emitters better than the diffraction limit [17]. In this approach, we consider a molecular system where a control state can be excited from the ground state with the control laser as is depicted in Fig. 16. As was assumed for the scheme of Fig. 1, the control state possesses a longer population lifetime vs. decay to the ground state. However, we assume that the control state $|4\rangle$ does not have a rapid population exchange with state $|2\rangle$, such that the $R_{41} + R_{42}$ corresponds to a lifetime in the order of a nanosecond [26]. The dephasing rates between states are assumed to be of the order of picoseconds [25–27]. All the laser pulse durations are set to 7 ps $(1/e^2)$ except for the control laser field, which we consider as being constant during the

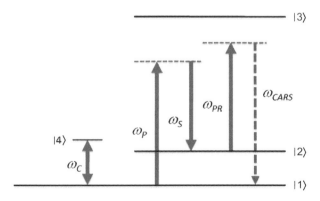

Fig. 16 Energy level diagram for obtaining subdiffraction resolution in CARS microscopy. The control laser excites Rabi oscillations at the $|1\rangle$–$|4\rangle$-transition which induces intensity-dependent sidebands in the CARS signal

Nonlinear Optics Approaches Towards Subdiffraction Resolution in CARS Imaging 319

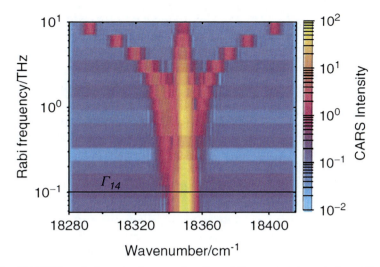

Fig. 17 CARS emission spectra as a function of the control-laser-induced Rabi frequency, Ω_R. Note that the sidebands are notable only when the Rabi frequency is greater than the decoherence rate, Γ_{14}, as indicated by the *black line*. The position of the sidebands relative to the central peak is equal to Ω_R

interval of CARS emission. The numerical calculations extend over 30 ps in steps of 0.1 as.

To understand how the control laser modifies the CARS emission spectrum, the amplitude of the control field and, hence, its Rabi frequency (Ω_R) was varied in steps, while the dephasing (decoherence) rate, Γ_{14}, is held fixed at 0.1 THz. As shown in Fig. 17, the spectrum of the CARS emission is single peaked when $\Omega_R < \Gamma_{14}$. However, once the Rabi frequency is larger than the dephasing rate, the spectrum of the CARS emission splits, showing two symmetrical sidebands. These sidebands have a spacing from the CARS carrier frequency which coincides with Ω_R, given by the generalized Rabi frequency [16]:

$$\Omega_R = \sqrt{\frac{E_C^2 \mu_{14}^2}{\hbar^2} + \Delta_{14}^2} \qquad (11)$$

In this expression, $E_C = \sqrt{(2 I_C/\varepsilon_0)}$ is the control laser's electric field, and μ_{14} and Δ_{14} are the dipole moment at the $|1\rangle$–$|4\rangle$-transition, and a possible detuning of the control field from that transition. Examination of the ground state population as a function of time showed that the Rabi oscillations periodically deplete the ground state, which modulates the amplitude of the vibrational coherence. As a result, the amplitude of the CARS emission is modulated, creating two Rabi sidebands in the CARS emission spectrum. The splitting, which is large enough to overcome the decoherence, is several THz, which should be large enough to be distinguishable in a real CARS experiment using standard spectrum analyzers.

The frequency of the sidebands depends on the local intensity of the control laser beam, I_C. When applying a Gaussian (TEM$_{00}$) control beam, I_C and thus also the local values of Ω_R in (11) varies in a well-known Gaussian fashion throughout the control beam focus in the sample [34]. In spectroscopy, such spatial distribution of Rabi frequencies is usually considered an undesirable effect that limits spectral resolution. Here, however, we show that the effect can be used to obtain subdiffraction-limited resolution.

The radial distance between the emitter and the center of the control beam can be calculated by measuring the exact frequency of the sidebands, which we refer to as Rabi labeling of the emitter position. The absolute position of an emitter location can be calculated by trilateration. Consider, as a two-dimensional example, an emitter located at the Cartesian coordinates x_1 and y_1. The frequency of the CARS sidebands depends on the local intensity of the control laser, which depends on both, x_1 and y_1, such that neither of these coordinates cannot be determined directly. Instead, the sideband frequency corresponds to the distance to the center of the beam, $r_1 = \sqrt{(x_1^2 + y_1^2)}$, i.e., a range of possibilities in the shape of a ring about the beam center. To obtain additional information about the emitter position, the control laser is scanned over a known distance dx, along the x-axis (thus to $x_1 - dx$ and y_1), which brings the emitter to another radial distance, $r_2 = \sqrt{[(x_1 - dx)^2 + y_1^2]}$, where another Rabi sideband frequency is measured. Then the two square root expressions can be solved for x_1. Analogously, a known control beam displacement in the y-direction, dy, can be used to obtain y_1 as well.

The resolution of such an imaging process will depend on how well the sidebands can be resolved and how strongly the local Rabi frequency varies in the location of the emitter. In the following, we show that the described method would enable a subdiffraction-limited resolution.

Consider the pump, Stokes, probe, and control laser all centered on the same location and focused to their respective diffraction-limited Gaussian-shaped intensity profiles. The distribution of emitters then generates a distribution of Rabi sidebands on the CARS signal. The range of the Rabi sidebands is given by $\Omega_{max} = |2\pi\mu_{14}E_0/h|$ in the beam center to $\Omega_{min} = (1/e)\cdot|2\pi\mu_{14}E_0/h|$ at a distance w_0, where E_0 is the maximum of the control laser field amplitude in the beam center and w_0 is the distance where the field is smaller by $1/e$. The radial resolution then depends on how accurately the spectral content of the Rabi sidebands are determined. Given a Rabi frequency, Ω, corresponding to a distance, r, from the beam center and that the Rabi sideband frequency can be measurement with an accuracy, $d\Omega$, it can be shown that the radial resolution at that distance is dr, with

Nonlinear Optics Approaches Towards Subdiffraction Resolution in CARS Imaging

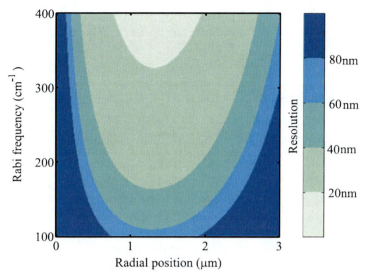

Fig. 18 Spatial resolution of CARS Rabi labeled microscopy calculated from (12) vs. the Rabi frequency and the spatial distance of the emitter from the center of the control beam

$$dr = \frac{w_0^2}{2r + dr} \ln\left(\frac{\Omega}{\Omega - d\Omega}\right) \quad (12)$$

Equation (12) has to be evaluated numerically, but it can be seen from the argument of the logarithm that dr becomes smaller (the resolution increases) with increasing Rabi frequency. Also, according to the $1/r$ dependence, the resolution improves with increasing distance from beam center. For a given control laser intensity, the resolution reaches its maximum value where the steepest intensity slope of the control laser beam is found, which is approximately at the distance equal to the Gaussian beam radius. This is illustrated in Fig. 18, where we have evaluated the resolution given by (12) for a range of Rabi wavenumbers and distances from the beam center. As the limit for the obtainable spectral resolution, $d\Omega$, we use a value of 3 cm^{-1}, which is a representative value for the spectral width of typical vibrational resonances in liquid samples. The figure shows that an appreciable improvement over diffraction-limited CARS microscopy is possible.

For a specific implementation of a microscope with a numerical aperture of NA = 1.2 and the laser wavelengths that were used in our calculations, the diffraction-limited resolution is 171 nm. If the control laser with a wavelength of λ = 3.3 μm and focused to a diffraction-limited spot induces Rabi oscillations with a frequency of 100 cm^{-1}, the spatial resolution attains its smallest value of 65 nm at a distance of 1.2 μm from beam center. From the data presented in [35], we estimate that 100 cm^{-1} Rabi oscillations can, indeed, be generated with a control laser intensity of about 500 MW cm^{-2}.

8 Conclusions

We have used a density matrix model of a four-level system to investigate several routes with the potential to beat the diffraction limit in CARS microscopy. In general, two classes of approaches were identified. The first can be called an incoherent approach and is based on the saturation of transitions achieved with a control beam at high intensities. This type of approach can be applied in various manners, depending on the specific level structure and damping parameters offered by the sample under inspection. In this sense, these approaches resemble STED from fluorescence microscopy, which has been rather successful with samples that offer fluorescent transitions, such as through chemical labeling. To achieve CARS suppression, one or more control lasers are used to prepopulate a vibrational state, or to partially deplete the ground state. Using typical parameters for molecular transitions probed by CARS microscopy, we showed that depletion factors of 99 % and above might be obtainable. This corresponds to improving the resolution to values in the order of $\sim\lambda/(20\text{ NA})$, where λ is the wavelength of the probe beam and NA the imaging numerical aperture. The required pulse area for saturation corresponds to intensities that span a range from 100 MW cm^{-2} to 200 GW cm^{-2}, which seems tolerable in terms of low photo damage, since the molecules remain in their electronic ground state. We also give references to sample molecules that may be suitable for basic proof-of-principle experiments. In an example for a level system that seems quite generally available, we observe that similarly high-resolution factors might be achievable. However, the saturation in such systems leaves a nonsaturable background that can be as high as 50 %. In such case, the images need to be recorded with a correspondingly higher signal-to-noise ratio—through longer averaging times—such that the noise level is on the order of a percent, and the high-resolution images need to be retrieved by subtracting a resolution-limited reference image. CARS suppression can also be achieved with very small population changes, i.e., by removing less than a percent from the ground state. In this case, the control laser is chosen to be resonant with a transition from the upper vibrational state. For control transitions with longer coherence times, we have shown that resonantly driving the excitation of the control state results in intensity and, thus, spatially dependent sidebands in the CARS emission spectrum. Accurately measuring the resulting Rabi splitting in the CARS spectrum allows objects to be resolved below the diffraction limit. We estimate from typical experimental parameters that a resolution in the order of 65 nm may be achievable. We also note that our calculations show that any other emission process involving the ground state, such as CSRS, also generates Rabi sidebands, provided the decoherence rate is slower than the Rabi frequency. Hence, the Rabi labeling may be applicable to other microscopy techniques as well.

References

1. Hell SW, Wichmann J (1994) Breaking the diffraction resolution limit by stimulated-emission—stimulated-emission-depletion fluorescence microscopy. Opt Lett 19(11): 780–782
2. Klar TA et al (2000) Fluorescence microscopy with diffraction resolution barrier broken by stimulated emission. Proc Natl Acad Sci U S A 97(15):8206–8210
3. Betzig E et al (2006) Imaging intracellular fluorescent proteins at nanometer resolution. Science 313(5793):1642–1645
4. Huang B et al (2008) Whole-cell 3D STORM reveals interactions between cellular structures with nanometer-scale resolution. Nat Methods 5(12):1047–1052
5. Rust MJ, Bates M, Zhuang XW (2006) Sub-diffraction-limit imaging by stochastic optical reconstruction microscopy (STORM). Nat Methods 3(10):793–795
6. van de Linde S et al (2011) Direct stochastic optical reconstruction microscopy with standard fluorescent probes. Nat Protoc 6(7): 991–1009
7. Zenobi R (2008) Analytical tools for the nano world. Anal Bioanal Chem 390(1):215–221
8. Rittweger E et al (2009) STED microscopy reveals crystal colour centres with nanometric resolution. Nat Photonics 3(3):144–147
9. Wildanger D, Maze JR, Hell SW (2011) Diffraction unlimited all-optical recording of electron spin resonances. Phys Rev Lett 107(1):017601
10. Dyba M, Hell SW (2002) Focal spots of size lambda/23 open up far-field florescence microscopy at 33nm axial resolution. Phys Rev Lett 88(16):163901
11. Andrews JR, Hochstrasser RM, Trommsdorff HP (1981) Vibrational transitions in excited-states of molecules using coherent Stokes Raman-spectroscopy—application to ferrocytochrome-C. Chem Phys 62(1–2):87–101
12. Fu Y et al (2006) Characterization of photodamage in coherent anti-Stokes Raman scattering microscopy. Opt Express 14(9): 3942–3951
13. Lin CY et al (2011) Picosecond spectral coherent anti-Stokes Raman scattering imaging with principal component analysis of meibomian glands. J Biomed Opt 16(2):021104
14. Nikolaenko A, Krishnamachari VV, Potma EO (2009) Interferometric switching of coherent anti-Stokes Raman scattering signals in microscopy. Phys Rev A 79(1):13823
15. Raghunathan V, Potma EO (2010) Multiplicative and subtractive focal volume engineering in coherent Raman microscopy. J Opt Soc Am A 27(11):2365–2374
16. Milonni PW, Eberly JH (2010) Lasers physics. Wiley, Hoboken, NJ, xiv, 830p
17. Beeker WP et al (2010) Spatially dependent Rabi oscillations: an approach to sub-diffraction-limited coherent anti-Stokes Raman-scattering microscopy. Phys Rev A 81(1):012507
18. Beeker WP et al (2009) A route to sub-diffraction-limited CARS microscopy. Opt Express 17(25):22632–22638
19. Cleff C et al (2013) Stimulated-emission pumping enabling sub-diffraction-limited spatial resolution in coherent anti-Stokes Raman scattering microscopy. Phys Rev A 87(3): 033830(1)–033830(9)
20. Gardner JR et al (1993) Suboptical wavelength position measurement of moving atoms using optical-fields. Phys Rev Lett 70(22): 3404–3407
21. Shen YR (1984) Principles of nonlinear optics. Wiley, New York
22. Deak JC et al (2000) Ultrafast infrared-Raman studies of vibrational energy redistribution in polyatomic liquids. J Raman Spectrosc 31(4): 263–274
23. Golonzka O et al (2001) Coupling and orientation between anharmonic vibrations characterized with two-dimensional infrared vibrational echo spectroscopy. J Chem Phys 115(23):10814–10828
24. Laubereau A et al (1978) Vibrational population lifetimes of polyatomic-molecules in liquids. Chem Phys 31(3):335–344
25. Asbury JB et al (2003) Hydrogen bond dynamics probed with ultrafast infrared heterodyne-

detected multidimensional vibrational stimulated echoes. Phys Rev Lett 91(23):237402
26. de Vivie-Riedle R, Troppmann U (2007) Femtosecond lasers for quantum information technology. Chem Rev 107(11):5082–5100
27. Wurzer AJ et al (1999) Comprehensive measurement of the S-1 azulene relaxation dynamics and observation of vibrational wavepacket motion. Chem Phys Lett 299(3–4):296–302
28. Brion E et al (2007) Universal quantum computation in a neutral-atom decoherence-free subspace. Phys Rev A 75(3):032328
29. Hein B, Willig KI, Hell SW (2008) Stimulated emission depletion (STED) nanoscopy of a fluorescent protein-labeled organelle inside a living cell. Proc Natl Acad Sci U S A 105(38):14271–14276
30. Ventalon C et al (2004) Coherent vibrational climbing in carboxyhemoglobin. Proc Natl Acad Sci U S A 101(36):13216–13220
31. Cleff C et al (2012) Ground-state depletion for subdiffraction-limited spatial resolution in coherent anti-Stokes Raman scattering microscopy. Phys Rev A 86(2):023825(1)–023825(11)
32. Offerhaus HL (2011) Private communication
33. Boller KJ, Imamoglu A, Harris SE (1991) Observation of electromagnetically induced transparency. Phys Rev Lett 66(20):2593–2596
34. Schouwink P et al (2002) Dependence of Rabi-splitting on the spatial position of the optically active layer in organic microcavities in the strong coupling regime. Chem Phys 285(1):113–120
35. Witte T et al (2004) Femtosecond infrared coherent excitation of liquid phase vibrational population distributions (v > 5). Chem Phys Lett 392(1–3):156–161
36. Graener H, Laubereau A (1987) Ultrafast vibrational-energy transfer of polyethylene investigated with picosecond laser-pulses. Chem Phys Lett 133(5):378–380
37. Laubereau A, Kaiser W (1978) Vibrational dynamics of liquids and solids investigated by picosecond light-pulses. Rev Mod Phys 50(3):607–665
38. Fendt A, Fischer SF, Kaiser W (1981) Vibrational lifetime and Fermi resonance in polyatomic-molecules. Chem Phys 57(1–2):55–64
39. Graener H, Laubereau A (1982) New results on vibrational population decay in simple liquids. Appl Phys B 29(3):213–218
40. Okamoto H, Yoshihara K (1991) Femtosecond time-resolved coherent Raman-scattering from beta-carotene in solution—ultrahigh frequency (11-Thz) beating phenomenon and subpicosecond vibrational-relaxation. Chem Phys Lett 177(6):568–572
41. Ambroseo JR, Hochstrasser RM (1988) Pathways of relaxation of the N-H stretching vibration of pyrrole in liquids. J Chem Phys 89(9):5956–5957
42. Heilweil EJ, Cavanagh RR, Stephenson JC (1987) Population relaxation of Co(V=1) vibrations in solution phase metal-carbonyl-complexes. Chem Phys Lett 134(2):181–188
43. Tokmakoff A, Sauter B, Fayer MD (1994) Temperature-dependent vibrational-relaxation in polyatomic liquids—picosecond infrared pump-probe experiments. J Chem Phys 100(12):9035–9043

Chapter 13

Photooxidation Microscopy: Bridging the Gap Between Fluorescence and Electron Microscopy

Annette Denker and Silvio O. Rizzoli

Abstract

Eighty years after its development, electron microscopy still represents the gold standard in terms of resolution. A major disadvantage is, however, the requirement for fixed specimens—especially in view of the numerous live fluorescence microscopy methods that have been developed during the last few decades. This drawback can be largely compensated by combining both microscopy techniques, live imaging and electron microscopy, by transforming a fluorescent signal into one that can be visualized in the electron microscope. This can be achieved by employing photooxidation. This procedure uses the production of reactive oxygen species by excited fluorescent dyes to oxidize the substrate diaminobenzidine, which in turn forms an electron-dense precipitate in the immediate proximity of the dye. In this chapter, we explain the photooxidation protocol in detail, focusing mainly on FM dyes as markers of membrane trafficking, especially in synaptic physiology. We also discuss the use of numerous other labels for photooxidation applications. We conclude that this approach is applicable to a wide variety of cellular targets and processes and therefore has a great potential in linking diffraction-limited light imaging to the high resolution of electron microscopy.

Key words Correlative light and electron microscopy, Diaminobenzidine, FM (styryl) dyes, Genetically encoded tags, Synaptic vesicle recycling

1 Introduction

Fluorescence microscopy has proven to be an invaluable tool for visualizing the structure and function of biological substrates. To provide only one example, the introduction of genetically encoded fluorescent tags (such as green fluorescent protein, GFP) revolutionized the whole field of cell and molecular biology, as it allowed investigators to monitor the localization and, importantly, also the dynamics of virtually all cellular proteins. However, two major drawbacks are intrinsically associated with fluorescence microscopy: first, only fluorescently labeled substrates can be imaged, which limits the detection of multiple different structures simultaneously; second, light microscopy is inherently diffraction limited, resulting in a maximal resolution of about half the wavelength of light (as explained in Chap. 1). Even with the many recently developed super-resolution fluorescence microscopy techniques presented in the rest of this volume, in practical terms most fluorescence microscopes still function in the range of a few tens of nanometers [1, 2].

Electron microscopy (EM), on the other hand, provides unsurpassed resolution, down to the Ångström range, and reveals ultrastructural information unavailable in conventional light microscopy. In addition, in EM there is no need to introduce specific labels for every structure of interest, as different organelles can generally be identified by their morphological appearance simply via heavy metal labeling. One of the major disadvantages of EM, however, is the strict requirement for fixed samples, rendering live imaging of cellular processes impossible. Furthermore, the detailed investigation of protein localization is aggravated by difficulties associated with immunogold labeling, with many antibodies being unable to bind resin-embedded epitopes with high affinity and specificity.

Both of these difficulties have been partially addressed by combining live fluorescence microscopy with electron microscopy, using the photooxidation technique. Here, organelles or proteins can be specifically labeled by different methods (including GFP tagging), and the structures can be imaged in living cells. The cells are afterwards fixed and the fluorescent signal is transformed into an electron-dense precipitate, which is easily observed in electron microscopy. In this chapter, we explain the photooxidation technique, focusing on its application to membrane trafficking at the synapse.

1.1 The Precursor of Photooxidation: Horseradish Peroxidase (HRP) as an Endocytic Marker

An important theme in synaptic physiology has been the morphology of the organelles involved in synaptic vesicle recycling and metabolism. The synaptic vesicles are small membranous organelles (~40–50 nm in diameter) that contain neurotransmitter molecules. When a neuron is activated, the vesicles collapse into the plasma membrane of the cell and release their neurotransmitter contents onto the neighboring cells, thereby activating or silencing them. After release, the protein and lipid components of the vesicles are gathered by a complex protein machinery and a new vesicle forms from the plasma membrane, completing what is termed synaptic vesicle recycling [3].

The first attempts to investigate the morphology of the organelles involved in synaptic vesicle recycling have used inherently dense molecules, such as the polysaccharide dextran. Synapses (such as nerve–muscle preparations, also termed neuromuscular junctions or NMJs) are bathed in dextran solution, which is taken up by the newly formed vesicles. This can be observed in EM as small "dots" of dextran inside the lumens of the recycled vesicles [4].

However, the density of dextran particles is not particularly high, making the labeled vesicles hard to identify. Therefore, an alternative technique has been employed since the early 1970s, using the hemoprotein horseradish peroxidase (HRP) as tracer. Due to its intrinsic slight electron opacity, the enzyme could in principle be directly visualized by EM [5]. However, this feature is not sufficient to reliably track and identify small amounts of tracer, and a method that takes advantage of HRP's enzymatic activity was therefore developed. This was based on the use of the HRP-catalyzed transfer of two electrons from a donor substrate such as 3,3'-diaminobenzidine (DAB) to hydrogen peroxide. The oxidized DAB polymerizes and forms a dark brown precipitate insoluble in water or ethanol [6], which can be easily visualized by conventional transmitted light microscopy. This method was first introduced by Graham and Karnovsky [7], who found that the DAB precipitate was also electron opaque and osmiophilic, allowing for easy identification of newly recycled synaptic vesicles in EM [4, 8].

Using the HRP-induced oxidation of DAB, Ceccarelli and coworkers showed that vesicles are reformed from the axolemma after fusion and neurotransmitter release at the frog neuromuscular junction [4, 9]. Heuser and Reese simultaneously performed a seminal study on the pathway of vesicle recycling in the same system [8] and the technique quickly spread to other preparations, including the lobster NMJ [10], fetal mammalian spinal cord neurons [11], and frog photoreceptors [12].

However, one major caveat was still associated with the procedure: HRP can promote the peroxidation of membrane lipids, thereby possibly interfering with membrane function [13]. Indeed, distortions of membrane structure and increased vesicle release have been reported after HRP application [8]. Also, the technique could not be combined with live imaging, therefore providing only snapshots of synaptic physiology.

1.2 Combining Light and Electron Microscopy: Photooxidation

Such a combination was highly desirable, as it would have provided the opportunity to combine the advantages of both light and electron microscopy (live imaging and superb ultrastructural resolution, respectively). The first attempt to perform such an experiment used the same principle as in the HRP experiments: preparations were soaked with DAB, with its oxidation providing the necessary EM marker. Instead of using HRP, Maranto irradiated lucifer-yellow-filled neurons of the leech in the presence of DAB in 1982 [14]. This procedure resulted in bleaching of the lucifer-yellow-associated fluorescence and in the formation of an electron-dense DAB precipitate (same as in the HRP-catalyzed reaction). However, huge amounts of dye were necessary to trigger precipitate formation. The technique was then further developed and extended to a variety of different dyes and to weaker initial fluorescence labeling by Sandell and Masland [15].

The reaction mechanism of this photooxidation reaction (sometimes also referred to as photoconversion) seems to be very similar to the HRP approach, with DAB likely being oxidized and thereby precipitated. Oxidation is probably mediated by highly reactive singlet oxygen, which is produced as the photoexcited fluorescent molecules undergo the phenomenon of intersystem crossing [16] (see also Fig. 2a). Such a reaction mechanism would also explain the requirement for molecular oxygen [15] as detailed below.

Here, we explain the technical basics underlying the photooxidation reaction as employed in synaptic preparations. We will shortly discuss the further handling of the samples after the photooxidation reaction as well as the EM processing, including comments on heavy metal staining. The chapter contains a troubleshooting section dealing with problems associated with background staining (especially in mitochondria) and with wrongly chosen illumination times. The last part of the chapter presents some of the results obtained by using the photooxidation technique (with special emphasis on studies of synaptic physiology) and provides an outlook on new developments, including new genetically encoded tags.

2 The Photooxidation Methodology

2.1 Staining Nerve Terminals with FM Dyes

The most important application for photooxidation in the field of synaptic physiology has been its combination with FM (styryl) dyes as sensitive markers for membrane retrieval. FM dyes share a general structure, which is explained in Fig. 1a for the widely used FM 1–43. A hydrophobic part, which usually consists of two aliphatic chains, allows for their reversible insertion into membranes. The length of these chains determines the hydrophobicity of the dye and thus its affinity for membranes. The dye molecule also contains a hydrophilic head group (with two positive charges), which prevents penetration through the membranes. These two moieties are connected by aromatic rings and one or more double bonds that determine the dye's spectral properties. Importantly, the dyes are several hundred fold more fluorescent in the lipid environment of the membrane than in water, which renders the dye molecules essentially invisible in solution.

Figure 1b displays a typical FM dye labeling experiment for monitoring vesicle recycling in a presynaptic terminal. Upon dissecting the preparation of interest (for instance, an NMJ), dye-containing buffer is added. The dye diffuses into the preparation and inserts into the synaptic plasma membrane, where its fluorescence increases substantially. When vesicles fuse with the plasma membrane (for instance, triggered by electrical stimulation) and undergo recycling, they will take up the dye. After vesicle reformation, the vesicles remain labeled, while the extracellular space can be

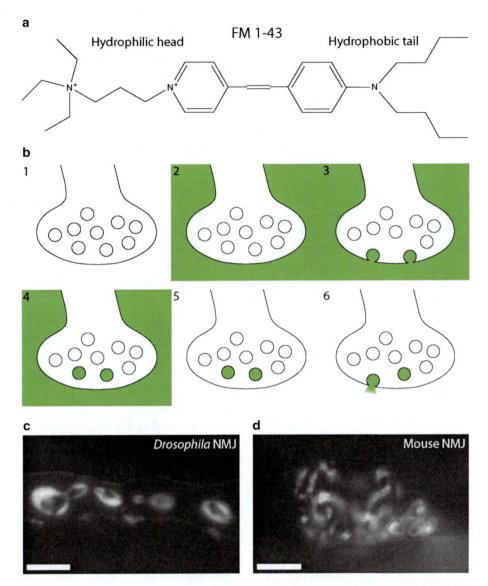

Fig. 1 FM dye characteristics and staining procedure. (**a**) The structure of FM 1–43. As each FM dye, the molecule contains a hydrophilic head group and a hydrophobic tail group connected via aromatic rings and double bonds. (**b**) (1) A presynaptic terminal filled with synaptic vesicles. (2) The FM dye is added to the preparation and inserts into the terminal membrane. (3) Vesicle fusion is stimulated, resulting in (4) dye uptake into the recycling vesicles. (5) The preparation may then be washed to remove excess dye. (6) Further stimulation in the absence of FM dye triggers a new round of vesicle recycling and destaining of previously labeled vesicles. (**c**) *Drosophila* NMJ labeled by FM 1–43 uptake. Size bar: 5 μm. (**d**) Mouse NMJ stained by FM 1–43 labeling. Size bar: 10 μm

washed to remove excess dye. Furthermore, the labeled vesicles can be destained, if a second round of stimulation is triggered in absence of the dye, resulting in the loss of label from the preparation. Example images of synapses stained with FM dyes are displayed in

Fig. 1c, d. Several types of imaging experiments can be devised, with different combinations of staining/destaining, testing various synaptic characteristics such as vesicle pool properties (i.e., vesicle populations with different release abilities [17]).

2.2 Photooxidation of FM Dyes

Protocol (a detailed video protocol is provided in [18]):

(a) After FM dye labeling and fluorescence imaging, the preparations should be fixed. We generally employ 2.5 % glutaraldehyde fixation (for 45–90 min), as formaldehyde has been reported to increase spontaneous release frequency [19]. The temperature for fixation also requires careful consideration; whereas the fixation procedure is faster at room temperature, fixation on ice decreases the amount of vesicles fusing spontaneously while fixation is not yet complete. We therefore generally combine the two approaches by first fixing on ice for ~30 min, followed by further fixation at room temperature to speed up the process.

(b) Fixation is followed by thorough washing in phosphate buffered saline (PBS). Like all subsequent steps until fluorescence illumination, this washing is performed at 4 °C.

(c) The sample is then removed from the dissection dish (it is relatively robust after glutaraldehyde fixation) and washed for ~20 min in 100 mM NH_4Cl (in PBS). This step will neutralize free aldehyde groups of the remaining glutaraldehyde and it also decreases glutaraldehyde autofluorescence. If this step is omitted or shortened, residual free glutaraldehyde will later react with DAB and form a transparent crystalline precipitate, which is easily observed by visual inspection. In our hands, samples that display such precipitate formation should be discarded, as photooxidation is rarely successful in this case. Should precipitate formation be observed, it is advisable to prolong the NH_4Cl washing or to add an additional step of washing with 100 mM glycine for the next preparations processed.

(d) The sample is transferred to PBS and washed thoroughly. It is then pinned into a new dish and exposed to a second round of PBS washing. These washing steps should eliminate all residual free fixative.

(e) The sample is incubated for 30–60 min at 4 °C with (filtered) 1.5 mg/mL DAB (in PBS). The DAB solution should either be prepared fresh or stored for not more than a few hours at 4 °C.

(f) Before illumination, the DAB solution used for the incubation of the preparation is replaced with fresh solution. The sample is then placed under the microscope objective. A 20× objective provides a good compromise between the size of the illumination field (and thereby also the size of the photooxidation spot) and the illumination time (which increases with lower magnification).

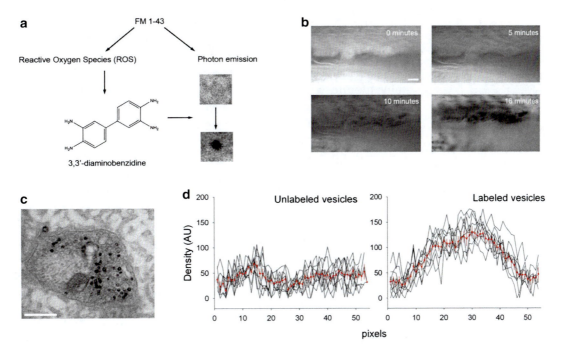

Fig. 2 The photooxidation reaction. (**a**) The principle of the photooxidation reaction. Illumination results in photon emission and in the production of ROS. When preparations are simultaneously incubated with DAB, the ROS will oxidize the DAB, causing it to form an electron-dense precipitate, which can be visualized by EM. As the ROS are highly reactive and short-lived, this reaction only occurs in the immediate proximity of the labeled structures (synaptic vesicles in this case). (**b**) The progress of the photooxidation reaction in a *Drosophila* NMJ. At the beginning of the reaction, the FM fluorescence is bleached. Shortly afterwards, the brown DAB precipitate appears, marking the beginning of successful photooxidation. The reaction should not be stopped at this point but be continued for another 5–10 min to ensure successful and complete DAB oxidation. At this point, synapses have turned dark brown and some photooxidation product is observed on neighboring tissues such as the muscle surface (for instance, due to glutaraldehyde autofluorescence). Note that these images were taken with a 63× objective to better visualize the process (we generally employ a 20× objective). Size bar: 2 μm. (**c**) A successfully labeled and photooxidized *Drosophila* NMJ. Size bar: 500 nm. (**d**) Labeled and unlabeled vesicles can be reliably distinguished. Line scans were performed through ten unlabeled and ten labeled vesicles from the synapse displayed in (**c**). The *red line* indicates the average density. The profiles of labeled and unlabeled vesicles are clearly quite different, both in density levels and in profile shape, due to the fact that labeled vesicles have a densely stained core, whereas unlabeled vesicles display a clear core. Modified with permission from [22]

A dry objective can be used (not immersed in the solution on upright microscope setups) to avoid objective contamination with DAB.

The photooxidation reaction, as described in Fig. 2a, will begin upon the start of illumination. The FM dye is excited and emits photons; at the same time, it produces reactive oxygen species (ROS). As the DAB is membrane permeant and can penetrate into the preparation, it will reach the labeled vesicles and be oxidized, resulting in its precipitation, as described above. Importantly, this reaction only takes place in the immediate proximity of the dye (in our example, the vesicular lumen) as the ROS are highly reactive and have a very short lifetime.

The progress of the photooxidation reaction should be carefully monitored. As indicated in Fig. 2b, the illumination first results in the bleaching of the FM dye fluorescence. A few minutes after bleaching is complete, the preparation assumes a brown color. We generally continue the reaction for another 5–10 min to ensure complete photooxidation. At this time, the nerve terminals have turned completely dark brown (clearly visible on top of the residual glutaraldehyde autofluorescence of the background) and some photooxidation product can also be observed on neighboring tissues such as the muscle surface. This is probably due to DAB precipitation induced by the tissue or fixative autofluorescence. Successful precipitate formation can also be easily observed by transmission light. As will be discussed below, the exact time point of stopping the reaction (by simply switching the lamp off) needs to be chosen carefully and depends to a large extent on light intensity and on properties of the preparation. In our hands, illumination between 30 and 45 min often gave optimal results (using a 20× objective). When photooxidation is not completed after 1 h, the sample can generally be discarded—no further oxidation is to be expected. This can be due to poor DAB penetration or old, not intense enough fluorescence lamps (also refer to Sect. 2.4.1 for discussion of low light intensity).

2.3 EM Processing and Imaging of Photooxidized Preparations

Protocol:

(a) When photooxidation is complete, the DAB can be washed off and replaced with PBS. The photooxidation spot is easily identified under a normal dissection microscope, which allows it to be cut from the rest of the preparation (this ensures that photooxidized synapses are quickly found in electron microscopy). The cut tissue can be stored at 4 °C in PBS for up to 2 days.

(b) The samples are then postfixed and stained with 1 % osmium tetroxide (in filtered PBS). Generally, about 150–200 μL per sample are sufficient. Incubation in osmium tetroxide should be performed at room temperature for 45–60 min. If penetration into the tissue might be limiting, this time can be increased to 1.5 or even 2 h (in the latter case the incubation should be done at 4 °C to avoid osmium tetroxide precipitation) and the osmium concentration can be increased up to ~2 %. The osmium incubation should be performed under a ventilated hood, using gloves and eye protection.

(c) The osmium is removed by thorough washing with filtered PBS (at least three times; at room temperature, as all subsequent steps).

(d) The samples are dehydrated according to the following scheme:

30 % Ethanol	5 min
50 % Ethanol	5 min
70 % Ethanol	5 min
90 % Ethanol	10 min
95 % Ethanol	10 min
3× 100 % Ethanol	10 min
1:1 Ethanol:propylene oxide	10 min
3× Propylene oxide	10 min
1:1 Propylene oxide:epon	10–12 h (under constant agitation)
For samples in which osmium tetroxide penetration is not an issue (such as the *Drosophila* NMJ), the last three steps involving propylene oxide may be omitted. Instead, samples should be transferred into a 1:1 mixture of ethanol:epon resin and incubated for 10–12 h under constant agitation.	

(e) The propylene oxide:epon or ethanol:epon mixture is removed and replaced with fresh 100 % epon resin. The samples are kept in open vials for ~8 h to allow for complete evaporation of the organic solvent.

(f) The samples are then incubated in moulds in fresh resin for 24–48 h at 60 °C to allow for epon polymerization.

(g) The samples are cut into thin or ultrathin sections on a microtome. It is important to note that no further staining is necessary, i.e., no uranyl acetate or lead citrate stains. They may actually reduce the signal-to-noise ratio by diminishing the recognition of the DAB product in EM. We find the use of the osmium stain alone to be sufficient for visualizing the sample's ultrastructure, at the same time allowing for reliable identification of the photooxidation product.

(h) The samples may now be imaged by EM. An example image of a successfully labeled and photooxidized synapse is provided in Fig. 2c.

As the photooxidation reaction proceeds in an all-or-none manner, labeled and unlabeled vesicles can generally be readily distinguished by visual inspection, but the ease of defining the labeling state of a vesicle depends to some extent on the clarity of the cytoplasm. However, even within the rather dense cytoplasm of the *Drosophila* preparation, we find that labeled and unlabeled vesicles can be reliably distinguished as indicated by performing line scans through vesicles designated as labeled or unlabeled, respectively (Fig. 2d): the average densities and shapes of the two profiles differ

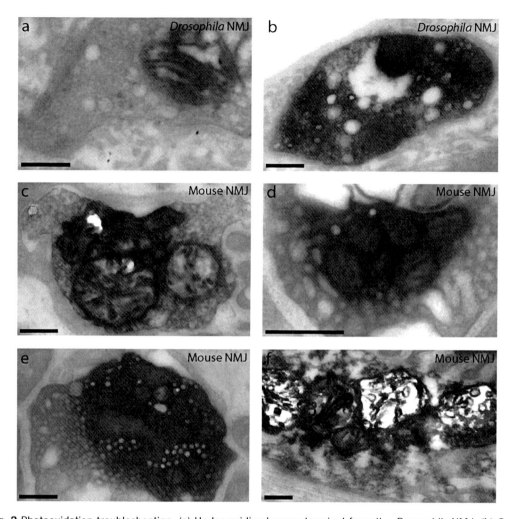

Fig. 3 Photooxidation troubleshooting. (**a**) Under-oxidized nerve terminal from the *Drosophila* NMJ. (**b**) Over-oxidized nerve terminal from the *Drosophila* NMJ. Note dark cytosol. (**c–f**) Different stages of mitochondrial distortions due to accumulation of DAB precipitate in the mouse NMJ. Mitochondria first start to swell (**c**), and then burst, releasing the DAB content into the cytosol (**d**). This results in the "negative" image of dark cytosol and clear vesicles typical for over-oxidized samples (**e**) (compare also (**b**)). Finally, mitochondria having spilled most of their contents can even appear "empty," while the morphology of the entire synapse is extremely poor (**f**). Size bars are 300 nm for all images

substantially, which is expected as unlabeled vesicles present a clear core (lumen), whereas labeled vesicles have a densely stained core.

2.4 Troubleshooting

2.4.1 Over- and Under-Oxidation

One of the difficulties of the photooxidation technique is the occurrence of under- or over-oxidized terminals as displayed in Fig. 3a, b, respectively.

In under-oxidized synapses, no reaction product is observed. This can be due to several factors, such as limited DAB penetration, too low light intensity, and too short illumination times (i.e., the reaction was stopped too early). The limited penetration of DAB into the preparation is especially evident for very thick samples (>300 μm): for these tissues we often observed successful photooxidation in surface layers (on both sides of the preparation), but more central structures did not reveal precipitate formation. For most muscle preparations, where the terminals are found on the muscle surface, the limited DAB penetration is generally not a major drawback of the technique. Photooxidation can however also be limited by too low light intensity, which can be increased by using a high intensity xenon or mercury lamp (possibly combined with a back mirror to collect back-scattered light) and a high numerical aperture objective. The illumination spot should be well focused on the structure of interest. The most common error resulting in under-oxidized synapses is the choice of too short illumination times. It is important to keep in mind that completely oxidized synapses are generally not found immediately after the disappearance of the FM fluorescence, and not even after the initial appearance of the dark DAB precipitate, but ~5–10 min later.

On the other hand, illuminating the samples too long is also not an option, as this easily results in over-oxidation. Over-oxidized terminals display a "negative" image of what one expects to see in a well-oxidized sample (compare Figs. 2c and 3b): the cytosol is now dark, due to DAB precipitate spilling out of damaged organelles, and unlabeled organelles (which exclude the dark cytosol) appear clear. All potentially labeled structures are obscured by the very dark staining of the cytosol. Furthermore, over-oxidized samples often display a heavily distorted ultrastructure, with many vacuolar structures. As described below, these problems mainly result from the photooxidation of mitochondria.

In summary, the time window in which to stop the photooxidation reaction is quite narrow. It is also important to mention that experience with illumination times from one specific preparation cannot necessarily be directly applied to another preparation. For instance, we found mouse NMJs to be much more susceptible to over-oxidation than all other NMJs we investigated (including insects, frog, fish, and chicken). Therefore, it might be advisable to test different illumination times, when experience with the photooxidation of a specific preparation is limited.

2.4.2 DAB Precipitation Within Mitochondria

Interestingly, FM-labeled structures are generally not the first to oxidize DAB. Instead, mitochondria often start to precipitate DAB even in synapses which were illuminated for too short time intervals to observe any labeled synaptic vesicles (note, for instance, mitochondrion in Fig. 3a). This precipitation is in our hands dependent on illumination and is not observed outside the photooxidation spot, but it does not depend on the presence of the dye. Whereas the photooxidation of the mitochondria can be used as an indication for successful DAB penetration and sufficient light intensity, it also represents the main cause for the deleterious effects of over-oxidation: when the mitochondria accumulate too high amounts of DAB precipitate, they start to swell, distorting the synaptic ultrastructure. When illumination is continued further, the mitochondria will eventually burst and spill their dark DAB precipitate content into the cytoplasm, resulting in the electron-dense cytoplasm described above. The process is displayed in Fig. 3c–f for the mouse NMJ.

Different mechanisms have been proposed to account for the high susceptibility of mitochondria to over-oxidation, and in both cases the mitochondrial respiratory chain plays a special role: first, cytochromes display strong autofluorescence and will therefore also produce ROS and oxidize the DAB under illumination. Second, respiratory enzymes may display residual enzymatic activity even after fixation, also resulting in the production of ROS. In line with the second hypothesis, addition of potassium cyanide, which is an inhibitor of the mitochondrial respiratory chain, reduces background DAB precipitation in the mitochondria [16, 20, 21]. The fact that illumination is required for the appearance of precipitation in the mitochondria (see above) argues for a combination of the two mechanisms.

If illumination times are chosen carefully, spilling of mitochondrial contents can generally be avoided for most synapses. Because of its high toxicity, potassium cyanide should be regarded as a last resort, rather than the default solution to over-oxidation.

3 Applications of the Photooxidation Technique to Biological Questions

3.1 Studies on Synaptic Physiology

As displayed in Fig. 4, the photooxidation technique has been established in a vast variety of synaptic preparations. These include adult and developing NMJ and central nervous system (CNS) synapses from vertebrates and invertebrates. We have used the technique for the comparative investigation of distinct vesicle pools [22, 23]. The photooxidation technology allows several parameters to be measured, which could not have been gathered from (diffraction-limited) light microscopy, including vesicle positions, morphologies, and numbers.

One important question in the vesicle pool field has been how many of the vesicles are actually involved in recycling under defined stimulation paradigms. For most stimulation conditions and preparations tested, the percentage of labeled (i.e., recycling) vesicles was found to be surprisingly small. For instance, Harata and colleagues found only 10–20 % of the vesicles in hippocampal neurons labeled after 10 Hz stimulation for 2 min or after high potassium application for 90 s (note that such stimulation had been reported to cause the complete turnover of the functional vesicle population [24]). Similar numbers of labeled vesicles were found in the frog NMJ after 30 Hz stimulation for 10 s [25]. For the calyx of Held synapse in the rat auditory pathway, the pool of recycling vesicles was found to be even smaller: only ~5 % of the vesicles were found

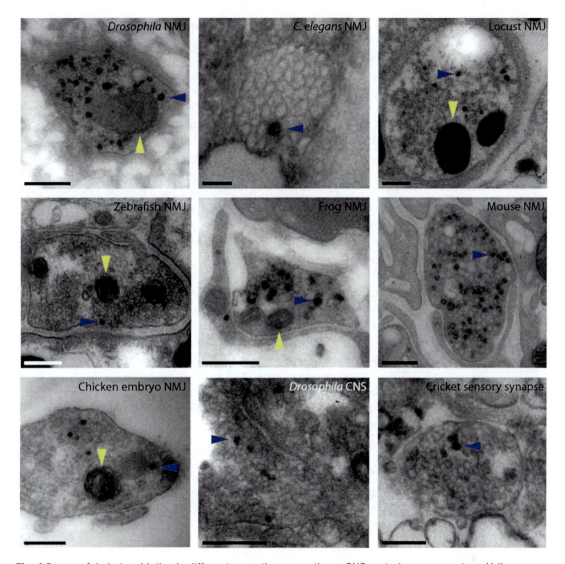

Fig. 4 Successful photooxidation in different synaptic preparations. *CNS* central nervous system. *Yellow arrowheads* indicate mitochondria (refer to Sect. 2.4.2 for an explanation of mitochondrial staining). *Blue arrowheads* indicate example labeled vesicles. Size bars are generally 300 nm (*C. elegans* NMJ: 100 nm; *Drosophila* CNS and cricket sensory synapse from the optic lobe: 200 nm)

labeled after stimulation at 5 Hz for 20 min or at 20 Hz for 5 min. Interestingly, high potassium stimulation for 15 min resulted in the recruitment of many more vesicles (~40 %; probably from the so-called reserve vesicle pool) [26].

The studies cited above also indicated that the recycling pool vesicles seem to be intermixed with the reserve vesicles in all these preparations ([24–26]; also refer to [27]). Similar results have been presented for the *Drosophila* NMJ [22]. This observation bears several important implications: first, the recycling vesicles need to be highly mobile to be able to quickly reach the active zones.

Second, since there is no clear difference in vesicle positions between the recycling and reserve vesicles (i.e., the recycling vesicles are not found closer to the membrane than the reserve ones), molecular tags must exist which distinguish between recycling and reserve vesicles and make the recycling vesicles more active than the reserve ones.

A new concept which was recently introduced in the vesicle pool field was the idea of a "super-pool," suggesting that synaptic vesicle recycling is not necessarily restricted to one nerve terminal or bouton, but that vesicles can be exchanged between boutons ([27]; also refer to [17]). Darcy and colleagues bleached individual labeled synapses in hippocampal neurons, allowed for recovery over several minutes and then photooxidized the preparation. Consequently, all labeled vesicles in the bleached synapses had been imported from unbleached terminals during the recovery period. The newly acquired vesicles were found to be thoroughly intermixed in spatial terms with the native vesicle population of the host synapse—with photooxidation again providing information which would be impossible to gather by conventional imaging means.

The photooxidation technique has proven useful in visualizing the morphology of the vesicle recycling process, by fixation at different time points after stimulation. Using this method, Richards and collaborators could demonstrate, for example, that strong stimulation triggers the formation of membrane infoldings and cisternae which then bud off new vesicles in the frog NMJ [28]. Similar results have also been found for the rat calyx of Held [26]. Recently, Hoopmann and coworkers used photooxidation to suggest that the vesicles of the so-called readily releasable pool (RRP; those vesicles which are docked at the membrane and primed for release) are recycled via endosomal sorting: the vesicles were rapidly retrieved from the membrane, and then fused to slightly larger endosomes (and possibly to each other), from which new vesicles were formed [29].

3.2 Applications of the Photooxidation Technique to Other Dyes and Systems

Photooxidation has also been used with other dyes or fluorescent proteins, although the FM dyes still represent the most successful application of the technique. Because of the much greater sensitivity of photooxidation when compared to immunogold labeling, one would of course like to extend it beyond markers of membrane uptake—with the major problem being to find dyes that provide both high fluorescence and a substantial singlet oxygen quantum yield.

One dye which has been found to provide an acceptable combination of these two characteristics is eosin, a fluorescein isothiocyanate derivative containing four bromine atoms which increase the molecule's probability to undergo intersystem crossing and thereby create singlet oxygen [16] (compare Sect. 1.2). Clearly, a higher singlet oxygen quantum yield must be accompanied by higher resistance of the dye molecule itself to these reactive species to prevent self-destruction (as would, for instance, occur in the genetically encoded phototoxic "Killer Red" [30]). Typically, the reaction should be performed at low temperatures (<5 °C), which increases the saturation point of oxygen within the solution. Low temperatures were also reported to minimize both the spontaneous oxidation of DAB and the diffusion of the DAB precipitate from the reaction site. This allowed for imaging of cytosolic proteins and structures such as microtubules.

However, eosin cannot be genetically encoded and therefore needs to be applied as a fusion product with target-specific macromolecules such as antibodies. The approach therefore has the same limitations as conventional immunostaining (such as limited diffusion and the strict requirement for permeabilization to label intracellular structures). It was therefore desirable to develop a genetically encoded tag, which allowed for correlative light and electron microscopy. Such an approach was presented by Gaietta and colleagues who fused a protein of interest (here the gap junction component connexin) to a small tetracysteine tag and incubated the cells with the membrane-permeant biarsenical ReAsH dye, which displays a strong increase in fluorescence upon binding to the tetracysteine motif [31]. The authors used this method successfully to investigate protein trafficking. However, the technique has several major drawbacks, such as a relatively modest singlet oxygen quantum yield [32] and the requirement for using 1,2-ethanedithiol (EDT) as an antidote to minimize cell toxicity (which is due to the binding of ReAsH to endogenous thiols) [31].

To extend the application of correlative light and electron microscopy to various cellular targets, several attempts have been made to render the photooxidation method compatible with GFP, which represents the most popular genetically encoded fluorescent tag in cell biology. Unfortunately, this has been only applied successfully in a few studies. For instance, Monosov and coworkers fused GFP to a peroxisomal targeting signal in yeast and then monitored successful targeting by using photooxidation [21]. The method was further developed and named GRAB (GFP recognition after bleaching) by Grabenbauer et al. [20]. Again, it was found to be essential to increase the oxygen concentration in the DAB solution (by bubbling pure oxygen gas and reducing the temperature). This is probably due to the β-barrel structure of the GFP molecule, which shields the chromophore within from oxygen and thereby minimizes the singlet oxygen production. As GFP photooxidation results in a relatively low signal, reduction of any background DAB precipitation was also found to be essential. This was achieved by potassium cyanide application (as described in Sect. 2.4.2) and by quenching of aldehyde-associated autofluorescence by application of ammonium chloride and glycine (as indicated in Sect. 2.2) and sodium borohydrate.

Nevertheless, the use of GFP for photooxidation remains challenging, mainly due to its low singlet oxygen quantum yield. Furthermore, GFP has the inherent disadvantage of being a relatively big tag of ~25 kDa, which can influence the localization and function of the protein of interest. Recently, a much smaller genetically encoded tag with considerably increased singlet oxygen quantum yield has been reported: the fluorescent flavoprotein miniSOG (mini Singlet Oxygen Generator) [32]. MiniSOG is about half the size of GFP and could be successfully tagged to such diverse cellular proteins as α-actinin, histones, connexins, and SynCAMs. It therefore is a very promising tool for applying correlative microscopy to a plethora of cellular targets.

4 Conclusion

Although photooxidation has proven most successful for studying the recycling of synaptic vesicles and synaptic physiology, it is now being employed for a wide variety of biological questions. This is mainly due to the development of new genetically encoded photooxidation-compatible tags, which extend the use of this technique to several other proteins and subcellular compartments.

Importantly, other technologies between light and electron microscopy are rapidly emerging, allowing for new practical applications and near-nanometer resolution in both fixed and live specimens. For example, array tomography and nano-resolution fluorescence electron microscopy are two recently developed techniques that are sure to provide substantial advances in biology in years to come [33, 34].

References

1. Donnert G, Keller J, Medda R et al (2006) Macromolecular-scale resolution in biological fluorescence microscopy. Proc Natl Acad Sci U S A 103(31):11440–11445
2. Wildanger D, Medda R, Kastrup L et al (2009) A compact STED microscope providing 3D nanoscale resolution. J Microsc 236(1):35–43
3. Sudhof TC (2004) The synaptic vesicle cycle. Annu Rev Neurosci 27:509–547
4. Ceccarelli B, Hurlbut WP, Mauro A (1973) Turnover of transmitter and synaptic vesicles at the frog neuromuscular junction. J Cell Biol 57(2):499–524
5. Novikoff AB (1963) Lysosomes in the physiology and pathology of cells: contributions of staining methods. In: de Rueck AVS, Cameron MP (eds) Ciba foundation symposium-lysosomes. Little, Brown and Company, Boston
6. Sambrook J, Russell DW (2001) Molecular cloning: a laboratory manual. Cold Spring Harbor Laboratory Press, Cold Spring Harbor
7. Graham RC Jr, Karnovsky MJ (1966) The early stages of absorption of injected horseradish peroxidase in the proximal tubules of mouse kidney: ultrastructural cytochemistry by a new technique. J Histochem Cytochem 14(4):291–302
8. Heuser JE, Reese TS (1973) Evidence for recycling of synaptic vesicle membrane during transmitter release at the frog neuromuscular junction. J Cell Biol 57(2):315–344
9. Ceccarelli B, Hurlbut WP, Mauro A (1972) Depletion of vesicles from frog neuromuscular junctions by prolonged tetanic stimulation. J Cell Biol 54(1):30–38
10. Holtzman E, Freeman AR, Kashner LA (1971) Stimulation-dependent alterations in peroxidase

uptake at lobster neuromuscular junctions. Science 173(998):733–736
11. Teichberg S, Holtzman E, Crain SM et al (1975) Circulation and turnover of synaptic vesicle membrane in cultured fetal mammalian spinal cord neurons. J Cell Biol 67(1):215–230
12. Schacher S, Holtzman E, Hood DC (1976) Synaptic activity of frog retinal photoreceptors. A peroxidase uptake study. J Cell Biol 70(1):178–192
13. Gennaro JF Jr, Nastuk WL, Rutherford DT (1978) Reversible depletion of synaptic vesicles induced by application of high external potassium to the frog neuromuscular junction. J Physiol 280:237–247
14. Maranto AR (1982) Neuronal mapping: a photooxidation reaction makes Lucifer yellow useful for electron microscopy. Science 217(4563):953–955
15. Sandell JH, Masland RH (1988) Photoconversion of some fluorescent markers to a diaminobenzidine product. J Histochem Cytochem 36(5):555–559
16. Deerinck TJ, Martone ME, Lev-Ram V et al (1994) Fluorescence photooxidation with eosin: a method for high resolution immunolocalization and in situ hybridization detection for light and electron microscopy. J Cell Biol 126(4):901–910
17. Denker A, Rizzoli SO (2010) Synaptic vesicle pools: an update. Front Syn Neurosci 2:135. doi:10.3389/fnsyn.2010.00135
18. Opazo F, Rizzoli SO (2010) Studying synaptic vesicle pools using photoconversion of styryl dyes. J Vis Exp 36. http://www.jove.com/details.php?id=1790. doi:10.3791/1790
19. Smith JE, Reese TS (1980) Use of aldehyde fixatives to determine the rate of synaptic transmitter release. J Exp Biol 89:19–29
20. Grabenbauer M, Geerts WJ, Fernadez-Rodriguez J et al (2005) Correlative microscopy and electron tomography of GFP through photooxidation. Nat Methods 2(11):857–862
21. Monosov EZ, Wenzel TJ, Luers GH et al (1996) Labeling of peroxisomes with green fluorescent protein in living P. pastoris cells. J Histochem Cytochem 44(6):581–589
22. Denker A, Krohnert K, Rizzoli SO (2009) Revisiting synaptic vesicle pool localization in the Drosophila neuromuscular junction. J Physiol 587(Pt 12):2919–2926
23. Denker A, Bethani I, Krohnert K et al (2011) A small pool of vesicles maintains synaptic activity in vivo. Proc Natl Acad Sci U S A 108(41):17177–17182
24. Harata N, Ryan TA, Smith SJ et al (2001) Visualizing recycling synaptic vesicles in hippocampal neurons by FM 1-43 photoconversion. Proc Natl Acad Sci U S A 98(22):12748–12753
25. Rizzoli SO, Betz WJ (2004) The structural organization of the readily releasable pool of synaptic vesicles. Science 303(5666):2037–2039
26. de Lange RP, de Roos AD, Borst JG (2003) Two modes of vesicle recycling in the rat calyx of Held. J Neurosci 23(31):10164–10173
27. Darcy KJ, Staras K, Collinson LM et al (2006) Constitutive sharing of recycling synaptic vesicles between presynaptic boutons. Nat Neurosci 9(3):315–321
28. Richards DA, Guatimosim C, Rizzoli SO et al (2003) Synaptic vesicle pools at the frog neuromuscular junction. Neuron 39(3):529–541
29. Hoopmann P, Punge A, Barysch SV et al (2010) Endosomal sorting of readily releasable synaptic vesicles. Proc Natl Acad Sci U S A 107(44):19055–19060
30. Bulina ME, Chudakov DM, Britanova OV et al (2006) A genetically encoded photosensitizer. Nat Biotechnol 24(1):95–99
31. Gaietta G, Deerinck TJ, Adams SR et al (2002) Multicolor and electron microscopic imaging of connexin trafficking. Science 296(5567):503–507
32. Shu X, Lev-Ram V, Deerinck TJ et al (2011) A genetically encoded tag for correlated light and electron microscopy of intact cells, tissues, and organisms. PLoS Biol 9(4):e1001041
33. Micheva KD, Smith SJ (2007) Array tomography: a new tool for imaging the molecular architecture and ultrastructure of neural circuits. Neuron 55(1):25–36
34. Watanabe S, Punge A, Hollopeter G, Willig KI, Hobson RJ, Davis MW, Hell SW, Jorgensen EM (2011) Protein localization in electron micrographs using fluorescence nanoscopy. Nat Methods 8(1):80–84

Chapter 14

Requirements for Samples in Super-Resolution Fluorescence Microscopy

Marko Lampe and Wernher Fouquet

Abstract

The preparation of samples and the choice of appropriate labeling techniques have become instrumental for the development of light microscopy techniques with increasingly high resolution. Both localization microscopy and STED approaches require fluorophores with specific features, including high photostability, specific excitation–emission spectra, and selective switching of single molecules to "on" and "off" state. Additionally, at higher resolutions the limits of conventional immunostaining often become apparent, as clearly exemplified by rather fragmented stainings of continuous cellular components such as microtubules and membranous organelles. Hence, the correct exploitation of fluorescent probes is of crucial importance for successful super-resolution imaging.

Here, the most prominent techniques related to super-resolution imaging are briefly explained, followed by a more detailed technical description of fluorophores and embedding media that are required for such imaging procedures. Some relevant aspects of troubleshooting the preparation of super-resolution samples are included to offer practical support for such experiment.

Key words Sample preparation, STED, GSDIM, Localization microscopy, Immunofluorescence

1 Introduction

Light diffraction-independent fluorescence-imaging methods that have been developed over the past decades are being implemented into a continuously growing number of biological, biomedical, and material research studies and are quickly gaining popularity among the scientific community. These technologies, with an effective resolving power of few tens of nanometers, are seen as a promising way to bridge the gap between regular light microscopy and high resolving imaging procedures such as electron and near-field microscopy [1–3].

The experimental results of microscopy applications do not exclusively depend on the technical capabilities of the microscope used. Some of the factors that contribute to the interpretation of the data include the quality and the density of the staining, the correct choice of the dyes and the selection of appropriate embedding media. In this chapter, we resume some knowledge regarding sample preparation and image acquisition as a brief guide for super-resolution fluorescence imaging techniques.

Fluorescence super-resolution techniques are commonly divided in three categories depending on the photophysical processes and the technology that allow to achieve super-resolution:

- *Stimulated emission depletion microscopy (STED)*: targeted readout of small amounts of reversible switching fluorescent dyes by superimposing two distinct light sources and intensity profiles, which influence the photophysical state of the fluorophores [4, 5].

- *Localization microscopy (including GSDIM, STORM, and PALM)*: procedures related to stochastic readout of single fluorescent molecules followed by mathematical reconstruction of a subdiffraction image on the basis of the determined localization of each molecule [2, 6–8].

- *Structured illumination microscopy (SIM)*: methods that rely on projected light patterns (Moiré patterns) to reveal high-order spatial information followed by Fourier-related transformations to achieve images with subdiffraction resolution [9, 10].

Each one of these techniques has specific and sometimes demanding sample preparation procedures. We will first discuss the common aspects of sample preparation, and a more detailed description of sample preparation for STED and GSDIM/dSTORM will follow. SIM is the least demanding of the techniques, and any (bright) sample prepared for conventional fluorescence confocal imaging should provide a good starting point for SIM imaging.

2 General Remarks

Generating good-quality raw data is extremely important, since additional image processing steps like denoising and deconvolution will rarely increase the amount of information that can be extracted from one image, besides contributing to distort the information that is present [11]. Thus, during experimental design, it is important to carefully consider which kind of information should be gathered with the specific approach that has been chosen. As an example, if small structures need to be segmented and counted, it is important that the signal-to-noise ratio (SNR) allows to reliably separate the desired structures. Alternatively in a time lapse experiment, one has to make sure that the frame rate is high enough to temporally resolve the desired event. SNR might be compromised

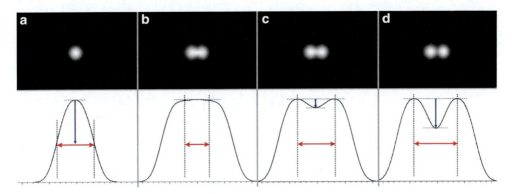

Fig. 1 Different methods for determining the resolution. (**a**) Full-width half-maximum (FWHM) as a measure of resolution (*red arrow*). (**b–d**) Peak to peak measurements (*red arrows*) with different stringencies, depending on the respective location of the Airy disk maxima. (**b**) Sparrow's criterion [17] given by the formation of a plateau between two single PSFs. (**c**) Dawes' limit [18], characterized by 5 % dip-in of the maximum value of the peaks and (**d**) Rayleigh's criterion with a roughly 20 % dip-in which corresponds to the distance at which the first peak falls at the position of the first dark fringe of the second PSF [13, 19]

by insufficient acquisition speed. Thus, dim structural information may be lost in short frame intervals and, vice-versa, fast moving objects will appear blurred or distorted at long exposure times and slow scanning speeds.

2.1 The Microscope

It may sound trivial, but it is important to keep in mind that the practical experience at the microscope allows to get familiar with the setup to understand its limitations by adjusting the imaging parameters and obtaining increasing image quality. Practical experience will also help to troubleshoot a series of technical problems and to recognize artifacts that might be encountered while working with the microscope.

In super-resolution microscopy, the resolution itself is obviously one of the most important parameters to deal with. There are several methods to determine the resolving power of microscopes. Among them a popular approach comprises the analysis of the spatial intensity distribution or point spread function (PSF), which is determined by the diffraction pattern of light of the imaging system (also called Airy disk [12]). This can be overcome by imaging very small fluorescent objects (e.g., small beads, single fluorescent molecules, or even background of antibody stainings) smaller than the resolution of the setup, whose intensity profile depends on the resolving power of the microscope. By convention in this particular case, the resolution (or resolving power of the microscope) is defined as the diameter of the PSF at the half maximum of its peak (full width, half maximum, FWHM, Fig. 1a, [13, 14]). Using the FWHM as a measure of resolution is a straightforward method, and in principle it is easy to be calculated for almost any imaging system. The only requirement for this technique is the size of the structure used for the determination of the PSF (that should be smaller than the microscope's resolution).

Not only a more precise but also more elaborate family of methods takes into account the distance between two structures that are found at a very close proximity. In these cases, the resolving power is defined as either the distance at which the two maxima form a plateau (Sparrow's criterion, Fig. 1b) or the distance between the two maxima at which the saddle (dip-in) between the two peaks reaches a certain fraction of the maxima (Fig. 1c, d). Finding random structures or regions in the imaged samples that intrinsically have the necessary characteristics for this analysis is often difficult since the two maxima need to be equal and at a precise distance. Efforts are currently made to create "nanorulers," e.g., small objects based on self-assembling DNA Origami technology [15], labeled with fluorescent dyes and placed at defined distances with nanometer precision [16].

A closely related issue to the resolving power of the microscope is the identification of artifacts and the reliability of the acquired information. Understanding the imaging artifacts that may arise is crucial as they may influence the scientific conclusions of experimental observations. Artifacts are as diverse as the number of available microscopes. They may be of physical (optical aberrations, mechanic instabilities, thermal drifts, etc.) or theoretical nature (wrong imaging parameters, excessive image processing or the application of the inappropriate statistical analysis). To avoid artifact it is important to:

- Check the performance of the microscope, preferably with a known reference sample that is suited for this purpose with the setup in use.
- Perform the experiments in a controlled environment at a constant temperature to avoid the detrimental effects of thermal drifts on both mechanical and optical components. Good mechanical isolation is of great importance, since the effect of sample movement is more obvious in high-resolution imaging.
- Avoid image manipulations that are not strictly necessary.

2.2 Fluorescence Labeling

Currently, a wide range of fluorescent markers have been developed and commercialized to fulfill most of the requirements of super-resolution techniques. While in SIM microscopy the photophysical properties of dyes are not particularly relevant, provided that the SNR is sufficiently high, some localization microscopy techniques require dyes with specific properties that can be further adjusted by controlling the embedding medium/environment (see specific sections below).

In conventional fluorescence imaging, the fluorescent tag is often associated with an antibody or a high affinity ligand that in turn recognizes the structure of interest, resulting in lower costs and higher flexibility of fluorescent labeling.

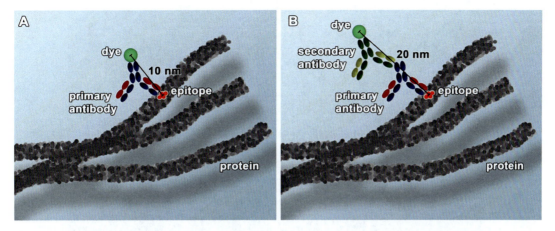

Fig. 2 Distances between epitope and fluorescent dye are relevant in super-resolution. Schematic representation of fluorescent labeling through antibody technology. In the examples, the fluorescent molecules are either directly bound to the primary antibody (**a**) or to the secondary antibody (**b**). Secondary antibodies are regularly used in order to potentiate the fluorescent signal by increasing the amount of antibodies that bind to the same epitope. Inevitably, the distance between the epitope and dye increases with the number of antibodies used

While in diffraction limited imaging this detail has no incisive implication, in super-resolution applications the distance between the fluorescent tag and the actual probe has to be taken into account. Commonly the structure of interest is labeled with a pair of primary and secondary antibodies, where the primary antibody recognizes the target protein and is responsible for the labeling specificity, while the secondary antibody is conjugated with fluorescent dye and binds to the primary antibody (Fig. 2b). It is not trivial to predict the exact distance between the epitope and the dye for all antibody pairs since fluorescent molecules are randomly bound to the entire secondary antibody. In any case, since both antibodies are macromolecules with sizes of ~5–10 nm [20], the distance between the protein of interest and the dye is in a radius of ~20 nm. For this reason the protein of interest can be localized with a precision of ~20 nm, even when the nominal resolving power of the super-resolution microscope is higher. To avoid this loss of resolving precision in super-resolution imaging, it is necessary to use smaller tags, including fluorescent primary antibodies (Fig. 2a), SNAP/CLIP protein tags, fluorescently tagged proteins, F_{ab} fragments [21], and nanobodies [22], which minimize the distance between the structure of interest and the fluorescent label.

2.3 Fluorescent Proteins and Live Imaging

The implementation of fluorescent proteins into life sciences has revolutionized the approach to Biology, allowing to follow in vivo processes in real time [23–25]. Super-resolution live imaging is gaining territory in biological applications, but is still mostly limited to methods that have adequate acquisition times and take advantage of fluorescent proteins. Compared to modern organic

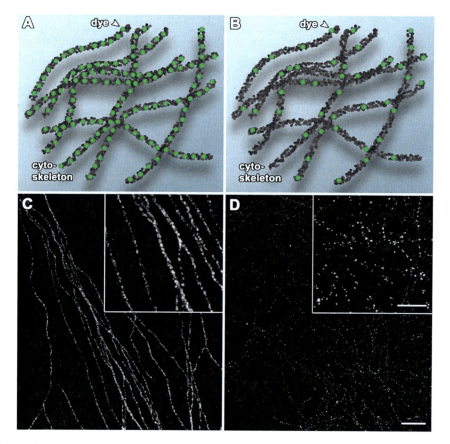

Fig. 3 Effects of low labeling density on super-resolution images. Representation of microtubule fibrils with high (**a**) and low (**b**) labeling density, demonstrating the loss of information due to the lack of fluorescent dyes. The same phenomenon can be observed in real GSDIM images (**c, d**). While in (**c**) the microtubule structure is properly depicted, in (**d**) very little information can be obtained. Scale bars 1 μm and 500 nm. Images acquired with the Leica SR GSD microscope

fluorescent dyes, the genetically encoded markers are commonly less photostable and may require more conservative imaging conditions [25].

2.4 Sample Preparation

A good quality of the sample in super-resolution imaging is important since the resolution itself is limited by label density and distribution (Fig. 3). There is no gold standard for sample preparation, as every specimen has specific requirements. A given fluorescent marker will probably act differently in cultured rodent neurons than in intact *C. elegans* tissue because the environment (oxygen availability, redox state, and buffering agents) has a significant influence on the dyes. Unpredictably, the characteristics of dyes may be advantageous for one method but inadequate for another. For example, long dark states of fluorophores are important for GSDIM but should be avoided in STED.

It is important to understand the requirements of the dyes and labeling procedures for each method and optimize the protocol accordingly. For this, it is advisable to start with a simple established protocol and improve it step by step. A good staining can be defined by the following properties: first, the antibodies (primary and secondary) should label the structure of interest with low cross-reactivity and minimal unspecific binding. This is normally easy to verify when the localization of the protein is known, but may require elaborate biochemical analyses otherwise. Second, the fluorescent signal should be strong enough to produce images with proper SNR. Third and most importantly, the label density must be higher than the microscope's resolution. Using a single fiber as an example (see Fig. 3) if a dye only binds at every 150–200 nm in a wide-field image, the fiber may look continuously stained, but with higher resolution the signal will result to give rise to several small spots. The observer is not able to probe the fiber's integrity between the fluorescent spots. In both conventional and super-resolution microscopy, the label density follows rules depending on the sampling rates that are related to the Nyquist criterion [14]. Adjusting labeling density can be as easy as increasing the antibody concentration in solution or prolonging the incubation. Sometimes, however, antibodies and epitope availability have intrinsic limitations that hamper high density staining. Solving intrinsic label density issues requires a trial and error approach and might be achieved by optimizing fixation and staining procedures.

Once staining procedures have been optimized, it is advisable to follow them as strictly as possible to ensure experiment reproducibility. The preservation of a specific cellular structure is strongly dependent on the fixation method used [26]. Fixation should be optimized for the structure of interest and the antibody used. A common immunofluorescence protocol for cell culture takes advantage of paraformaldehyde (PFA) fixation (see Box 1).

As previously mentioned, the mounting medium influences the dye properties and should be adjusted to the requirements of the specific technique. We describe here five mounting media that are commonly used in super-resolution approaches and can be easily prepared. Some of them have a specific influence on dye photophysics as described below.

Aqueous Embedding Media
- *MEA in PBS*
  PBS containing β-mercaptoethylamine (MEA; Sigma-Aldrich, # 30070) also known as Cysteamine. MEA concentration can be adjusted to keep dyes stay in long-lived dark states. It is used at concentrations ranging from 10 to 100 mM. Diluted MEA solutions should be prepared right before use.

Box 1 Fixation with Paraformaldehyde with Preextraction

Reagents

- PBS
- 2 % Paraformaldehyde (PFA) in PBS
- Triton X-100
- Bovine Serum Albumin (BSA)
- Fetal Bovine Serum (FBS)

Procedure

All steps are performed at room temperature.
1. Aspirate the cell culture medium
2. Rinse coverslips with 3× with PBS
3. Pre-fix the cells with 2 % PFA in PBS for 20 s
4. Rinse 1× with PBS
5. Preextract cells with 0.1 % Triton in PBS for 3 min
6. Rinse 1× with PBS
7. Fix cells with 2 % PFA in PBS for 15 min
8. Rinse 3× times with PBS
9. Wash 3× 5 min with PBS
10. Permeabilize cells with 0.1 % Triton in PBS for 10 min
11. Rinse 3× with PBS
12. Block 1 h at room temperature with 2 % BSA and 2 % FBS in PBS
13. Incubate with primary antibody 1 h at RT or overnight at 4 °C
14. Wash 3× 5 min with PBS
15. Secondary 1 h at RT
16. Wash 3× 5 min

Steps 1–6 are not essential when the staining quality sufficient. Larger specimen and whole mounts may also skip steps 1–6.

Tips

1. Use #1.5 coverslips (0.170 mm thick), as microscope objectives are corrected for this thickness.
2. Adjust incubation times according to sample needs:
   - *Blocking agents*: influences background and specificity of antibodies.
   - *Detergents*: longer incubation (or higher concentrations) may increase dye penetration and epitope accessibility. Excessive exposure may drastically reduce tissue preservation.
   - *Antibodies*: increased antibody exposure increases the penetration depth and staining quality. Increased concentrations may ameliorate not only labeling density but also unspecific background.
   - *Washing*: longer washing periods decrease background levels and unspecific staining.
3. Store samples at 4 °C in PBS, glycerol-based or hardening mounting medium, avoiding light exposure. Note that sample quality might degrade over time. For longer periods specimen can also be stored at −20 °C, if kept in glycerol-based mounting media. In order to avoid sample degradation, it is advisable to image samples within 24 h.
4. Washing steps and antibody incubation times might be prolonged, according to the sample's thickness and amount of background.

Alternatively, stock solutions can be prepared and stored at −20 °C and thawed right before use. Frozen aliquots can be stored for several weeks. Molecules containing thiols, such as MEA, are frequently used for localization microscopy related to the GSDIM and dSTORM techniques [7, 27].

- *Glucose-oxidase mix (GLOX)*
  PBS containing 10 % (w/v) glucose, 0.5 mg/mL glucose oxidase (e.g., Sigma-Aldrich G2133), and 40 µg/mL catalase (e.g., Sigma-Aldrich C1345) adjusted to pH 7.4. It is possible to prepare a glucose stock solution and mix the reagents freshly before mounting the coverslip. As for MEA, GLOX is widely used to modulate the duration of dark states and the two reagents can be combined (see [27] for details). The combination of MEA- with GLOX-Buffer is highly recommended and often provides superior results than other aqueous buffers alone. Concentrations of the buffer components should be carefully adjusted. Further information can be found in [28, 29].

- *Glycerol*
  By combining different amounts of water (or PBS) and glycerol, the refractive index (RI) can be precisely adjusted between 1.33 and 1.47. Additionally, this mounting medium is easy to prepare, suitable for long sample storage, and works well with some GSDIM, STED, and SIM dyes.

Hardening Embedding Media

- *Polyvinyl alcohol*
  Solution of 1 % polyvinyl alcohol (PVA; MW 25,000, 88 mol% hydrolyzed; Polysciences, #02975-500) can be prepared in PBS (adjusted to pH 7.4). A 1–2 µm thick film is applied, normally with the aid of a spin coater device [7]. After the evaporation of the water, the sample can be stored for several months. This medium is commonly used for certain dyes not only in localization microscopy (oxygen deprivation) but may also work well in STED and SIM. This embedding medium requires rather thin samples and, as for all media that have a refractive index (RI) similar to glass, they are not suited for TIRF.

- *Mowiol*
  A working solution can be easily prepared with 6 g of glycerol (analytical grade), 2.4 g of Mowiol powder (Calbiochem # 475904), 6 mL of double distilled water, and 12 mL of 0.2 M Tris buffer with pH 8. The solution should be stirred for approximately 4 h and rest for 2 h. Before storage Mowiol should be warmed for 10 min at 50 °C (water bath) and centrifuged for 15 min at $5,000 \times g$. The supernatant can be kept at −20 °C for storage. Mowiol is the medium of choice for STED images, and also works with some GSDIM dyes and SIM.

Common antifades, e.g., diazabicyclo-octane (DABCO) or *n*-propyl gallate (NPG) [30, 31], might cause significant changes to the photophysical properties of dyes and can be used in STED and SIM imaging. Antifading reagents can influence the photophysical behavior of dyes for GSDIM dramatically and often decrease image quality. Therefore, antifading reagents should not be used for GSDIM without being carefully tested.

In three-dimensional acquisition approaches, the use of embedding media and clearing solutions to adjust the refractive index to the immersion medium and to the objective will contribute to increase tissue penetration and ameliorate image quality [32].

3 STED

Under the name of *reversible saturable optical fluorescence transitions* (RESOLFT) [33] are grouped some of the most prominent high-resolution techniques that are used to circumvent the light diffraction barrier. In these applications, fluorophores are transitioned between emitting "on" and not emitting "off" states by irradiating them with light of a given wavelength. In the targeted readout mode, images are obtained by scanning the focal spot across the sample while overlaying two PSFs, one wavelength (focal spot) exciting the dye and one (doughnut) for fluorescence emission depletion. Typically, lateral resolutions around 40–90 nm are obtained. The axial resolution is dependent on the method applied (special PSF, or in combination with 4Pi) and may reach 50 nm. The nature of the "on/off" transitions can be achieved by different approaches including stimulated emission (STED, [4]), ground-state depletion (GSD, [34]), as well as by other conformational molecular changes, or other methods in which fluorophores have different behavior depending on the light irradiation method [35]. Slowly switching molecules, like photoactivatable proteins (Dronpa and Padron, [33, 36]), require low light intensities for switching, but longer acquisition times, as this method relies on a high number of switching cycles for image acquisition. Fast processes, such as STED, need correspondingly stronger light sources, but imaging is also significantly faster [33]. Because of the great popularity of STED amongst other scanning RESOLFT techniques, this chapter will be mainly concerned in presenting applicative knowledge for this method.

3.1 Available Fluorophores

In STED microscopy, the perfect matching between fluorescent markers and excitation and depletion laser lines is of crucial importance. The process of stimulated emission is most effective at the spectral range of the fluorophore's emission maximum [37]. Unfortunately, irradiating the marker at such wavelengths often leads to additional anti-Stoke's excitation (commonly referred to as reexcitation), which competes with simulated emission process and

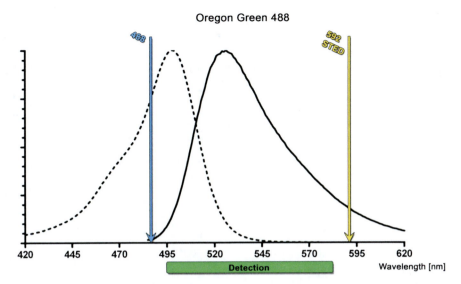

Fig. 4 Requirements of a dye for STED microscopy. Absorption (*dotted line*) and emission (*continuous line*) spectrum of Oregon Green 488. Excitation line (*blue arrow*) and depletion line (*yellow arrow*) are selected in such a way that the strong 592 nm irradiation cannot excite the dye (anti-Stoke's). The excitation spectrum is narrow enough in order not to absorb the depletion beam, but there is a sufficiently high probability for stimulated emission to occur at 592 nm and enough space for fluorescence detection (*green bar*)

thus spoils the resolution improvement. In order to avoid anti-Stoke's excitation by the depletion beam, a configuration is needed in which the STED light still lays inside the emission spectrum of the dye, without reexciting it. This can be achieved by choosing fluorophores with narrow excitation spectra, with long "tails" in the emission spectrum, or with long Stoke's shifts (e.g., Oregon Green 488, see Fig. 4). Moreover, the ideal dye should feature high brightness and photostability under the given imaging conditions.

These limitations may sound extremely stringent at first, but there are already numerous dyes that were found to perform properly (see Table 1). The list of STED compatible fluorescent markers is not restricted to organic dyes. A number of genetically encoded fluorescent proteins have been reported to work with commercially available STED systems (Table 2).

Recently, much effort has been invested to discover and develop new fluorescent markers that show a significantly increased performance. Some promising, but yet not available solutions were reported with fluorogen-activating proteins (FAP) [38] and nitrogen vacancies in diamonds [39].

Describing the localization of a single protein/marker is insufficient for a wide number of imaging experiments in life sciences. There is a growing demand for multicolor imaging beyond light diffraction, which allows the investigation of biomolecular arrangements on nanometer scale. For two-color (2C) STED, the rules for conventional fluorescence microscopy apply. Ideally, to acquire

Table 1
Fluorophores reported to work in STED microscopy

Fluorophore	Excitation (nm)	Depletion (nm)	Provider
ATTO 425	440	532	ATTO-TEC
BD Horizon V500[a]	458	592	Becton and Dickinson
Cascade Yellow[b]	458	592	Life Technologies
Pacific Orange[b]	458	592	Life Technologies
Abberior STAR 440SX[a]	458	592	Abberior
ATTO 488[b]	488	592	ATTO-TEC
Abberior STAR 488[b]	488	592	Abberior
Alexa Fluor 488[b]	488	592	Life Technologies
Chromeo 488[b]	488	592	Active Motif
Chromeo 505[a]	488	592	Active Motif
Oregon Green 488[a]	488	592	Life Technologies
FITC[b]	488	592	Life Technologies
DY-495[b]	488	592	Dyomics
DyLight 488[b]	488	592	Pierce Technology
ATTO 532	488	600	ATTO-TEC
Abberior CAGE 500	500	600	Abberior
Abberior STAR 512	512	605	Abberior
Oregon Green 514	514	592	Life Technologies
Alexa Fluor 532	514	660	Life Technologies
ATTO 565	532	640	ATTO-TEC
Abberior STAR 470SX[b]	532	750	Abberior
Chromeo 494[a]	532	750	Active Motif
Mega 520[a]	532	750	Sigma Aldrich
Abberior CAGE 552	552	655	Abberior
ATTO 590	570	690	ATTO-TEC
Alexa Fluor 594[b]	570	690	Life Technologies
ATTO 594	570	690	ATTO-TEC
DyLight 594	570	690	Pierce Technology
ATTO 633	630	735	ATTO-TEC
ATTO 647N[a]	635	750	ATTO-TEC
Alexa Fluor 647	635	750	Life Technologies
ATTO 655[b]	635	760	ATTO-TEC
Abberior STAR 635[a]	640	750	Abberior
ATTO 665[b]	640	775	ATTO-TEC

Values are based on reported experiments and published works (see http://www.4pi.de)
[a]Recommended dyes for commercial setups
[b]Dyes that were reported to work with commercial STED microscopes

Table 2
Fluorescent proteins reported to work with commercially available STED microscopes

Fluorescent proteins	Excitation (nm)	Depletion (nm)
eGFP	488	575
Citrine	488	592
EmGFP	488	592
eYFP	514	592
Venus	514	592

Table 3
Selected 2C STED dye combinations for systems using 592-nm depletion lasers

Continuous wave; excitation: 458 and 514 nm; depletion: 592 nm					
Dye1			**Dye2**		
Name	**Excitation**	**Emission**	**Name**	**Excitation**	**Emission**
BD Horizon V500	458	460–510	Oregon Green 488/Chromeo 505	514	520–580
Abberior STAR 440SX	458	465–515	Oregon Green 488/Chromeo 505	514	520–580

images in two colors, fluorophores should have distinct excitation/emission spectra. To date, there are several ways to realize 2C STED microscopy. To reduce costs and optimize efficiency most commonly, a single depletion wavelength is used for multicolor experiments. This is achieved in practice by combining a classic dye with standard Stoke's shift with a dye showing a large Stoke's shift (see Tables 3 and 4). In this case, the fluorophores must show similar STED performances for the depletion wavelength of choice, which means that the two fluorophores feature partially overlapping emission spectra (see Fig. 5). This simple STED setup avoids the bleaching of fluorophores with strong depleting light at improper wavelengths and allows several consecutive acquisitions of the same field of view [40].

There are also ways to combine two depleting lasers, avoiding cross-interference between spectra [41, 42]. In this approaches, the exciting and depleting wavelengths are optimized for the fluorophores, which increases the overall performance, but renders the microscope more complex.

Table 4
Selected 2C STED dye combinations for systems using multiphoton titanium–sapphire laser for fluorescence depletion

Pulsed; excitation: 531 and 640 nm; depletion: 745–775 nm					
Dye1			**Dye2**		
Name	Excitation	Emission	Name	Excitation	Emission
Chromeo 494	531	550–605	ATTO 647N	640	665–705
Alexa 532	531	550–605	Abberior STAR 635	640	665–705
Mega 520	531	665–705	ATTO 665	640	665–705

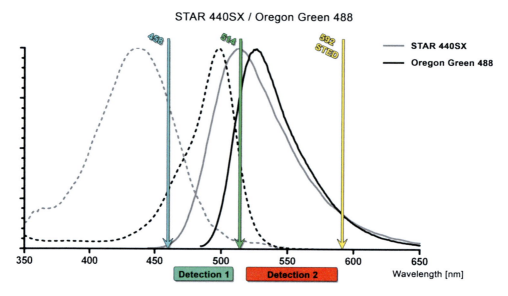

Fig. 5 Example of fluorophores with good separated excitation peaks and partial overlapping emission spectra suited for 2C STED. Excitation of STAR 440SX at 458 nm (*blue arrow*), and Oregon Green 488 at 514 nm (*green arrow*). The *yellow arrow* depicts the 592-nm depletion line. Signal acquisition in green (Detection 1) and red (Detection 2) requires to be sequential to avoid cross-talk between channels. Suboptimal excitation (at 514 nm) further helps to avoid cross-talk, as anti-Stokes excitation is very weak for STAR 440SX at 514 nm. Samples with single labels would be further recommended as controls

One example for 2C STED is the combination of Oregon Green 488 and STAR 440SX. By exciting Oregon Green 488 with 514 nm, STAR 440SX is not cross-excited and thus the signal of Oregon Green 488 can be separated from STAR 440SX. On the other hand side, almost no Oregon Green 488 fluorescence falls into the STAR 440SX detection window (Fig. 5). Images can be acquired with minimal crosstalk and therefore in most cases no postprocessing is required.

Dye Selection for Dual-Color STED

In general the optimal STED dye features:

- Excitability with the available lines
- Some emission at depletion wavelengths (high STED efficiency)
- Appropriate fluorescence lifetimes
- Very little anti-Stoke's excitation at the depletion wavelength (if any)
- High quantum yield
- High photostability under STED conditions

Dye pairs for dual-color STED should further show:

- Little cross-excitation
- Differences in the emission spectra
- Good STED efficiency at the same wavelength
- No significant absorption (bleaching) at different STED wavelengths

The performance of fluorophores for STED is mainly determined by the *saturation intensity* (I_{SAT}) [43], which describes the light intensity needed to completely saturate stimulated emission and thus quench all fluorescence. I_{SAT} can be obtained by plotting the dye intensity against the depleting laser power. I_{SAT} depends on a number of parameters including the depletion wavelength and the fluorescence lifetime. How strongly the lifetime influences I_{SAT} depends on how the depleting light is presented to the molecule through either continuous or pulsed laser sources. Pulsed lasers concentrate all energy into very small intervals and therefore achieve very high peak powers, compared to the nominal average light intensity. Highest STED efficiency is obtained by carefully timing excitation and depletion pulses. With continuous wave (CW) laser sources, the energy is evenly spread over time and the peak power equals the mean power. Especially, in CW the fluorophore's lifetime (in the range of a few nanoseconds) plays an important role. It is less likely for fluorescent molecules with short lifetimes to be exposed to enough photons while being in an excited state than for molecules with longer lifetimes. In pulsed systems, because of the brevity of the pulse, the majority of photons are presented to the fluorophore while it is in an excited state [44, 45].

3.2 Sample Preparation/ Mounting Medium

STED microscopy is fully compatible with standard staining and fixation protocols, but needs fluorophores suited for this method (see above). Staying as close as possible to conventional sample preparation protocols will most likely give the best results. Intuitively, the best results are obtained from bright samples with high contrast. As STED is a technology based on confocal

Table 5
Mounting media reported to work with STED microscopy

Mounting media	RI	Remarks
TDE	1.514	
ProLong Gold	1.46	Hardening
Mowiol ± 2.5 % DABCO	1.5	Hardening
86 % glycerol + 4 % NPG	1.452	
86 % glycerol + 2.5 % DABCO	1.452	

microscopy, any sample optimized for confocal microscopy approaches is expected to give good results in STED. Due to the inherent optical sectioning capabilities of confocal microscopes relatively thick preparations can be investigated. Moreover, the generation of special STED PSFs and the implementation of the 4Pi principle further increase the axial resolution [46].

It is of cardinal importance to check if the sample is absorbing the depleting light since this might hamper the emission depletion process. Some tissues can be problematic, such as chloroplasts, pigmented cells in zebrafish, and blood vessels in rodent brain slices. Such issues might be overcome, by applying specific clearing procedures, by using different mounting media or genetically modified animals, as in the case of pigment-free zebrafish [47].

STED has few limitations regarding mounting media, mainly related to the absorbance at the STED depletion wavelength. When the mounting medium absorbs in the depletion wavelength, the heat produced can destroy the sample. Additionally, refractive index matching helps when deep sample penetration is required. Table 5 resumes the embedding media that were found to work well with STED. Note that some embedding media may affect the performance of some fluorophores. Thiodiglycol (TDE), for example, is not compatible with a series of blue/green dyes and with some fluorescent proteins (for a list of dyes working efficiently in TDE see [48]).

3.3 Tips and Tricks

Since stimulation emission is a very fast switching process, high densities of photons are required to make sure that every excited fluorophore is "emission depleted." As long as these strong light intensities are not absorbed, no harm is done to the sample. Interestingly, several fluorophores bleach easily when simultaneously exposed to regular values of exciting and depleting light, but do not suffer photodamage when exposed to the same light intensities separately. One assumption is that dyes, as soon they reach higher energetic states (singlet and triplet), are prone to absorb the strong STED light, which leads to irreversible bleaching. This is avoided if molecules have enough time to relax back to the ground state.

A pulsed excitation followed by depletion is hence a better choice for conserving fluorescence, although acquisition speed can be faster with CW depletion lasers. Another way to give the necessary time for fluorophore relaxation is to increase the scan speed and decrease the pixel dwell time of the focal spot. Practically, it is more advantageous for many fluorophores to acquire the same amount of photons over several fast scans (by averaging or adding up the images), than just over one slow scan.

As already mentioned, 2C STED setups with one dye pair and a single depletion laser come at the compromise of a partial overlap of the emission spectra of the two dyes. Signal detection is very similar to regular fluorescent imaging with two dyes of similar fluorescence wavelength range. In order to maximize spectral separation, it might be useful to acquire images sequentially, while using detection ranges that are advantageous for the corresponding dye and even tolerating the choice of suboptimal excitation lines (see Fig. 5) to reduce cross excitation.

Another way to prevent unnecessary molecular stress and to preserve the specimen is to adapt the laser intensity according to the required resolution. If the desired information can be extracted at a lower resolution, it is not useful to work with full laser power. By reducing the depleting light intensities, the samples will get brighter, less excitation power will be needed, and more images can be taken from the same field of view. This is especially important for living samples.

4 Localization Microscopy

GSDIM, STORM, PALM, and affiliated techniques achieve super-resolution by spatiotemporal reconstruction of single molecule localization [2, 6–8]. These methods take advantage of the ability of fluorophores to remain in two distinct molecular states: an excitable (on) state, and a dark (off) state. Different from STED, where fluorescence is detected by a scanning microscope, localization microscopy determines the position of photon bursts derived from a single fluorophore. The system uses a camera and therefore can record randomly occurring events with a large field of view. To assemble a high-resolution image, the majority of fluorophores inside this field of view need to be in a dark state, while a small number of molecules spontaneously (or induced, depending on the technique) switches to an "on" state. These single molecule photon bursts (often termed "events") are localized and their coordinates are integrated into a matrix. Subsequently, the emitting molecules are switched back to a nonfluorescent state. This procedure is repeated several thousand times and all the coordinates of the molecules are placed into the same matrix, which finally represents the acquired super-resolution image. Determining

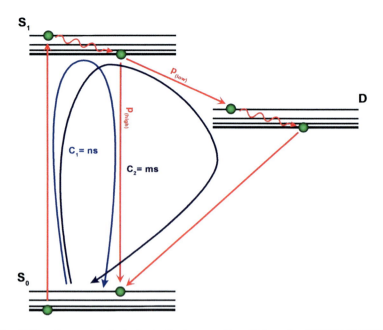

Fig. 6 Requirements of photophysical properties of dyes for GSDIM/dSTORM. Simplified Jablonski diagram showing distinct energetic states of a fluorophore in GSDIM/dSTORM. After being driven from the ground state (S_0) to the excited state (S_1), the fluorophore has two options: either to relax back to the ground state by releasing a photon (cycle 1, C_1) or via a nonradiative (no fluorescence emission) relaxation over a long-lived dark state D (cycle 2, C_2). Suitable properties for good dyes would consist of (1) very fast C_1 cycles (nanoseconds) and rather slow C_2 cycles (milliseconds to seconds); (2) as little irreversible photobleaching as possible, and (3) rather low probabilities for the dye to pass into the dark state. Short C_1 cycles, good photostability, and low probabilities to dark state conversion lead to events that are fast and bright (many photons are emitted, before the molecule is turned off), whereas very long dark states ensure that the majority of fluorescent molecules are kept in the "off" state

the center of a PSF of single emitting molecules, and consequently the localization of molecules, can be done with up to single digit nanometer precision, leading to resolutions in the range of 20 nm laterally [2, 6–8] and 50 nm axially [49]. The localization precision for each molecule depends on the number of photons detected during the acquisition [50].

4.1 Available Fluorophores

The requirements for fluorophores in localization microscopy can be summarized into three basic prerequisites (see Fig. 6). First, molecules need to be switched between emitting and nonemitting states by influence of (continuous) illumination with at least one light source of a given wavelength. Second, the equilibrium between the "on" and the "off" state should be shifted towards the "off" state. For this, the majority of the fluorescent molecules need to naturally stay in or be driven into rather long-lived dark states to enable single molecule detection and temporal reconstruction of the field of view. Third, when molecules are in the "on" state, they should emit as many photons as possible in a short timeframe ("burst") to avoid spreading the signal over multiple frames since this reduces the SNR (see Fig. 6). This delicate balance between "on" and "off" states is mainly determined by illumination intensity and environmental factors in direct proximity to the fluorophore in the case of dSTORM/GSDIM [7, 28].

Ideally, the number of events per frame should be high enough to minimize acquisition time and low enough to still detect nonoverlapping single molecules. Achieving the right number of events per frame is comparably easy when using photoswitchable molecules (PALM) [2], which require near-UV light (e.g., 405 nm) to be switched to an "on" conformation. The number of events can be increased or decreased by modulating the 405 nm light. For PALM, the environment has only a mild influence on the switching rate of the fluorophores. In dSTORM and GSDIM [7, 28], which make use of long-lived dark states to keep the majority of fluorophores from emitting light, the surrounding medium is of vital importance for maintaining the correct number of events per frame. In fact, many additives are introduced into the medium (such as oxygen scavengers, thiol-containing solutions, and antioxidants) to keep fluorescent molecules in long-lived dark states (see Table 6) [7, 27, 28, 51].

The benefit from using the more elaborate procedure for GSDIM/dSTORM compared to PALM is in general a higher photon count per event and therefore a higher localization precision. Furthermore, by using an antibody sandwich, more fluorophores are brought to structure of interest leading to a higher staining density with fluorophores. Both measures can significantly improve structure visualization and enhance the practically obtained resolution.

The composition of the embedding medium, the interaction between the sample and the molecular microenvironment has a huge impact on switching rates. Dyes can also display varying performance depending on the age of the sample, e.g., fluorophores that have a good performance in freshly mounted samples, may not look as good after months, weeks, or even hours. Finding the proper environment for each dye and combining several dyes for multicolor imaging is sometimes challenging and requires some optimization.

In Table 6, we provide some combinations of dyes and embedding that were reported to have a good performance.

Performing multicolor imaging can be addressed in different ways. Depending on the inherent setup that is used, the different color paths either need to be split on different camera chips or acquired sequentially. For simultaneous detection, the emission of different fluorophores is normally separated by dichroic mirrors.

Table 6
Relation of suitable dye/embedding medium combinations for single and multicolor imaging

Laser	Fluorophore	Aqueous media		Hardening media	
		MEA in PBS	Glucose-oxidase	PVA	Mowiol
488	Alexa Fluor 488	+	+	+	
	ATTO 488	+	+	+	+
	Chromeo 488	+		+	
	Oregon Green 488	+		+	
	Chromeo 505	+			+
532	ATTO 520	+		+	
	Alexa Fluor 532	+	+	+	+
	ATTO 532		+	+	
	Alexa Fluor 546			+	
	Chromeo 546			+	
	Rhodamine 6G	+	+	+	+
	ATTO 565		+	+	
	Alexa Fluor 568	+	+		
642	ATTO 633	+			
	Alexa Fluor 647	+			
	ATTO 647N	+			+
	ATTO 655	+	+		+
	Alexa Fluor 680	+	+		
	Alexa Fluor 700	+			+

Summary of current experience, tested by the authors (Giulio Simonutti, personal communication) and extracted from published reports [27]. These are empirical results and can serve as a first guidance. Dye/medium combinations that are more likely to work are displayed with "+." For more details regarding quantifications of dyes, please refer to Zhuang et al. Protocols for mounting media can be found in Sect. 2 of this chapter or in [52]

Thus, as in regular fluorescence microscopy, the most straightforward approach would be to choose dyes with distinct excitation wavelengths that can be completely separated over the excitation/emission spectrum (e.g., ATTO 488 and Alexa Fluor 647, or Alexa Fluor 532 and Alexa Fluor 647). Another, more elegant way of addressing multicolor imaging benefits from the use of single molecule detection. Assuming that single molecules of a certain type have a singular spectral signature, the nature of the molecule can be precisely determined by the ratio of photons detected on each side of the dichroic mirror. Both channels are finally added back to one single event and attributed to the dye of the corresponding signature. In this manner, dyes that are spectrally close and excited by the same light source can be distinguished (e.g., Alexa Fluor 488, ATTO 532, and Cy3) [53]. Table 6 can be used as a guide for the choice of a common mounting medium when multicolor imaging is required.

Performing live-imaging experiments with localization microscopy is rather complex. There is a wide range of fluorescent proteins that work with localization microscopy, but the long acquisition time typically necessary to have super-resolution localization of molecules (around 2–10 min) is incompatible with the time course of most of the biological processes of interests. For this reason at present, the acquisition of super-resolution images may be limited to structures showing slow dynamics [54]. Furthermore, the increased exposure to light compared to conventional microscopy might exert strong phototoxic effects and should be considered accordingly. Single particle tracking (PAINT and uPAINT), on the other hand, can be effectively achieved [55].

4.2 Sample Preparation/Mounting Medium

At resolutions typically around 30 nm, high labeling density is necessary to achieve a continuous representation of the labeled structure in localization microscopy. The optimization of the immunostaining should always start with a rather simple protocol and should be performed as in regular fluorescence microscopy, following the recommendations of Sect. 2. We have chosen a standard immunofluorescent staining, in which the structure of interest is marked with a regular antibody sandwich, to describe some common issues related to this super-resolution method (see Figs. 2 and 3 and "tips and tricks" below). It is important to note that since localization microscopy can reach the highest resolutions among super-resolution techniques, it is also the method in which poor staining quality can have the strongest effects. As previously discussed, the use of fluorescently labeled primary antibodies and F_{ab} fragments increases the localization accuracy of the fluorescent signal with respect to the structure of interest. In general, the signals obtained from directly labeled antibodies and F_{ab} fragments are somewhat weaker. For this reason, a compromise between number of localized events and the structure representation should be achieved, as both parameters contribute to the final image quality.

The majority of localization microscopes are based on TIRF technology. Therefore, almost every setup can automatically benefit from the greatly improved axial resolution derived by the limited excitation triggered by TIRF, since only fluorescent molecules that are in close vicinity to the coverglass are excited (100–250 nm) [56]. However, the TIRF mode can only be used with samples that are imaged in aqueous media, as the RI mismatch is needed for creating the evanescent fields. Two of the most frequently used aqueous mounting media (MEA and GLOX, see Table 6) are described in Sect. 2. For aqueous solution, it is advisable to seal off the imaging chamber to avoid gas (oxygen) exchange in order to stabilize the desired dye characteristics.

Since sample deterioration can occur quickly (hours to weeks) in aqueous media, Glycerol as well as hardening media like PVA and Mowiol (see Sect. 2) could be used as an alternative to preserve the samples for significantly longer periods.

4.3 Tips and Tricks

The most important aspect of stochastic single molecule localization based images is to understand the dynamics of the blinking events and adjust the imaging parameters accordingly, in order to avoid artifacts and obtain high-quality images.

A discontinuous representation of a structure, as shown in Fig. 3, might be caused by either low labeling density or by poor performance of the fluorophore. Therefore, it is important to check the recommended combination of dye/embedding medium and make sure that the dye is performing as previously recorded. If the event number decreases more quickly than expected, this may be a hint of poor dye performance. Preparing new solution or changing the embedding media might solve the problem. The staining should always show a good SNR. Low SNR might be due to poor labeling density. Increasing the antibody concentration and adjusting the fixation procedures might help in this case. Hardening embedding media, like Mowiol need enough time to cure, as dyes may perform differently if the embedding medium is not dry.

Freely diffusing dyes that increase the unspecific background should be limited. Filtering all used solutions might help reducing the background as well. During our tests some antibodies showed strong background noise only in aqueous media, which implies that the dye is disassociating from the antibody during image acquisition. The super-resolution image should reflect the overall structure observed in the wide-field image. Artifacts might occur if the registered localizations are based on two or more simultaneous events. In this case, one could adjust the laser power and embedding solutions accordingly. So-called "hot pixels" appear with certain frequency and might be either due to high background, or molecules that cannot be bleached, nor driven into "off" states. These events may be removed from the event list later, but the detection of other events in direct proximity may be compromised.

As image acquisition in localization microscopy takes long times, samples might undergo drift. Movements may be seen as smeared structures (towards the same direction) that otherwise should be sharp. These movements are very often derived from temperature fluctuations and vibrations of the room. To avoid this, the microscope should not be exposed to direct airflow (e.g., open doors and air-conditioning), the sample should be properly fixed on the sample holder and the vibration dampers should work correctly. Fiducial markers may additionally help to align the image during image reconstruction and processing. The proper attachment of the fiducial markers to a given sample is important since their movement should only reflect drift and no other motions (e.g., the displacement of a membrane to which they are attached). Furthermore, using fiducials often requires additional steps during sample preparation and further optimization of fiducial attachment and concentration.

5 Conclusion

Super-resolution cellular imaging is rapidly becoming an essential area of modern cellular studies. Obtaining images that reliably reflect the microarchitecture of cells at the macromolecular level require optimization of sample preparation procedures. In this chapter, we have discussed a possible workflow for the optimization of these procedures including the choice of the appropriate fixation methods, labeling tools, mounting media, and possible dye combinations for multicolor imaging. It should be emphasized that super-resolution microscopy also requires particular attention to image processing and quantification. For example, while in confocal microscopy colocalization studies were performed by overlaying both signals and measuring the overlap between the two channels, the same staining may not show any type of overlap at higher resolutions. With less/no overlap between channels, the proximity relation between the now much smaller structures starts to play a bigger role. A careful combination of experimental procedures, super-resolution microscopy techniques, and image analysis tools is essential for further implementing super-resolution microscopy in cell biology research.

References

1. Watanabe S, Punge A, Hollopeter G et al (2011) Protein localization in electron micrographs using fluorescence nanoscopy. Nat Methods 8:80–84. doi:10.1038/nmeth.1537
2. Betzig E, Patterson GH, Sougrat R et al (2006) Imaging intracellular fluorescent proteins at nanometer resolution. Science 313:1642–1645. doi:10.1126/science.1127344
3. Schermelleh L, Heintzmann R, Leonhardt H (2010) A guide to super-resolution fluorescence microscopy. J Cell Biol 190:165–175. doi:10.1083/jcb.201002018
4. Hell SW, Wichmann J (1994) Breaking the diffraction resolution limit by stimulated emission: stimulated-emission-depletion fluorescence microscopy. Opt Lett 19:780–782
5. Dyba M, Jakobs S, Hell SW (2003) Immunofluorescence stimulated emission depletion microscopy. Nat Biotechnol 21:1303–1304. doi:10.1038/nbt897
6. Hess ST, Girirajan TPK, Mason MD (2006) Ultra-high resolution imaging by fluorescence photoactivation localization microscopy. Biophys J 91:4258–4272. doi:10.1529/biophysj.106.091116
7. Fölling J, Bossi M, Bock H et al (2008) Fluorescence nanoscopy by ground-state depletion and single-molecule return. Nat Methods 5:943–945. doi:10.1038/nmeth.1257
8. Rust MJ, Bates M, Zhuang X (2006) Sub-diffraction-limit imaging by stochastic optical reconstruction microscopy (STORM). Nat Methods 3:793–795. doi:10.1038/nmeth929
9. Heintzmann R (1999) Laterally modulated excitation microscopy: improvement of resolution by using a diffraction grating. Proc SPIE 3568:185–196
10. Gustafsson MG (2000) Surpassing the lateral resolution limit by a factor of two using structured illumination microscopy. J Microsc 198:82–87
11. Swedlow JR (2003) Quantitative fluorescence microscopy and image deconvolution. Methods Cell Biol 72:349–367
12. Airy GB (1835) On the diffraction of an object-glass with circular aperture. Trans Camb Philos Soc 5:283–291
13. Born M, Wolf E, Bhatia AB (1999) Principles of optics: electromagnetic theory of propagation, interference and diffraction of light, seventh edition, ISBN: 978-0-521-64222-4 (Print), page 436–443, Cambridge University Press
14. Pawley JB (1995) Handbook of biological confocal microscopy, second edition, ISBN: 978-0-387-25921-5 (Print) 978-0-387-45524-2 (Online), page 128–129, Springer Science+Business Media, Inc.

15. Barish RD, Schulman R, Rothemund PWK, Winfree E (2009) An information-bearing seed for nucleating algorithmic self-assembly. Proc Natl Acad Sci U S A 106:6054–6059. doi:10.1073/pnas.0808736106
16. Steinhauer C, Jungmann R, Sobey TL et al (2009) DNA origami as a nanoscopic ruler for super-resolution microscopy. Angew Chem Int Ed Engl 48:8870–8873. doi:10.1002/anie.200903308
17. Sparrow CM (1916) On spectroscopic resolving power. Astrophys J 44:76–86
18. Dawes WR (1867) Catalogue of micrometrical measurements of double stars. Mon Not R Astron Soc 27:217–238
19. Barakat R (1965) Rayleigh wavefront criterion. J Opt Soc Am 55:572–573
20. Ban N, Escobar C, Garcia R et al (1994) Crystal structure of an idiotype-anti-idiotype Fab complex. Proc Natl Acad Sci U S A 91:1604–1608
21. Brown JK, Pemberton AD, Wright SH, Miller HRP (2004) Primary antibody-Fab fragment complexes: a flexible alternative to traditional direct and indirect immunolabeling techniques. J Histochem Cytochem 52:1219–1230. doi:10.1369/jhc.3A6200.2004
22. Ries J, Kaplan C, Platonova E et al (2012) A simple, versatile method for GFP-based super-resolution microscopy via nanobodies. Nat Methods 9:582–584. doi:10.1038/nmeth.1991
23. Westphal V, Rizzoli SO, Lauterbach MA et al (2008) Video-rate far-field optical nanoscopy dissects synaptic vesicle movement. Science 320:246–249. doi:10.1126/science.1154228
24. Rankin BR, Hell SW (2009) STED microscopy with a MHz pulsed stimulated-Raman-scattering source. Opt Express 17:15679–15684
25. Nägerl UV, Bonhoeffer T (2010) Imaging living synapses at the nanoscale by STED microscopy. J Neurosci 30:9341–9346. doi:10.1523/JNEUROSCI.0990-10.2010
26. Wheatley SP, Wang YL (1998) Indirect immunofluorescence microscopy in cultured cells. Methods Cell Biol 57:313–332
27. Dempsey GT, Vaughan JC, Chen KH et al (2011) Evaluation of fluorophores for optimal performance in localization-based super-resolution imaging. Nat Methods 1–14. doi:10.1038/nmeth.1768
28. Heilemann M, van de Linde S, Schüttpelz M et al (2008) Subdiffraction-resolution fluorescence imaging with conventional fluorescent probes. Angew Chem Int Ed Engl 47:6172–6176. doi:10.1002/anie.200802376
29. Van De Linde S, Endesfelder U, Mukherjee A et al (2009) Multicolor photoswitching microscopy for subdiffraction-resolution fluorescence imaging. Photochem Photobiol Sci Off J Eur Photochem Assoc Eur Soc Photobiol 8:465–469
30. Johnson GD, Davidson RS, McNamee KC et al (1982) Fading of immunofluorescence during microscopy: a study of the phenomenon and its remedy. J Immunol Methods 55:231–242. doi:10.1016/0022-1759(82)90035-7
31. Ono M, Murakami T, Kudo A et al (2001) Quantitative comparison of anti-fading mounting media for confocal laser scanning microscopy. J Histochem Cytochem 49:305–312
32. Diaspro A, Federici F, Robello M (2002) Influence of refractive-index mismatch in high-resolution three-dimensional confocal microscopy. Appl Opt 41:685–690
33. Hofmann M, Eggeling C, Jakobs S, Hell SW (2005) Breaking the diffraction barrier in fluorescence microscopy at low light intensities by using reversibly photoswitchable proteins. Proc Natl Acad Sci U S A 102:17565–17569. doi:10.1073/pnas.0506010102
34. Hell SW, Kroug M (1995) Ground-state-depletion fluorscence microscopy: a concept for breaking the diffraction resolution limit. Appl Phys B Lasers Opt 60:495–497. doi:10.1007/BF01081333
35. Heintzmann R, Jovin TM, Cremer C (2002) Saturated patterned excitation microscopy—a concept for optical resolution improvement. J Opt Soc Am A 19:1599. doi:10.1364/JOSAA.19.001599
36. Andresen M, Stiel AC, Fölling J et al (2008) Photoswitchable fluorescent proteins enable monochromatic multilabel imaging and dual color fluorescence nanoscopy. Nat Biotechnol 26:1035–1040. doi:10.1038/nbt.1493
37. Rittweger E, Rankin BR, Westphal V, Hell SW (2007) Fluorescence depletion mechanisms in super-resolving STED microscopy. Chem Phys Lett 442:483–487. doi:10.1016/j.cplett.2007.06.017
38. Fitzpatrick JAJ, Yan Q, Sieber JJ et al (2009) STED nanoscopy in living cells using Fluorogen Activating Proteins. Bioconjug Chem 20:1843–1847. doi:10.1021/bc900249e
39. Han KY, Willig KI, Rittweger E et al (2009) Three-dimensional stimulated emission depletion microscopy of nitrogen-vacancy centers in diamond using continuous-wave light. Nano Lett 9:3323–3329. doi:10.1021/nl901597v
40. Moneron G, Medda R, Hein B et al (2010) Fast STED microscopy with continuous wave

41. Willig KI, Harke B, Medda R, Hell SW (2007) STED microscopy with continuous wave beams. Nat Methods 4:915–918. doi:10.1038/nmeth1108
42. Donnert G, Keller J, Wurm CA et al (2007) Two-color far-field fluorescence nanoscopy. Biophys J 92:L67–L69. doi:10.1529/biophysj.107.104497
43. Westphal V, Hell S (2005) Nanoscale resolution in the focal plane of an optical microscope. Phys Rev Lett. doi:10.1103/PhysRevLett.94.143903
44. Vicidomini G, Moneron G, Han KY et al (2011) sharper low-power sted nanoscopy by time gating. Nat Methods 8:1–5. doi:10.1038/nMeth.1624
45. Moffitt JR, Osseforth C, Michaelis J (2011) Time-gating improves the spatial resolution of STED microscopy. Opt Express 19:4242–4254
46. Dyba M, Hell S (2002) Focal spots of size $\lambda/23$ open up far-field florescence microscopy at 33 nm axial resolution. Phys Rev Lett. doi:10.1103/PhysRevLett.88.163901
47. Kelsh RN, Brand M, Jiang YJ et al (1996) Zebrafish pigmentation mutations and the processes of neural crest development. Development 123:369–389
48. Staudt T, Lang MC, Medda R et al (2007) 2, 2 0-Thiodiethanol: a new water soluble mounting medium for high resolution optical microscopy. Microsc Res Tech 9:1–9. doi:10.1002/jemt
49. Huang B, Wang W, Bates M, Zhuang X (2008) Three-dimensional super-resolution imaging by stochastic optical reconstruction microscopy. Science 319:810–813. doi:10.1126/science.1153529
50. Thompson RE, Larson DR, Webb WW (2002) Precise nanometer localization analysis for individual fluorescent probes. Biophys J 82:2775–2783. doi:10.1016/S0006-3495(02)75618-X
51. Vogelsang J, Cordes T, Forthmann C et al (2009) Controlling the fluorescence of ordinary oxazine dyes for single-molecule switching and superresolution microscopy. Proc Natl Acad Sci U S A 106:8107–8112. doi:10.1073/pnas.0811875106
52. Toomre D, Bewersdorf J (2010) A new wave of cellular imaging. Annu Rev Cell Dev Biol 26:285–314. doi:10.1146/annurev-cellbio-100109-104048
53. Testa I, Wurm C a, Medda R et al (2010) Multicolor fluorescence nanoscopy in fixed and living cells by exciting conventional fluorophores with a single wavelength. Biophys J 99:2686–2694. doi:10.1016/j.bpj.2010.08.012
54. Jones SA, Shim S-H, He J, Zhuang X (2011) Fast, three-dimensional super-resolution imaging of live cells. Nat Methods 8:499–508. doi:10.1038/nmeth.1605
55. Giannone G, Hosy E, Levet F et al (2010) Dynamic superresolution imaging of endogenous proteins on living cells at ultra-high density. Biophys J 99:1303–1310. doi:10.1016/j.bpj.2010.06.005
56. Axelrod D (1981) Cell-substrate contacts illuminated by total internal reflection fluorescence. J Cell Biol 89:141–145

Chapter 15

Probing Biological Samples in High-Resolution Microscopy: Making Sense of Spots

Felipe Opazo

Abstract

In recent years, microscopy techniques have reached high sensitivities and excellent resolutions, far beyond the diffraction limit. However, images of biological specimens obtained with super-resolution instruments have the tendency of being dominated by spots. The quality or faithfulness of the observed structure depends in great manner on the labeling density achieved by affinity probes. To obtain the required high labeling densities, several problems still need to be addressed. Prevalent staining methodologies are mainly based on antibodies. Due to their large size (~10–15 nm, ~150 kDa), antibodies penetrate poorly into biological samples, find only few epitopes and position the fluorophores far from the intended targets (relative to the resolutions currently achieved). These problems drastically limit imaging efforts, irrespective of the quality of the microscopes. Recently, RNA-based affinity probes (termed aptamers, ~15 kDa) and camelid-derived small single-chain antibodies (termed nanobodies, ~13 kDa) are offering a possible solution. Both of these probes have an improved imaging performance, not only for diffraction-unlimited microscopy techniques but also for conventional microscopy. The effort to develop new affinity probes of smaller dimensions should go together with improving the preservation of biological samples. The extraordinary amount of details obtained by super-resolution microscopy also enhances the detection of artifacts that fixatives and detergents cause to cellular structures. Therefore, thorough optimizations of current methodologies of sample preparation are necessary to achieve stainings capable of representing genuine biological structures and accurate protein distributions. Eventually, new fixation and permeabilization procedures able to retain more faithfully the biological structure under investigation need to be developed in the light of the current microscopy technologies.

Key words Affinity probes, Antibodies, Aptamers, Nanobodies, VHH, Super-resolution

1 Introduction

Light microscopy has been limited to a lateral resolution of about 200–300 nm, the so-called diffraction limit of light already described by Ernst Abbe in 1873. Light microscopy has recently been revolutionized by the introduction of super-resolution imaging [1, 2], with several techniques going beyond the diffraction limit and thus providing images with higher amount of details. These include, among others, *stimulated emission depletion*

(STED), *photo-activated localization microscopy* (PALM), *stochastic optical reconstruction microscopy* (STORM), *structural illumination microscopy* (SIM), and *ground state depletion microscopy followed by individual molecule return* (GSDIM) (see Chaps. 2, 5, and 7 & [3]). Several of these advanced microscopy techniques are now commercially available, which is significantly expanding their accessibility.

Despite the various limitations and challenges, these new visualization possibilities have changed the way biologists approach cellular research. The different features of each super-resolution technique give biologists the opportunity to pursue questions that have not been possible to answer with conventional diffraction limited light microscopy. Cell biologists have used mainly two methodologies to detect their proteins of interest (POI), immunostainings and the fusion of the POI to fluorescent proteins. Both approaches have advantages and limitations.

The antibody stainings applied by most biologists today are suboptimal, since current methods were developed with and for conventional microscopy, thus lacking the precision required for super-resolution microscopy. Typically, POIs are identified by primary antibodies, which are then detected with fluorescently coupled secondary antibodies. The primary–secondary antibody complex can be up to 30 nm long. This dimension is an obvious problem when current techniques are able to deliver resolutions of 9 nm [4]. Therefore, it has been repeatedly noted that antibodies are not optimal tools for this type of work [5–8].

Since antibodies are still the preferred affinity tool for staining biological samples, the quality of super-resolution images is currently limited by the low accuracy of immunostainings regardless of all technical breakthroughs in the field of optics.

An alternative to antibody staining would be the overexpression of fluorescent proteins. Normally, the gene coding for a protein of interest (POI) is fused genetically to the sequence coding for a fluorescent protein in a vector plasmid. The introduction of such plasmid in a cell (with liposome transfection or viral infection) will eventually produce the POI tagged with the desired fluorescent protein. The super-resolution technique PALM is the only one that principally requires the expression of photo-switchable fluorescent proteins [9]. However, several fluorescent proteins have been developed for techniques that conventionally use antibody stainings [10, 11].

Fluorescent proteins have generated invaluable information in cell biology investigations. However, it is necessary to bear in mind that their use is normally associated with the overexpression of a POI fused to a fluorescent protein. Unfortunately, overexpression might lead to undesired protein aggregation, mislocalization, and loss or gain of function of the POI [12–14]. Careful assessments need to be carried out to exclude any potential malfunction or mislocalization of the overexpressed POI, a practice that has not been broadly adopted by the field of cell biology. Additionally, the overexpression of a POI fused to a fluorescent protein is typically performed in cells containing the endogenous nontagged protein of interest, which results in a mixture of fluorescent and "invisible" POI leading to misinterpretations and mislocalization. Finally, the expression of a foreign gene cannot be performed when studying human samples like biopsies. This limits the whole field of disease diagnostics to use only affinity probes and their associated techniques.

In my view, the ideal probes for labeling biological samples should be (1) sufficiently small to penetrate easily and to find all potential epitopes, (2) with high binding affinities to an endogenously expressed target to ensure a good signal-to-noise ratio avoiding the need for overexpression, and (3) bright to provide assays with high sensitivity. In the following sections, I will discuss in detail the affinity probes that are commonly used today for advanced microscopy and the new emerging generation of small probes capable of finding more epitopes and delivering high-density stainings.

2 Antibodies as Affinity Probes for Biological Stainings

Much has been improved since the first immunostainings were performed in 1941 [15]. Currently, antibody stainings are the most used affinity probes for the detection of POI in biological samples. Their success has been mainly achieved due to their high specificity and the simplicity to couple them with organic fluorophores.

The rapid improvements of super-resolution microscopy techniques during the last few years brought them to a level where biological samples can be imaged with ease. The super-resolution microscopy community adopted antibody-staining methodologies from the first biological staining attempts.

The first immunostained sample obtained with super-resolution microscopy was performed with an antibody against tubulin and fluorescently coupled secondary antibodies revealing the filaments of microtubuli in eukaryotic cells [16]. Subsequently, a seminal work studying the fate of synaptic proteins upon synaptic vesicle recycling [17] demonstrated that advanced microscopy combined with immunostainings could have major implications on biological questions. Several new super-resolution techniques have then followed and almost all of them (except PALM) have been using primary and secondary antibodies as affinity probes for stainings [17–20]. Only PALM relies principally in the overexpression of photo-switchable fluorescent proteins [21].

The antibody labeling of cytoskeletal structures became a gold standard as a biological staining to demonstrate the resolution power of newly appearing techniques. This is not surprising since

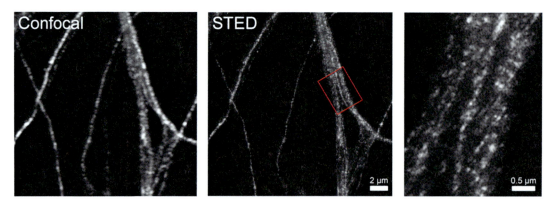

Fig. 1 Spotty pattern staining of continuous structures. Microtubules stained with an antibody against tubulin and detected with fluorescently coupled secondaries. When using advanced microscopy techniques (in this case STED) at a sufficiently high zoom (*right most panel*), the spotty pattern of stained microtubules becomes evident

very good antibodies for such targets are readily available. Cytoskeletal filaments normally contain many copies of the target protein (typically a homopolymer) ensuring a good signal and even if the staining is poor it results in structures that are easy to be recognized (filaments). Therefore, the immunostainings of cytoskeletal structures are a good choice to demonstrate the proof of principle and optical advantages of the microscopy technique under study.

However, irrespective of the super-resolution technique that has been used during the last ~10 years, immunostained samples typically result in images dominated by spots. In this respect, a common and to some extent accepted artifact regularly results in discontinuous stainings of continuous filamentous structures such as microtubuli or neurofilaments (Fig. 1 and see for example [22–24]).

Although some "spotty" pattern images obtained with super-resolution have been clearly demonstrated to be faithful representations of the biological processes investigated (see for review [25]), the interpretation of structures other than continued filaments can be challenging. The fact that the antibody stainings regularly result in a "spotty" pattern is nevertheless a concern. It is important to acknowledge that the "spotty" pattern artifact is mainly, if not only, relevant in super-resolution microscopy imaging and does not affect diffraction limited microscopy techniques.

Which could be the reason why spots dominate super-resolution images? Some of the problems might originate during sample preparation since fixation and permeabilization are known to alter cellular structures (I will briefly discuss this in Sect. 6). Another possible explanation is molecular "clumping" induced by antibodies. This artificial clustering of target molecules has

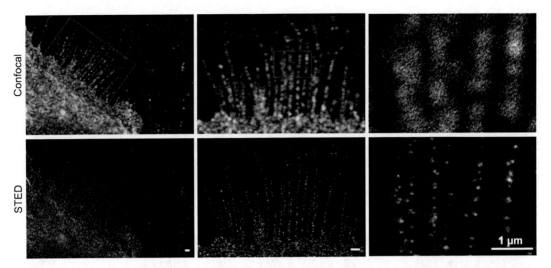

Fig. 2 Antibody-induced protein clustering. A431 cells stained for EGFR using primary/secondary conventional immunostaining. Three different zoom levels of confocal and STED images are presented. Note the regularly distributed spots in the highest zoom of the STED image (*bottom right panel*). Although the EGFR is described to form clusters, it is biologically improbably that all subunits are found only in clusters and no gradient or normal distribution can be observed in-between the spots. Scale bars represent 1 μm

principally been described during live-cell application due to the dual binding ability of antibodies (two binding pockets for every antibody). However, the immunostaining of formaldehyde fixed cells (where clumping is reduced but not fully avoided, see [26]) can also result in images dominated by "spotty" patterns due to antibody clustering (Fig. 2).

An additional explanation for obtaining mainly spots in advanced microscopy imaging might be the large size of antibodies. Normally, antibodies are composed of two heavy and two light polypeptide chains interconnected with disulfide bonds. Such immunoglobulins typically have a molecular weight of ~150 kDa and ~12 nm in length [27]. In a conventional immunostaining, the primary/secondary package has ~30 nm of length considering only one secondary per primary, which is the most conservative scenario. Therefore, it is reasonable to hypothesize that the large size of antibodies would directly affect the labeling density by hampering the recognition of all potential epitopes. Moreover, it is well known that their large size reduces the overall penetration in biological specimens; a fact that additionally limits the amount of recognizable epitopes. Therefore, smaller fluorescence probes with high affinity and selectivity are clearly needed in super-resolution microscopy applications (Fig. 3).

Fab fragments obtained by papain digestion of antibodies are an obvious choice, with an estimated size of ~9 nm and molecular weight of ~50 kDa [27]. However, our observations suggest that Fab

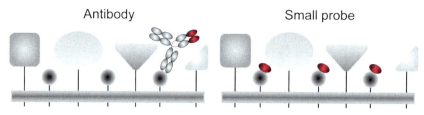

Fig. 3 Probe size can be a key determinant of the quality of staining. The scheme exemplifies a situation where the large size of the antibodies prevents them from reaching every epitope and a smaller probe would allow a higher labeling density than antibodies. Note that the cartoon shows only the fluorescently conjugated primary antibody. However, conventional immunostainings would have one or two fluorescently coupled secondary antibodies bound to the primary

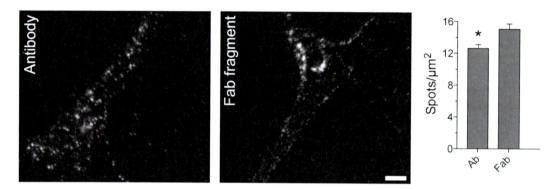

Fig. 4 Comparison between staining with full antibodies or Fab fragment. Cultures of hippocampal primary neurons were immunostained against syntaxin 1 using a full antibody or a Fab fragment. The *left most panel* (Antibody) shows a neuronal process stained using the monoclonal antibody followed by fluorescently coupled secondary antibodies. The *middle panel* (Fab fragment) was a similar neuronal culture stained with the Fab fragment derived from the same monoclonal antibody, directly coupled to a fluorophores. Scale bar represents 500 nm. The graph displays the quantification of the number of spots detected by the two stainings. While the antibody and Fab fragment images are relatively similar, the analysis demonstrates that the Fab fragments recognize ~20 % more spots than the antibodies. Error bars represent the SEM of four independent experiments. Significant differences were tested by student's *t*-tests, *$p<0.05$

fragments might still be too large for super-resolution applications. Fluorescently labeled Fab fragments provided images similar to those of the complete antibody increasing epitope recognition only by ~20 % (Fig. 4 and see [6]). This data suggest that the high price tag of fluorescently coupled Fab fragments can be only justified if "clumping" of full antibodies needs to be avoided.

Do other affinity tag exists that are capable of binding with similar strength and specificity but being smaller than Fab fragments? The answer is yes, and I will discuss these new fluorescent affinity probes that promise to significantly improve the structure recognition and overall image quality in the next sections of this chapter.

3 Camel Single Variable Domain Antibodies (VHH) or Nanobodies

The antigen recognition domain in conventional antibodies is only functional when the heavy (H) and the light (L) polypeptide chains are correctly assembled. Each chain contains a variable domain (VH and VL) that both together form the paratope responsible for the antigen recognition of antibodies. The Camelidae Family (which include camels, llamas, and alpacas among others) have naturally occurring heavy-chain antibodies in their immune system [28]. Such immunoglobulins are capable of binding their antigens without the need of pairing with a light chain. The functional variable domain derived from a camelid heavy-chain antibody is termed VHH or nanobody (Fig. 5). These nanobodies are considerably smaller than antibodies or antibody fragments with a typical molecular weight of ~15 kDa (see above in Sect. 2).

The process of nanobody selection involves several steps. After the immunization of a camelid, peripheral B-lymphocytes from blood samples need to be purified and cDNA libraries of the humoral immunoglubulins are generated. The variable domains responsible for the epitope recognition are amplified by conventional PCR and then typically cloned into a phagemid vector to perform phage-display experiments [29, 30].

Phage-display screening would normally begin with the production of a library of bacteriophages each displaying one different variable domain (nanobody) on their capsides. These chimeric phages are then exposed to the antigen (the same that was used for immunizing the animal) and only the phages binding such antigen will be retained, amplified, and incubated again with the antigen. This cycle is termed panning, and it is performed several times (3–6 iterations) to obtain specific antigen-binding nanobodies. The major advantage of a phage-display screening system is that the selected bacteriophages contain the genetic information for the

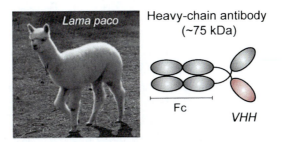

Fig. 5 Heavy-chain antibodies. The immune system of Alpacas (*Lama paco*) is capable of producing conventional immunoglobulins and the so-called heavy-chain antibodies. Nanobodies or VHH (represented as a whole domain in *red*) are the minimal functional sequence of the variable domain able to bind selectively to a target

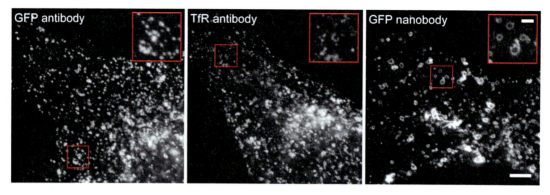

Fig. 6 Comparison of antibody versus nanobody staining. Baby hamster kidney (BHK) cells were transfected with human transferrin receptor (TfR) fused to GFP. Scale bars represent 2 μm for the main images and 1 μm in the zoomed insets

specific variable domain that it is bound to the antigen. After the last panning iteration, the DNA of selected phages is sequenced, and thereby the selected nanobodies can be finally cloned into protein expression vectors for large-scale production [29, 31].

The knowledge of camelids having single-chain antibodies and the know-how to produce nanobodies has been around for 2 decades [28]. However, the staining abilities of nanobodies for microscopy studies were never systematically investigated. Their laborious selection and the apparent lack of advantages over classical antibody stainings prevented the nanobody field to exploit them for conventional fluorescent microscopy. It took several years for the super-resolution microscopy field to notice that conventional antibody stainings are not accurate enough. With each improvement in resolution that has been made with these new techniques, it becomes more urgent and necessary to obtain smaller and brighter affinity probes.

Searching for such probes, Ries et al. [5] have recently demonstrated that a 13 kDa nanobody directed against GFP has a significant improvement (compared to normal antibodies) in the imaging precision achieved with localization-based super-resolution microscopy.

Our results obtained with the same nanobody used by Ries and colleagues cannot only confirm their observations but also add the fact that nanobody stainings provide a higher labeling density and accurate structure recognition when compared to conventional immunostaining (Fig. 6).

Nanobodies have the great potential of improving accuracy of biological stainings for all advanced microscopy techniques. Their small size and target specificity have improved the staining accuracy to a level that was not possible to achieve with conventional stainings [5]. Further efforts to develop new nanobodies still have to be made.

4 Aptamers

Aptamers are single-stranded DNA or RNA oligonucleotides capable of binding selectively to a target. They were introduced more than 20 years ago [32], but they have not been thoroughly evaluated as imaging tools. Aptamers are selected in vitro from oligonucleotide libraries through a process termed systematic evolution of ligands by exponential enrichment or SELEX (Fig. 7) [33]. Aptamers can be selected to bind a broad diversity of targets such as small molecules, purified proteins, or more recently whole cells—such as tumor-derived cells [34].

During the SELEX process, single-stranded RNA or DNA libraries containing $\sim 10^{15}$ different sequences are exposed to a particular target. After washing off the unbound oligonucleotides, target-bound aptamers are isolated and amplified. Each oligonucleotide sequence is flanked by known sequences allowing the amplification of the selected aptamers by using simple polymerase chain reactions. This "new" selected library is exposed again to the desired target. Several of such iterations (normally 6–15, with increasing stringency) ensure the selection of sequences able to bind strongly and with high specificity to the target molecule [35].

The secondary structure of DNA or RNA oligonucleotides depends primarily on their particular sequence. Moreover, the target specificity of aptamers is defined by their own sequence. Due to technical difficulties in automatic synthesis of oligonucleotides, aptamer libraries are currently restricted to sequences of ~100 nucleotides, which ultimately constrain the diversity of aptamers. Nevertheless, after successful SELEX procedures, the sequences of all "good binders" are aligned and conserved domains are

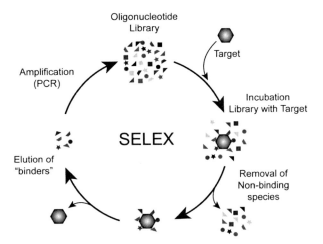

Fig. 7 Scheme of the necessary steps for aptamer selection performed by SELEX. Targets can be as diverse as ions, small molecules like ATP, proteins, viruses, or even whole cells

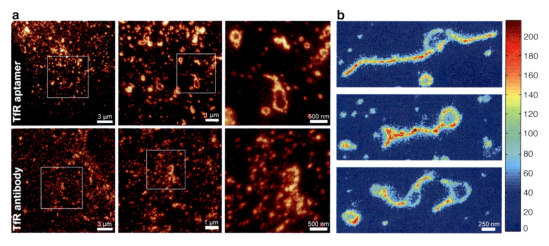

Fig. 8 Aptamers allow for substantially better structure recognition in STED microscopy. (**a**) STED images of cells stained with an aptamer selected against transferrin receptor (TfR aptamer, *upper panels*) and with a monoclonal TfR antibody (*lower panels*). Three zoom levels are indicated; scale bars, *left* to *right*, 3 μm, 1 μm, and 500 nm. Note that the TfR aptamer allows the visualization of the empty lumen of the endosome (*left panel*), while the antibody only reports one "spot" (*right panel*). (**b**) Aptamer staining allows the visualization of domains of TfnR within single endosomes. Note the enrichment of TfnR in the endosomal tubular structures, compared to the circular regions. Images are displayed using an intensity-dependent colormap (*right*). Scale bar, 250 nm

normally defined. Further trimming of sequences are performed to obtain shorter aptamers. Functional aptamers of lengths ranging from 15 to 100 nucleotides have been described in the literature. Aptamers are typically much smaller than antibodies with size estimations of 4–6 nm in length and ~15 kDa of weight for a ~50 nucleotide aptamer [36]. Their small size allows them to penetrate the samples easily and bind a higher number of target proteins. A recent example using a well-known aptamer against prostate-specific membrane antigen (PSMA) resulted in binding up to 100-fold more epitopes than the different antibodies tested [6]. This increase in detected epitopes gives a brighter, clearer signal and allows the identification of structures with much higher accuracy in super-resolution microscopy (Fig. 8). Moreover, the ability to find more epitopes makes fluorescence detection using aptamers significantly more sensitive for conventional microscopy, when compared to classical immunostainings [6].

Probes like nanobodies or aptamers are commonly directly coupled to a fluorescent molecule, which ultimately places the fluorophores closer to the intended target. This feature is of extreme importance for colocalization studies, especially when instruments can deliver resolutions of ~10 nm [4].

Aptamers have several additional advantages:

1. Once an aptamer sequence is selected, their production can be scaled and achieved by automatic oligonucleotide synthesizers (avoiding the use of animals).

2. DNA and RNA chemistry has been thoroughly investigated, and many modifications can be readily incorporated (biotins, fluorophores, gold particles, etc.). Special modifications in their nucleotides (typically a Fluor atom in the second carbon of the ribose sugar of the nucleotides) have been developed to avoid or retard the action of nucleases, thus increasing the life span of such aptamers in a cellular context or even in intravenous application [37].

3. Differently to regular antibodies, aptamers can have little or no immune response when injected in animals [38, 39].

4. Of particular interest is the fact that aptamers allow easy multicolor imaging limited only by the fluorophores, light source (excitation), and emission filters, but they do not depend on matching animal species between secondary and primary antibodies.

5. Theoretically, aptamers can be raised using molecular complexes as targets and select only sequences able to bind the complex but not the individual components of the complex. For a long time, this feature has been desired for antibodies, however with little success.

While many alternatives and modifications of the main SELEX procedure have been described [35], they all ensure the production of aptamers that can recognize proteins exposed on the surface of living cells, neglecting intracellular epitopes. This has in fact reduced the applicability of aptamers for imaging—although one should note that the main reason why aptamers have not been used in imaging is that most of them have simply not yet been tried in microscopy.

A second important reason that has also deterred aptamers from being embraced by the field of microscopy is their difficulty in detecting fixed proteins. Fixation is a major step in virtually every imaging protocol—but no methods have been established yet to select aptamers capable of recognizing their targets in already fixed tissue, since there was no real need for this until recently [6]. Interestingly, some years ago Zeng and colleagues demonstrated an exceptional case when an aptamer selected with classical SELEX (using native protein as a target) was capable of detecting its target in formalin-fixed samples [40].

To date, formaldehyde is probably the most common fixative used for immunocytochemistry and histochemistry. However, the fixative property of formaldehyde is based on the formation of methylene bridges resulting in intra- and intermolecular cross-linking by which the native structure of the protein is deformed [41]. It can be speculated that the aptamer binding to the target protein is strongly affected by the deformation of the protein due to fixation. This is also true, although to a lesser extent, for antibody epitope recognition. The drastic difference in aldehyde fixation sensibility between aptamers and antibodies might rely on the size/surface of the recognized epitope. Normally, the epitopes of antibodies range from 4 to 12 amino acids [42], but the oligonucleotide self-folded predicted structure of various aptamers suggested that their binding strength relies in several contacts to the target molecule, and as soon as one is missing the remaining contacts might not be sufficient enough to stabilize the interaction. Major concerns have been raised for decades concerning biological sample fixation, which I will briefly touch on Sect. 6 in this chapter. The appearance of super-resolution microscopy has created the necessity to develop new smaller staining molecules, and I would not be surprised if it also pushes forward new procedures for a better sample fixation and permeabilization.

Despite the obvious benefits that small aptamers can provide for the imaging quality obtained with super-resolution microscopy, it is also clear that major efforts need to be dedicated to generate aptamers able to work properly in fixed biological samples.

5 Other Small Probes for Microscopy

To find small molecules that bind selectively to a protein is a lengthy and laborious task. Nature, however, has been searching for such binders much longer time than scientists, and without binding specificity no biological complexity would be possible. A simple example of specific biological-relevant interactions is the system of receptors and ligands. Several natural ligands have been coupled to fluorescent molecules and used to principally investigate cellular endocytosis and organelle trafficking [43, 44]. Perhaps one of the most used proteins for this type of staining is the transferrin receptor in combination with its fluorescently labeled ligand transferrin [45, 46]. These kinds of stainings using the natural ligands would typically result in accurate distribution and localization of the receptor of interest, and their potential use for advance microscopy has recently been suggested [6]. However, only few proteins can be fluorescently tracked by this strategy, since not all proteins have a ligand (or known ligand) and most intracellular proteins cannot be easily followed by this approach.

Another important source of specific and typically strong binders is obtained from natural toxins or venoms. Several of them have been coupled to fluorophores and used for microscopy stainings. Some of the most prominent fluorescently coupled toxins are (a) alpha-bungarotoxin, which binds and inhibits nicotinic acetylcholine receptors causing muscular paralysis [47]; (b) Shiga toxin, which binds to the large subunit of eukaryotic ribosomes blocking protein synthesis [48]; and (c) tetradotoxin, a small molecule that binds and blocks sodium channels in neurons [49].

The need of smaller probes for advanced microscopy was once more made clear after Xu and colleagues obtained improved resolutions by using the small fungal toxin phalloidin. The small size of fluorescent phalloidin (~2 kDa) was able to detect filamentous actin with a high label density and thanks to its small size that places the fluorophore much closer to the target compared to the antibodies. These stainings resulted in STORM images [Chap. 2 (Super-resolution Microscopy: Principles, Techniques and Applications)] with an impressive lateral resolution of ~10 nm [4].

Should we stop using antibodies in microscopy? Certainly not! It is for instance clear that images from all conventional microscopy resolution are unaffected by the use of full antibodies, but care must be taken with antibody clumping effects and colocalization studies as mentioned above.

Additionally, some immunostainings of microtubules have been proved to work well also for some super-resolution microscopy images [50]. This is principally the case because the homopolymers of tubulin (microtubules) were fixed with glutaraldehyde, which is an excellent fixative to preserve structures (used principally in electron microscopy). Unfortunately, as I will briefly describe in the next section, very few antibodies are able to recognize their epitopes after cross-linking proteins with glutaraldehyde. Moreover, depending on the biological question that is pursued, some of the localization-based techniques like dSTORM [Chap. 2 (Super-resolution Microscopy: Principles, Techniques and Applications)] have used mathematical routines to average entire cellular structures or organelles and get improved images by averaging out some poor protein distribution obtained after antibody stainings [51].

The extensive list of readily available antibodies and their vast number of applications makes antibodies far from being an obsolete tool. However, current advanced microscopy techniques are bringing antibodies to their limits due to their (now) relative large size. Therefore, efforts to develop new smaller probes need to be augmented in order to match the increasing improvement in resolution of the current and new light microscopy instruments.

6 Basic Consideration on Biological Samples

Even with an ideal probe the staining of biological samples is not trivial and might still display unfaithful representation depending on the structure preservation of the sample. Correct sample preparation is essential, and it is a problem that electron microscopists have been facing for decades. Now it has also become critical for light super-resolution techniques. I will briefly describe the main sources of the difficulties which arise upon sample fixation and permeabilization.

Protein fixation refers to the formation of a protein network by cross-linking neighbor polypeptides. The main aim of sample

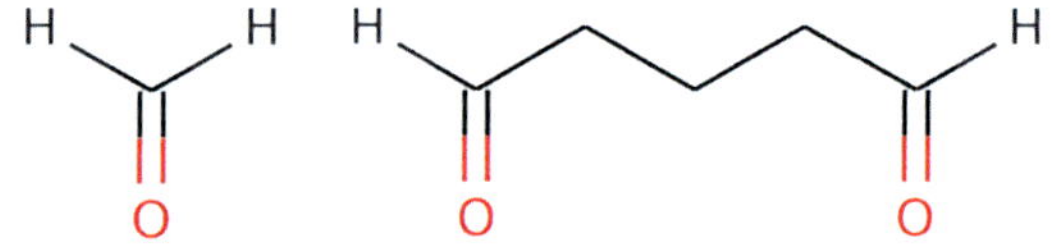

Fig. 9 Chemical structure of formaldehyde and glutaraldehyde

fixation is to have a "snapshot" of the specimen allowing the observation of cellular elements at a given time point. Formaldehyde and glutaraldehyde (Fig. 9) are the two major chemical fixatives currently used for microscopy preparations.

Formaldehyde is a small molecule that permeates fast through biological samples, and it contains only one reactive aldehyde group. Formaldehyde reacts first with the free amino group (NH_2) present in the side chain of lysines, and after this bond is formed it can further react with the nitrogen (NH) of any peptide bond resulting in methylene bridges (cross-linking) [41]. Typically, the formation of such methylene bridges is slow and can take weeks to reach saturation. On the other hand, carbohydrates, nucleic acids, and lipids are trapped between the protein matrix generated by formaldehyde, and they are typically not chemically modified unless formaldehyde fixation is prolonged for days.

Formaldehyde is especially used when immunostaining is the downstream application. Immunogenicity is typically well preserved after a mild formaldehyde fixation. However, the rather slow fixation speed results in deformations of cellular structures, similar to what occurs to an image that turns out blurred due to excessive exposure of moving objects. This typically results in a poor structure preservation effect that is enhanced when using the resolutions achieved by the current advanced microscopy technologies. Moreover, incomplete fixation facilitates the clumping of antigens by primary antibodies, and/or the primaries getting clumped by the secondary antibodies (Sect. 2, Fig. 2 [26]).

Glutaraldehyde contains two aldehyde groups connected by three carbons (Fig. 9), which confer a significantly higher reactivity than formaldehyde [52, 53]. Glutaraldehyde penetrates biological tissue slower than formaldehyde, but its cross-linking speed is several folds faster. The carbon linker between the aldehyde groups allows glutaraldehyde to interact with a higher degree of flexibility to more distant and even to closer reactive groups than formaldehyde.

Sabatini and colleagues introduced glutaraldehyde in 1963 as an improved fixative for electron microscopy due to its good structure preservation characteristics [54]. Since then, glutaraldehyde has become the standard fixative for electron microscopy preparations, but some difficulties remain when used for immunostainings. The dense matrix caused by the strong cross-linking ability of glutaraldehyde can affect the penetration of antibody molecules into the sample. Importantly, many epitopes are lost due to the

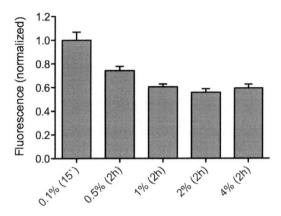

Fig. 10 Detergent extraction of membrane proteins after fixation. HeLa cells were fixed for 45 min with 4 % paraformaldehyde (in PBS) and immunostained against transferrin receptor (TfR). A short permeabilization of 15 min with low concentration of detergent (0.1 % of Triton X-100) was used as a reference. Already 1 % of Triton X-100 was capable of extracting ~40 % of TfR in the already fixed samples. Note that the reference value had already undergone a little extraction and that after 2 h with 1 % Triton X-100 extraction ceases

glutaraldehyde-induced protein alterations, resulting in very few antibodies able to recognize their target after glutaraldehyde fixation. Several attempts have been made to generate antibodies recognizing glutaraldehyde-fixed antigens and bypass the immunogenicity problems caused by glutaraldehyde fixation [55]. Another reason why glutaraldehyde is not commonly used in fluorescence microscopy is that samples fixed with this chemical tend to have strong background fluorescence.

A compromise can sometimes be achieved by stabilizing the tissue with the fast permeable formaldehyde and obtaining a good structure preservation with glutaraldehyde by mixing them both at empirically determined ratios [56]. Despite the true benefits in detail preservation by adding 0.1 % of glutaraldehyde to a 4 % paraformaldehyde solution, it is sufficient for most of the antibodies to partially or totally lose the ability to find their epitopes.

Sample permeabilization generally refers to the perforation of cellular membranes allowing access of macromolecules that otherwise could not cross such barriers. Typically, after protein fixation, mild detergents are used to create small apertures by extracting lipids from biological membranes. The detergent nature, its final concentration, and time of application are key variables for the final integrity of the remaining cellular elements. This destructive process is unfortunately necessary if antibodies or other affinity probes are used to detect cytosolic, organellar, or nuclear constituents.

In order to have an overview of the damage caused by detergents, I have followed the extraction of a transmembrane protein (TfR) after permeabilization. Interestingly, up to ~40 % of the TfR was extracted (Fig. 10). This evidence indicates that an optimization of the permeabilization step should be carefully performed, and that it needs to be taken into account for experimental conclusions.

Some of the most common detergents used for staining protocols are Tween 20, Triton X-100, digitonin, and saponin, each of them having different strengths and weaknesses [57]. Permeabilization steps can and should be avoided if only GFP-like moieties are followed, or if the staining dyes are cell permeable (like the DNA marker DAPI).

Interestingly, the aptamers (Sect. 4 above), but not the similar size nanobodies (Sect. 3 above), are able to enter a cell through the small fissures generated during the fixation step. Thus, aptamers lack the necessity of detergent-based permeabilization.

7 Final Remarks

The breakage of the diffraction limit of light brought a new perspective of the cellular elements and their interactions to cell biologists. However, it also brought interesting new challenges. Staining methodologies used successfully for conventional microscopy need to be revised if applied in current diffraction unlimited instruments. It has only recently been demonstrated that smaller monovalent probes (aptamers and nanobodies) can have an advantage over classical antibodies for super-resolution microscopy techniques. Unfortunately, only a couple of such probes are available today, different from the immense amount of readily available antibodies. Importantly, some antibody stainings have also been proven to give good and faithful representations of the structures investigated with super-resolution microscopes. In addition to choosing the correct affinity probe, the new level of details that the new instruments can deliver is forcing cell biologists to treat samples with much care to obtain accurate and meaningful results.

In view of the evidence, much effort still needs to be made to develop new small affinity tools, improving the biological sample preparation (fixation and permeabilization) and carefully optimizing existing antibodies to obtain a true representation of the element under investigation.

References

1. Hell SW (2007) Far-field optical nanoscopy. Science (New York, NY) 316:1153–1158
2. Hell SW, Wichmann J (1994) Breaking the diffraction resolution limit by stimulated emission: stimulated-emission-depletion fluorescence microscopy. Opt Lett 19: 780–782
3. Toomre D, Bewersdorf J (2010) A new wave of cellular imaging. Annu Rev Cell Dev Biol 26:285–314
4. Xu K, Babcock HP, Zhuang X (2012) Dual-objective STORM reveals three-dimensional filament organization in the actin cytoskeleton. Nat Methods 9:185–188
5. Ries J, Kaplan C, Platonova E, Eghlidi H, Ewers H (2012) A simple, versatile method for GFP-based super-resolution microscopy via nanobodies. Nat Methods 9:582–584
6. Opazo F, Levy M, Byrom M, Schäfer C, Geisler C, Groemer TW, Ellington AD, Rizzoli SO

(2012) Aptamers as potential tools for super-resolution microscopy. Nat Methods 9:938–939
7. McKinney SA, Murphy CS, Hazelwood KL, Davidson MW, Looger LL (2009) A bright and photostable photoconvertible fluorescent protein. Nat Methods 6:131–133
8. Lakadamyali M, Babcock H, Bates M, Zhuang X, Lichtman J (2012) 3D multicolor super-resolution imaging offers improved accuracy in neuron tracing. PLoS ONE 7:e30826
9. Henriques R, Griffiths C, Hesper Rego E, Mhlanga MM (2011) PALM and STORM: unlocking live-cell super-resolution. Biopolymers 95:322–331
10. Grotjohann T, Testa I, Leutenegger M, Bock H, Urban NT, Lavoie-Cardinal F, Willig KI, Eggeling C, Jakobs S, Hell SW (2011) Diffraction-unlimited all-optical imaging and writing with a photochromic GFP. Nature 478:204–208
11. Chang H, Zhang M, Ji W, Chen J, Zhang Y, Liu B, Lu J, Zhang J, Xu P, Xu T (2012) A unique series of reversibly switchable fluorescent proteins with beneficial properties for various applications. Proc Natl Acad Sci 109:4455–4460
12. Lisenbee CS, Karnik SK, Trelease RN (2003) Overexpression and mislocalization of a tail-anchored GFP redefines the identity of peroxisomal ER. Traffic (Copenhagen, Denmark) 4:491–501
13. Opazo F, Punge A, Bückers J, Hoopmann P, Kastrup L, Hell SW, Rizzoli SO (2010) Limited intermixing of synaptic vesicle components upon vesicle recycling. Traffic (Copenhagen, Denmark) 11:800–812
14. Rappoport JZ, Simon SM (2008) A functional GFP fusion for imaging clathrin-mediated endocytosis. Traffic (Copenhagen, Denmark) 9:1250–1255
15. Coons AH, Creech HJ, Jones RN (1941) Immunological properties of an antibody containing a fluorescent group. Exp Biol Med (Maywood) 47:200–202
16. Dyba M, Jakobs S, Hell SW (2003) Immunofluorescence stimulated emission depletion microscopy. Nat Biotechnol 21: 1303–1304
17. Willig KI, Rizzoli SO, Westphal V, Jahn R, Hell SW (2006) STED microscopy reveals that synaptotagmin remains clustered after synaptic vesicle exocytosis. Nature 440:935–939
18. Schermelleh L, Heintzmann R, Leonhardt H (2010) A guide to super-resolution fluorescence microscopy. J Cell Biol 190:165–175
19. Huang B, Wang W, Bates M, Zhuang X (2008) Three-dimensional super-resolution imaging by stochastic optical reconstruction microscopy. Science (New York, NY) 319:810–813
20. Testa I, Wurm CA, Medda R, Rothermel E, von Middendorf C, Fölling J, Jakobs S, Schönle A, Hell SW, Eggeling C (2010) Multicolor fluorescence nanoscopy in fixed and living cells by exciting conventional fluorophores with a single wavelength. Biophys J 99:2686–2694
21. Betzig E, Patterson GH, Sougrat R, Lindwasser OW, Olenych S, Bonifacino JS, Davidson MW, Lippincott-Schwartz J, Hess HF (2006) Imaging intracellular fluorescent proteins at nanometer resolution. Science (New York, NY) 313:1642–1645
22. Linde S, Kasper R, Heilemann M, Sauer M (2008) Photoswitching microscopy with standard fluorophores. Appl Phys B 93:725–731
23. Zhuang X (2009) Nano-imaging with Storm. Nat Photonics 3:365–367
24. Wildanger D, Medda R, Kastrup L, Hell SW (2009) A compact STED microscope providing 3D nanoscale resolution. J Microsc 236:35–43
25. Lang T, Rizzoli SO (2010) Membrane protein clusters at nanoscale resolution: more than pretty pictures. Physiology (Bethesda, MD) 25:116–124
26. Tanaka KAK, Suzuki KGN, Shirai YM, Shibutani ST, Miyahara MSH, Tsuboi H, Yahara M, Yoshimura A, Mayor S, Fujiwara TK, Kusumi A (2010) Membrane molecules mobile even after chemical fixation. Nat Methods 7:865–866
27. Kilár F, Simon I, Lakatos S, Vonderviszt F, Medgyesi GA, Závodszky P (1985) Conformation of human IgG subclasses in solution. Small-angle X-ray scattering and hydrodynamic studies. Eur J Biochem 147:17–25
28. Hamers-Casterman C, Atarhouch T, Muyldermans S, Robinson G, Hamers C, Songa EB, Bendahman N, Hamers R (1993) Naturally occurring antibodies devoid of light chains. Nature 363:446–448
29. Muyldermans S (2001) Single domain camel antibodies: current status. J Biotechnol 74: 277–302
30. Winter G, Griffiths AD, Hawkins RE, Hoogenboom HR (1994) Making antibodies by phage display technology. Annu Rev Immunol 12:433–455
31. Harmsen MM, De Haard HJ (2007) Properties, production, and applications of camelid single-domain antibody fragments. Appl Microbiol Biotechnol 77:13–22
32. Ellington AD, Szostak JW (1990) In vitro selection of RNA molecules that bind specific ligands. Nature 346:818–822

33. Tuerk C, Gold L (1990) Systematic evolution of ligands by exponential enrichment: RNA ligands to bacteriophage T4 DNA polymerase. Science (New York, NY) 249:505–510
34. Yan AC, Levy M (2009) Aptamers and aptamer targeted delivery. RNA Biol 6:316–320
35. Janas T, Janas T (2011) The selection of aptamers specific for membrane molecular targets. Cell Mol Biol Lett 16:25–39
36. Werner A, Konarev PV, Svergun DI, Hahn U (2009) Characterization of a fluorophore binding RNA aptamer by fluorescence correlation spectroscopy and small angle X-ray scattering. Anal Biochem 389:52–62
37. Chelliserrykattil J, Ellington AD (2004) Evolution of a T7 RNA polymerase variant that transcribes 2′-O-methyl RNA. Nat Biotechnol 22:1155–1160
38. Foy JW-D, Rittenhouse K, Modi M, Patel M (2007) Local tolerance and systemic safety of pegaptanib sodium in the dog and rabbit. J Ocul Pharmacol Ther 23:452–466
39. Ireson CR, Kelland LR (2006) Discovery and development of anticancer aptamers. Mol Cancer Ther 5:2957–2962
40. Zeng Z, Zhang P, Zhao N, Sheehan AM, Tung C-H, Chang C-C, Zu Y (2010) Using oligonucleotide aptamer probes for immunostaining of formalin-fixed and paraffin-embedded tissues. Mod Pathol 23:1553–1558
41. Hopwood D (1969) Fixatives and fixation: a review. Histochem J 1:323–360
42. Buus S, Rockberg J, Forsstr Oumlm BO, Nilsson P, Uhlén M, Schafer-Nielsen C (2012) High-resolution mapping of linear antibody epitopes using ultrahigh-density peptide microarrays. Mol Cell Proteom 11:1790–1800
43. Barysch SV, Aggarwal S, Jahn R, Rizzoli SO (2009) Sorting in early endosomes reveals connections to docking- and fusion-associated factors. Proc Natl Acad Sci 106:9697–9702
44. Rink J, Ghigo E, Kalaidzidis Y, Zerial M (2005) Rab conversion as a mechanism of progression from early to late endosomes. Cell 122:735–749
45. Wilner SE, Wengerter B, Maier K, de Lourdes Borba Magalhães M, Del Amo DS, Pai S, Opazo F, Rizzoli SO, Yan A, Levy M (2012) An RNA alternative to human transferrin: a new tool for targeting human cells. Mol Ther Nucleic Acids 1:e21
46. Jones SA, Shim S-H, He J, Zhuang X (2011) Fast, three-dimensional super-resolution imaging of live cells. Nat Methods 8:499–508
47. Kumari S, Borroni V, Chaudhry A, Chanda B, Massol R, Mayor S, Barrantes FJ (2008) Nicotinic acetylcholine receptor is internalized via a Rac-dependent, dynamin-independent endocytic pathway. J Cell Biol 181:1179–1193
48. Bujny MV, Popoff V, Johannes L, Cullen PJ (2007) The retromer component sorting nexin-1 is required for efficient retrograde transport of Shiga toxin from early endosome to the trans Golgi network. J Cell Sci 120: 2010–2021
49. Angelides KJ (1981) Fluorescent and photoactivatable fluorescent derivatives of tetrodotoxin to probe the sodium channel of excitable membranes. Biochemistry 20:4107–4118
50. Bates M, Huang B, Dempsey GT, Zhuang X (2007) Multicolor super-resolution imaging with photo-switchable fluorescent probes. Science (New York, NY) 317:1749–1753
51. Löschberger A, van de Linde S, Dabauvalle M-C, Rieger B, Heilemann M, Krohne G, Sauer M (2012) Super-resolution imaging visualizes the eightfold symmetry of gp210 proteins around the nuclear pore complex and resolves the central channel with nanometer resolution. J Cell Sci 125:570–575
52. Hopwood D (1967) Some aspects of fixation with glutaraldehyde. A biochemical and histochemical comparison of the effects of formaldehyde and glutaraldehyde fixation on various enzymes and glycogen, with a note on penetration of glutaraldehyde into liver. J Anat 101: 83–92
53. Hopwood D (1972) Theoretical and practical aspects of glutaraldehyde fixation. Histochem J 4:267–303
54. Sabatini DD, Bensch K, Barrnett RJ (1963) Cytochemistry and electron microscopy. The preservation of cellular ultrastructure and enzymatic activity by aldehyde fixation. JCB 17:19–58
55. Tagliaferro P, Tandler CJ, Ramos AJ, Pecci Saavedra J, Brusco A (1997) Immunofluorescence and glutaraldehyde fixation. A new procedure based on the Schiff-quenching method. J Neurosci Methods 77: 191–197
56. Reese TS, Karnovsky MJ (1967) Fine structural localization of a blood-brain barrier to exogenous peroxidase. JCB 34:207–217
57. Matsubayashi Y, Iwai L, Kawasaki H (2008) Fluorescent double-labeling with carbocyanine neuronal tracing and immunohistochemistry using a cholesterol-specific detergent digitonin. J Neurosci Methods 174:71–81

INDEX

A

Abbé, E. 6, 7, 9, 10, 27, 30, 41, 45, 116, 134–136, 158, 159, 259, 369
Abbe/Rayleigh criterion 5, 6, 9, 60, 62, 191
Abberior dyes 355, 356
Affinity
  probes 103, 371–374, 376, 383, 384
  tools 370, 384
AFM. *See* Atomic force microscopy (AFM)
Airy disk 4, 5, 9, 20, 345
Airy pattern 5, 60
Alpaca 375
Angular aperture...3–4
Antibody
  chains...375, 376
  variable domains
    primary 374
    secondary 374
    size 376
Anti-stoke's excitation 352, 353, 356, 357
APD. *See* Avalanche photo diodes (APD)
Aperture 3, 14, 42, 91, 116, 135, 170, 189, 260, 305, 335, 383
Aptamers 377–380, 384
Array tomography 340
Atomic force microscopy (AFM)
  applications... 236, 237
  biological applications..13
  principle..227–228
  sample preparation...234–235
  setups
    diffracting imaging (type II) 265
    Fresnel zone plate (FZP; type I) 260
    projection imaging (type III) 264
Atto dyes 114
Avalanche photo diodes (APD)..................... 56, 75, 77, 193

B

Back focal plane (BFP) 89, 146–148, 267
Binding.......................... 17, 80, 175, 179, 182, 183, 216, 231–233, 242, 246, 339, 349, 371, 373–375, 377–380
Bleaching................................. 32, 35, 82, 114, 162, 176–179, 294, 297, 327, 332, 358

C

Clusters................................... 27, 33, 47, 73–75., 173, 215, 216, 242–248, 372–373
Coherent anti-Stokes Raman (CARS)............. 207, 291–323
Colocalization.................................25, 36, 102, 107, 182–185, 203, 365, 378, 381
Confocal microscopy 20–22, 27, 28, 44, 57, 76, 78, 133–135, 153, 202, 358
Continuos excitation 54, 355
Contrast............................... 9–10, 20, 21, 24, 33, 50, 55, 81, 100, 115, 121, 122, 126, 133, 162, 168, 174, 176, 195, 203, 205, 208, 210–212, 215–217, 257–261, 264, 267–280, 282–285, 293, 357
Contrast cut-off distance 9, 20
Control laser 295–298, 300, 302, 304–309, 313–322
Correlative microscopy 285, 340
Crosslinking 203
cw STED Microscopy 54–56
Cytoskeleton...47, 103, 105–106, 203

D

Damping 294, 297, 299, 322
Data analysis in SMLM 113–129
Deconvolution 22, 58, 78, 95, 123, 143–145, 170, 171, 177, 180, 184, 300, 344
Density matrix formalism........................ 293, 297–299, 322
Density of labeling 61, 100
Dephasing 294, 296, 297, 300, 307, 313, 314, 316, 318, 319
Depletion beam 25, 45, 48, 50–54, 75–77, 304, 307, 308, 309, 313, 316, 322, 353, 355–359
Dextran..326, 327
Diffraction...................................... 2, 11, 41, 75, 87, 120, 135, 168, 189, 226, 258, 292, 325, 343, 369
  grating 6, 170, 266
  limit 4, 8, 10, 11, 14, 19, 20, 25, 26, 30, 31, 42, 45–48, 52, 75, 82, 94, 100, 125, 135, 158, 160, 168, 170, 183, 189, 190, 207, 226, 262, 295, 296, 300, 305, 308–316, 318, 320–322, 336, 347, 369, 370, 372, 384
  pattern 4–8, 42, 49, 59–61, 258, 264–266, 269, 272–274, 277, 282, 283, 345

Double resonant suppression 313–315
Dual-color 23, 81, 203, 357

E

Electron microscopy (EM) 28, 43, 88, 100, 101, 103, 174, 201, 259, 268, 270, 275, 277, 279, 285, 325–340
Embedding media/mounting media 174, 344, 349–352, 358, 364
Emission 10, 18, 45, 74, 88, 113, 134, 169, 209, 259, 292, 331, 352, 379
EM troubleshooting
   background staining 345
   cytochromes 117, 213, 336
   distorted ultrastructure 335
   EM processing
      dehydration 174
      embedding 28, 174, 176, 344, 346, 349, 351–352, 358, 361, 362, 364
      mitochondria 334
      organelle damage 335
      over-oxidation 335–336
Eosin 338, 339
Epitopes 174, 175, 179, 183, 245, 326, 347, 349, 371, 373, 374, 375, 378–383
Ewald sphere 147, 148, 152
Excitation 1, 18, 44, 74, 88, 115, 134, 169, 195, 249, 292, 331, 352, 379

F

Fab fragments 347, 363, 373, 374
Far-field microscopy 19, 26, 41–50, 75
Fast STED microscopy 56–57
Field of view (FOV) 76, 134, 163, 164, 265, 268, 270, 273, 313, 355, 359–361
Filaments 47, 82, 105, 106, 125, 215, 371, 372
FIONA. *See* Fluorescence imaging with one-nanometer accuracy (FIONA)
Fixation 330, 349, 350, 379
Fluorescence
   dyes/organic dyes 27, 34, 43, 45, 49, 50, 126, 174, 176, 353
   lifetime 45, 46, 54–55, 357
   microscopy 3, 8, 10, 100, 104, 128, 133–164, 202, 206, 211, 214–217, 322, 326, 343–365
Fluorescence imaging with one-nanometer accuracy (FIONA) 19
Fluorescent proteins
   for PALM 359
Fluorophore 77, 117, 123, 161–162, 176, 352–357, 360–363
FM dyes 77, 328–332, 338

Focus 3, 20, 21, 42, 44, 51–54, 91, 92, 94, 107, 116, 125, 134, 146–148, 152, 181, 191, 193, 200, 202, 216, 259, 261, 262, 263, 266, 280, 283, 305, 320, 321, 326
Force spectroscopy 16, 230–233
Formaldehyde 174, 175, 330, 373, 379, 382, 383
Fourier space
   weighted averaging in 142–143, 153
FOV. *See* Field of view (FOV)
Full with at half maximum (FWHM) 9, 22, 24, 33, 59–62, 119, 156, 157, 169, 191, 194, 200, 263, 310, 345
Fusion-tag 79

G

Gated STED Microscopy 55
Genetically encoded tags
   CLIP 80, 81, 347
   HALO 43, 80
   SNAP 32, 43, 81, 115, 117, 347
GFP. *See* Green fluorescent protein (GFP)
GFP recognition after bleaching (GRAB) 339
Glutaraldehyde
   autofluorescence 330–332
   quenching 175, 210, 211, 215, 339, 357
   temperature 174, 330, 332, 381
GRAB. *See* GFP recognition after bleaching (GRAB)
Grating 6, 7, 28, 29, 31, 134, 145–147, 151, 156, 161, 162, 170, 171, 173, 257, 266
Green fluorescent protein (GFP) 2, 23, 29, 33, 43, 61, 77, 79, 133, 153–155, 173, 176, 179, 325, 326, 339, 340, 355, 376, 384
Ground state depletion 45, 48, 50, 304, 306–313, 352, 370
Ground state depletion microscopy (GSDIM) 344, 348, 351, 352, 359–361, 370

H

Heavy metal labeling
   immunogold labeling 88, 326, 338
   live imaging 322
   osmium tetroxide labeling 332, 333
   ultramicrotomy,
Horseradish peroxidase (HRP) 326–328
Human samples 371

I

I^{5}M 10, 21, 24, 31, 34, 44, 150, 155
Image
   acquisition 9, 28, 31, 43, 49, 50, 61, 81–83, 94, 118, 128, 134, 139, 146, 160, 162, 168, 173, 177, 212, 236, 344, 348, 352, 356, 359, 364

formation 3, 6, 7, 118, 121, 139–145, 162, 191, 192, 269, 273
processing 26, 28, 99, 344, 346, 365
Immunostaining 43, 45, 169, 182, 184, 339, 363, 370–374, 376, 378, 381–383
Interference 6, 7, 21, 22, 24, 28–30, 34, 122, 146, 147, 149, 152, 155, 170, 172, 175, 257, 258, 266, 267, 270, 327, 355
In-vivo 79, 80, 116, 183, 203, 238, 244–245, 249, 347

J

Jablonski diagram 74, 360

K

Killer Red 339

L

Labeling
density 61, 90, 100, 173, 179–181, 348–350, 363, 364, 373, 374, 376
Leica TCS STED 75, 80, 81
Light diffraction 343, 352, 353
Live-cell imaging 19, 35, 49, 81–83, 94, 124, 125, 164, 173, 236, 259, 263
Live-cell marker 46, 82, 259
Live super-resolution 73–77, 81–83, 173, 347
Localization
microscopy 11, 21, 31, 33, 34, 83, 94, 98, 99, 114–119, 121, 123, 125, 180, 344, 346, 351, 359–364, 370
precision 31, 33, 50, 62, 90, 97, 99, 114, 116, 122, 129, 180, 360

M

Maxwell–Bloch equations 297
McCutchen aperture 147, 148, 150, 155
Microtubuli 128, 371, 372
Mini Singlet Oxygen Generator (miniSOG) 340
Moiré fringes 28–30
Mounting media 50, 174, 177, 349–351, 357–358, 362, 363, 365
Mowiol 351, 358, 362, 363, 364
Multicolor imaging 35, 76, 82, 83, 181, 353, 361, 362, 365

N

Nanobody 179, 347, 375–376, 378, 384
Nanoscopy 226, 259, 260
Near-field 13–19, 43
Near-field scanning optical microscopy (NSOM)
principle 13–36
probes 14, 189–190, 192–202, 204–212, 214–217

setup 14, 193, 194
tip enhanced near field optical microscopy (TENOM)
applications 207–208, 212–214, 216
optical antennas 208, 215–217
Noise 8–10, 16, 18, 20, 23, 30, 37, 56, 60–62, 77, 78, 81, 91, 92, 94, 95, 97, 98, 115, 116, 118–122, 129, 135, 142–145, 159, 160, 162, 174, 178, 179, 183, 202, 236, 269, 270, 271, 285, 300, 305, 308, 311–313, 322, 333, 344, 364, 371
NSOM. *See* Near field scanning optical microscopy (NSOM)
Numerical aperture 2–4, 7, 8, 23, 42, 91, 116, 122, 135, 151, 155, 156, 161, 162, 260, 266, 269, 305, 309, 321, 322, 335
Nyquist criterion 63, 349

O

Objective 2–8, 10, 20–22, 24, 29, 34, 42, 44, 56, 75, 89, 91–93, 116, 134, 135, 140, 146–153, 155, 156, 161, 162, 170, 171, 176, 177, 181, 193, 235, 264–268, 309–332, 335, 350, 352
Oligonucleotides 116, 377, 378
Optical
microscopy 2, 5, 13, 14, 16, 42, 43, 57, 63, 88, 168, 169, 189–218, 227, 264, 277, 292
nonlinearity 30
resolution 4, 115, 183–185, 189–191, 200, 203, 205, 209, 215, 216
sectioning 20, 21, 28, 46, 134, 147, 148, 150, 151, 153, 157, 168–170, 182, 358
unit 5
Optical transfer function (OTF) 136, 137, 139, 140, 142–145, 148–151, 153, 155–160, 177, 178, 191
Overexpression 370, 371

P

PALM. *See* Photoactivation localization microscopy (PALM)
Panning 375, 376
Paraformaldehyde (PFA) 174, 349, 350, 383
Patterned illumination 10, 30, 44, 138
Penetration (depth) 14, 46, 157, 176, 257, 328, 332–335, 350, 352, 358, 382
Permeabilization 173, 174, 339, 350, 372, 380, 381, 383–384
PFA. *See* Paraformaldehyde (PFA)
Phage display 375
Photoactivation 11, 21, 31, 33, 34, 49, 50, 79, 82, 87–107, 113, 114, 116, 126, 134, 292, 352
Photoactivation localization microscopy (PALM)
fluorescence photoactivation localization (fPALM) 11, 21, 31, 49, 113, 114

Photoactivation localization microscopy (PALM) (*cont.*)
  interferometric PALM (iPALM) 34, 35, 50, 93, 107
Photobleaching 15, 19, 30, 31, 46, 49, 57, 79, 81, 90, 92, 103, 107, 114, 115, 121, 126, 178–179, 211
Photon 2, 19, 45, 74, 88, 114, 159, 170, 189, 226, 258, 292, 331, 356
Photo-oxidation reaction
  bleaching 15, 19, 30, 31, 45, 50, 57, 79, 81, 90, 92, 103, 107, 115, 121, 126, 153, 178–179, 211, 294, 327
  DAB penetration 332, 334, 335
  DAB precipitation 327, 331, 332, 334–336, 339
  diaminobenzidine (DAB) 327–328, 330–336, 339, 352
  illumination field 328, 330–332, 334–336
  illumination time 57, 328, 330, 334–336
  intersystem crossing 328, 338
  light intensity 292, 293, 332, 334, 335, 357–359
Photostability 77, 79, 83, 114, 121, 124, 173, 179, 180, 203, 208, 217, 353, 357, 360
Photoswitching
  dyes/fluorophores 327–339, 352, 354
  mechanism 115, 116
  protein 30, 49, 68
Phototoxicity 49, 55, 82, 107, 173, 339, 363
4Pi microscopy 10, 21–24, 44, 155
Pinhole 20, 23, 44, 54, 133, 135, 146, 265, 272, 273
Pointillism 33, 34
Point spread function (PSF) 5, 8–10, 19–24, 26, 33, 53–55, 59, 60, 62, 82, 88, 93, 95–99, 115–125, 129, 139, 140, 144, 153, 156, 157, 177, 178, 191, 192, 345, 346, 352, 358, 360
Polarization 24, 26, 56, 106, 107, 146, 171, 198, 200, 201, 207, 209, 294–296, 299
Population 30, 48, 75, 78, 82, 114, 122–124, 185, 235, 292–297, 299–308, 313–319, 322, 330, 336, 338
Potassium cyanide
  respiratory chain 336
  under-oxidation 334
Principle of STED Microscopy 26
PSF. *See* Point spread function (PSF)
Pulsed excitation 54, 356, 359
Pulsed STED 54, 55

Q

Quantum yield 77, 176, 210, 211, 338–340, 357

R

Rabi splitting 296, 317, 322
Rayleigh's criterion 5–6, 9, 60, 62, 191, 345
Reactive oxygen species (ROS) 331, 332, 336
ReAsH 339
Refraction index 3
RESOLFT. *See* Reversible saturable optical fluorescence transitions (RESOLFT)
RESOLFT microscopy 48–50
Resolution 1–11, 13–36, 41–50, 52, 54–57, 59–63, 73–83, 87–88, 93–95, 100–107, 113–117, 120–129, 133–164, 168, 170, 173, 178–185, 189–191, 194, 199–218, 226–227, 231, 235–237, 257–269, 271–286, 291–323, 326, 340, 343–365, 369–384
Resolution in noisy images 134, 142
Resonant scanner/mirror 56, 59, 76
Reversible saturable optical fluorescence transitions (RESOLFT) 45, 352
RNA 174, 213, 377, 379
ROS. *See* Reactive oxygen species (ROS)

S

Sample preparation 2, 36, 89–95, 163, 173–178, 185, 234–235, 259, 268, 277, 285, 344, 348–352, 357–358, 363–365, 372, 381, 384
Saturation 10, 26, 29, 30, 49, 53, 54, 134, 159–162, 170, 175, 210, 259, 270, 292, 294, 297, 304, 305, 307–309, 322, 339, 357, 382
Scanning microscopy 15, 45, 359
Scanning speed 76, 345
Screening 189, 199, 375
Side lobes 22–24, 26
Signal-to-noise ratio (SNR) 10, 18, 20, 23, 30, 61, 77, 78, 81, 97–99, 116, 174, 202, 236, 305, 308, 312, 322, 333, 344, 346, 349, 361, 364, 371
SIM. *See* Structured illumination microscopy (SIM)
Single chain antibody 376
Single molecule localization
  software for SMLM 113, 126, 128, 129
Single-particle tracking 19, 33, 43, 82, 94, 106, 363
Singlet oxygen 338, 339, 340
SLM. *See* Spatial light modulator (SLM)
SNR. *See* Signal-to-noise ratio (SNR)
Sparrow criterion 9, 345, 346
Spatial light modulator (SLM) 145–147, 155, 164, 173
Spatial resolution 41–44, 48–50, 54–56, 59, 75, 107, 138, 180, 184, 275, 295, 300, 308, 311, 312, 321
Spontaneous emission 24, 51, 74–76, 293
Spot finding 118–119
Spot fitting 119–121
Staining 42, 43, 45, 61, 100, 114, 117, 172, 174, 180, 183, 202, 259, 260, 277, 279, 283, 328–335, 337, 344, 345, 349, 350, 357, 361, 363, 364, 365, 370–374, 376, 378, 380–382, 384
Stimulated emission depletion (STED)
  STED-FCS 57–59

Stochastic optical reconstruction (STORM)
    dSTORM 50, 88, 96, 99, 113–122, 124, 127, 344, 351, 360, 361, 381
Stokes shift 46, 52, 301
Structured illumination microscopy (SIM)
    channel alignment 181–183
    data acquisition 90, 95, 115, 140, 148–149, 153, 155–156, 231, 267
    image reconstruction 114, 115, 128, 141–142, 149–151, 153, 156, 270, 274, 364
    Moiré effect and patterns 28–30, 34, 135–136, 170, 172, 175, 328
    sample preparation 163, 173–178, 234–235, 268, 277, 344, 348–352, 357–358, 363, 372, 381, 384
    setups
        3D SIM 29, 30, 35, 154, 164, 168–177, 179, 180, 184, 185
        I^5-SIM 10, 24, 31, 155–158
        three beams 145–147, 149, 151–158, 170
        two beam 145–147, 149–153, 156–158, 160
Sub-diffraction 291–323, 344
Sub-diffraction microscopy 292, 295, 305, 318
Systematic evolution of ligands by exponential enrichment (SELEX) 377, 379

T

Temporal resolution 41, 56, 107, 231, 261, 344, 359, 361
Tetracysteine tag 339

Theory of image formation 7, 162
Three-dimensional (3D) imaging 8, 23, 27, 116, 120, 139, 168, 260, 268
Total internal reflection microscopy (TIRF) 13, 17–19, 35, 44, 47, 89–92, 95, 107, 116, 147, 151, 152, 156–158, 162, 351, 363
Two-color 25, 27, 29, 32, 34, 46, 56, 89, 92, 353, 355
Two-photon microscopy 23, 46, 107

V

VHH (llama VHH antibody) 375–376
Video-rate 28, 78, 83, 164, 293
Virtual volume super resolution microscopy (VVSRM) 93
Volume marker 79
Vortex-phase plate 75

W

Wavefront 21–23, 27, 263
Wavelength 2–8, 13, 14, 19, 22–24, 27, 28, 31, 41–43, 46, 52, 74, 76, 90, 91, 114, 134, 135, 147, 148, 150, 151, 155, 156, 161, 162, 168, 169, 173, 181, 189, 191, 197, 199, 207, 226, 257, 260, 262, 266, 283, 300, 305, 321, 322, 326, 352, 355, 357–360, 362
Wide-field 20, 24, 28, 35, 44, 49, 81–83, 88, 89, 91, 106, 114, 116–118, 121, 133, 134, 148–158, 160–164, 169–171, 177, 180, 184, 349, 364
Wiener filter deconvolution 143–144